MW01153870

MEMS Sensors: Design and Engineering Applications

MEMS Sensors: Design and Engineering Applications

Edited by
Holly Dunham

Published by States Academic Press,
109 South 5th Street,
Brooklyn, NY 11249, USA

ISBN: 978-1-63989-350-8

Cataloging-in-Publication Data

MEMS sensors : design and engineering applications / edited by Holly Dunham.
p. cm.
Includes bibliographical references and index.
ISBN 978-1-63989-350-8
1. Microelectromechanical systems. 2. Microelectromechanical systems--Design and construction.
3. Electromechanical devices. 4. Microtechnology. 5. Mechatronics. I. Dunham, Holly.
TK7875 .M46 2022
621.381--dc23

For information on all States Academic Press publications
visit our website at www.statesacademicpress.com

Contents

Preface

Every book is a source of knowledge and this one is no exception. The idea that led to the conceptualization of this book was the fact that the world is advancing rapidly; which makes it crucial to document the progress in every field. I am aware that a lot of data is already available, yet, there is a lot more to learn. Hence, I accepted the responsibility of editing this book and contributing my knowledge to the community.

MEMS, or micro electro-mechanical systems, is a technology associated with the manufacturing of microscale devices such as sensors, transducers, actuators, gears, pumps, switches, etc. These microscopic integrated devices combine electronic, electrical and mechanical elements. These elements work together, using microsystems technology, to carry out a single functional requirement. Sensors that are designed and manufactured using this technology are called MEMS sensors. Accelerometers, gyroscopes, magnetometers, pressure sensors, airflow sensors, microphones, temperature sensors, fuel sensors, impact sensors, etc., are some of the various MEMS sensors. They find applications in the automobile, chemical and pharmaceutical sectors, as well as, environmental and health sciences, computing and communications, and consumer products. There has been rapid progress in this field and its applications are finding their way across multiple industries. Different approaches, evaluations, methodologies and advanced studies on MEMS sensors have been included in this book. It is a vital tool for all researching or studying this field as it gives incredible insights into emerging trends and concepts.

While editing this book, I had multiple visions for it. Then I finally narrowed down to make every chapter a sole standing text explaining a particular topic, so that they can be used independently. However, the umbrella subject sinews them into a common theme. This makes the book a unique platform of knowledge.

I would like to give the major credit of this book to the experts from every corner of the world, who took the time to share their expertise with us. Also, I owe the completion of this book to the never-ending support of my family, who supported me throughout the project.

Editor

1

RF-MEMS Switches Designed for High-Performance Uniplanar Microwave and mm-Wave Circuits

Lluis Pradell, David Girbau, Miquel Ribó, Jasmina Casals-Terré, Antonio Lázaro, Adrián Contreras, Marco Antonio Llamas, Julio Heredia, Flavio Giacomozzi and Benno Margesin

Abstract

Radio frequency microelectromechanical system (RF-MEMS) switches have demonstrated superior electrical performance (lower loss and higher isolation) compared to semiconductor-based devices to implement reconfigurable microwave and millimeter (mm)-wave circuits. In this chapter, electrostatically actuated RF-MEMS switch configurations that can be easily integrated in uniplanar circuits are presented. The design procedure and fabrication process of RF-MEMS switch topologies able to control the propagating modes of multimodal uniplanar structures (those based on a combination of coplanar waveguide (CPW), coplanar stripline (CPS), and slotline) will be described in detail. Generalized electrical (multimodal) and mechanical models will be presented and applied to the switch design and simulation. The switch-simulated results are compared to measurements, confirming the expected performances. Using an integrated RF-MEMS surface micromachining process, high-performance multimodal reconfigurable circuits, such as phase switches and filters, are developed with the proposed switch configurations. The design and optimization of these circuits are discussed and the simulated results compared to measurements.

Keywords: RF-MEMS switches, MEMS fabrication technology, reconfigurable microwave circuits, uniplanar multimodal circuits, circuit simulation, electromechanical simulation, filters, phase switches

1. Introduction

Radio frequency microelectromechanical system (RF-MEMS) switches are aimed to perform the control function in tunable and reconfigurable RF/microwave and millimeter-wave (mm-wave) systems. Electrostatic actuation is often preferred to other actuation mechanisms like electrothermal [1, 2] and phase-change/phase-transition materials [3], due to its negligible current consumption, no requirement for external heating sources and integration capability with well-established technologies such as high-resistivity silicon [4–7], fused-quartz and glass substrates [8–11], or CMOS [12–15] and SiGe BiCMOS [16, 17] processes. The latter can provide totally integrated, efficient systems containing sensors, control electronics, and MEMS-reconfigurable RF communication circuits [18].

The mechanical and electrical design of the RF-MEMS switches has been comprehensively studied in the literature [19], and it highly depends on the circuit or transmission media in which it is to be integrated and the technology platform [20]. A number of solutions can be found, including integration in microstrip transmission lines [21], coplanar waveguides (CPWs) [4], coplanar striplines (CPSs) and slotlines [11], planar structures embedded in rectangular waveguides [22], and micromachined waveguides for sub-mm-wave frequencies [23]. Depending on the specific designs and dimensions, they can operate in the microwave and the mm-wave bands, at frequencies as high as 240 GHz as reported in [24] using BEOL in BiCMOS technology.

Series and parallel RF-MEMS switch topologies can be implemented, with either ohmic-contact [22] or capacitive-contact [4, 25]. While ohmic switches can operate in a very wide frequency band from DC to mm-waves featuring excellent OFF-state isolation and very low ON-state insertion loss, capacitive switches are frequency selective (being the center frequency defined by a series LC-resonant circuit) but their operation can be extended well beyond mm-wave frequencies by properly choosing the ON-state capacitance and the series inductance which depends on the membrane dimensions [24].

Mechanical topologies for RF-MEMS switches include bridge-type clamped-clamped or beam-suspension membranes and cantilever-type switches. Important switch parameters, such as the actuation voltage or the fabrication residual stress, are dependent on the particular selected topology [26–28]. Using three-dimensional (3D) mechanical simulation, the material physical properties are taken into account to a priori assess the behavior of the switch geometry (including the suspension type) in terms of initial membrane deformation, pull-in voltage, spring constant, mechanical resonant frequency, and transition times from OFF to ON states (and vice versa) [29]. Mechanical transients may produce bouncing phenomena [30–34] which degrade the RF behavior of the switch and can be studied more efficiently with energy models [35].

RF-MEMS switches featuring the above mechanical topologies are compatible with and can be conveniently integrated in uniplanar structures (CPW, CPS, and slotline) to perform a control function. In case of multimodal transmission lines like CPW, they can be used to selectively control the two CPW fundamental propagation modes (even and odd) [36]. To accurately analyze the interaction between modes in complex uniplanar structures (transitions, discontinuities), multimodal circuit models are derived from the application of the general multimodal theory [37–40]. Moreover, suitable equivalent circuits for both (ON/OFF) states of the switch can be obtained and integrated in the multimodal models. In this way, efficient and compact reconfigurable circuits for communication systems at microwave and mm-wave frequencies can be designed [6, 9–11].

In this chapter, a detailed study of RF-MEMS switches to be used in multimodal uniplanar circuits is presented. The switch electromechanical design considerations are explained in detail, and a number of switch configurations proposed, simulated mechanically, and fabricated using the FBK flexible technology platform [20]. The fabricated switches are measured, and the experimental results successfully compared to simulations, thus validating the design approach. An estimation of the RF behavior of the switches is obtained from 2.5 D electromagnetic simulation. The RF behavior after fabrication is assessed by measuring the switch transmission coefficient for both (ON/OFF) states. Equivalent circuit topologies are also proposed and the value of the circuit elements computed by fitting the simulated results to the measurements. The switch transmission coefficient is also used for the measurement of the switch hysteresis. The proposed switches are integrated into the microwave and mm-wave multimodal reconfigurable circuits to validate the multimodal design approach. Some examples of fabricated multimodal reconfigurable filters and phase switches using RF-MEMS switches with various mechanical topologies (bridge-type featuring ohmic contact and capacitive contact, and cantilever-type featuring ohmic contact) are presented.

This chapter is organized as follows. After this introduction, the multimodal circuits and models for uniplanar transitions and discontinuities are explained in Section 2. The RF-MEMS fabrication technology platform is described in Section 3. The electromechanical analysis derived from the energy approach is studied in Section 4. The fabricated switches are described in Section 5. The RF equivalent circuit for the switches is analyzed in Section 6. The reconfigurable multimodal microwave and mm-wave circuits are described in Section 7. The chapter ends with some conclusions.

2. Uniplanar lines and multimodal models for transitions and discontinuities

2.1. The slotline and the coplanar waveguide

The slotline and the CPW are uniplanar transmission lines. The slotline consists of two conductor strips on a dielectric substrate (**Figure 1(a)**). The CPW consists of three conductor strips on a dielectric substrate (**Figure 1(b)**). The slotline is a monomodal transmission line: it propagates only one fundamental quasi-transversal electromagnetic (TEM) mode, whose voltages and currents (both for the total voltage and current $V_s(z)$ and $I_s(z)$, and for the forward and backward propagating waves $V_s^+(z)$, $I_s^+(z)$, and $V_s^-(z)$, $I_s^-(z)$, respectively) are defined as in **Figure 2(a)** and can be circuitally modeled as an ideal transmission line (**Figure 2(b)**), with

$$\begin{aligned} V_S(z) &= V_S^+(z) + V_S^-(z) & V_S^+(z) &= V_S^+ e^{-j\beta_s z} & V_S^-(z) &= V_S^- e^{+j\beta_s z} \\ I_S(z) &= I_S^+(z) + I_S^-(z) & I_S^+(z) &= V_S^+(z)/Z_{0S} & I_S^-(z) &= -V_S^-(z)/Z_{0S} \end{aligned}$$

where Z_{0s} is the characteristic impedance of the slotline mode and β_s its phase constant.

The CPW is a multimodal transmission line: it can propagate two fundamental quasi-TEM modes simultaneously (the even and odd modes) whose voltages and currents are defined as in **Figure 3(a)**. The odd mode is often seen as spurious, and its propagation cut by means of air bridges (described subsequently). However, it can be used to design new kinds of compact uniplanar circuits. In a CPW section, the even and odd modes do not interact between them and therefore can be circuitally modeled as two independent ideal transmission lines, as shown in **Figure 3(b)**, with equations analogous to those of the slotline for either of them.

2.2. Multimodal models for CPW circuits

The even and odd modes behave differently when they encounter any transition or asymmetry, and there they may also interact between them. A multimodal model is a circuit model that makes the behavior of the different modes at a transition or asymmetry explicit. As an

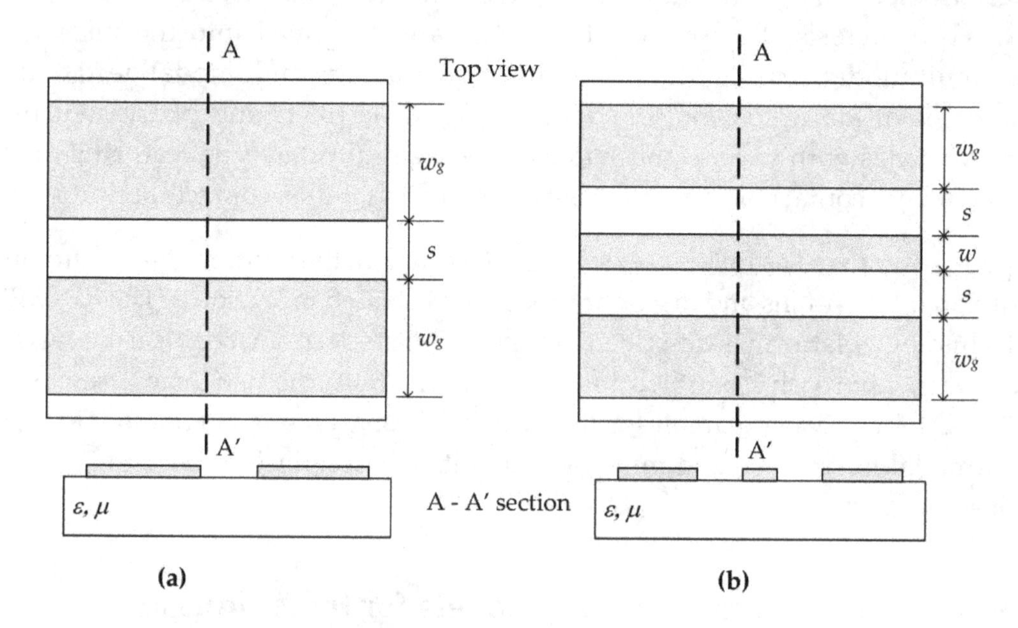

Figure 1. (a) Slotline. (b) CPW.

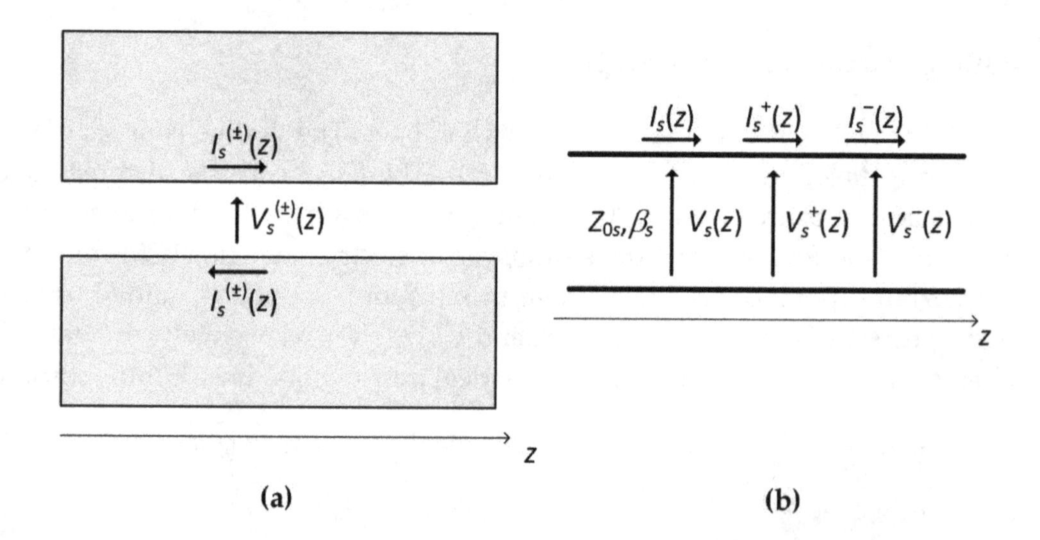

Figure 2. (a) Definitions of voltages and currents in a slotline. (b) Circuit model for a slotline section.

example, some simple multimodal models are presented subsequently; more complex ones are described in [6, 9–11, 38–40, 58].

2.2.1. Symmetric CPW-to-slotline transition

The layout of this transition is shown in **Figure 4(a)** (the depicted voltages and currents are the total ones for each mode, computed at the transition plane). Its behavior is easy to understand intuitively. At the transition, the odd mode transforms into the slotline mode and vice versa due to their similarity of voltage and current orientations (caused by the similarity of their electromagnetic fields). The even mode, however, is left in open circuit when the slotline begins since its current in the CPW central strip can flow no more. Therefore, the multimodal circuit model

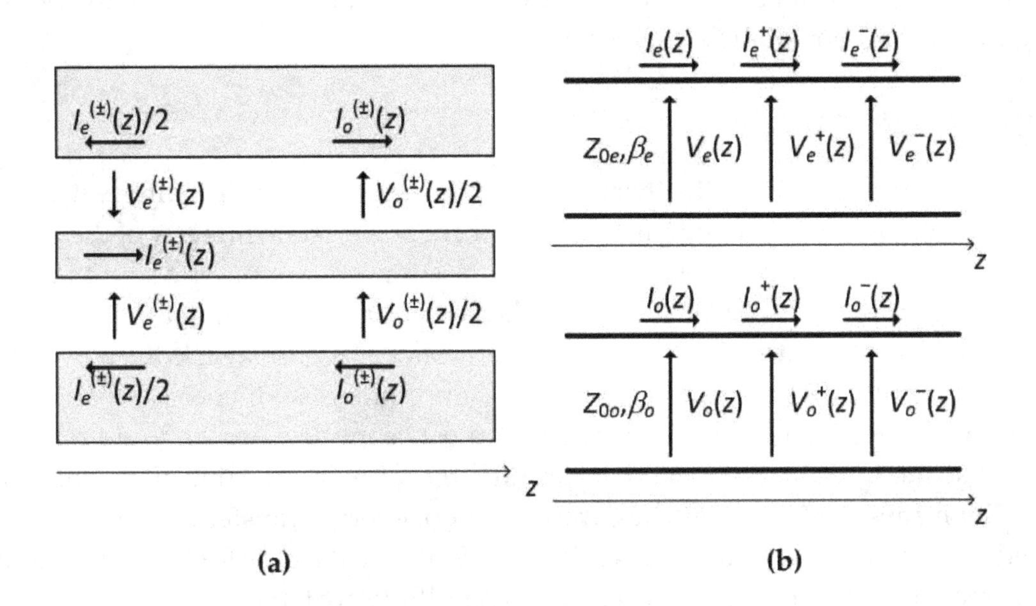

Figure 3. (a) Definitions of voltages and currents in a CPW. (b) Circuit model for a CPW section.

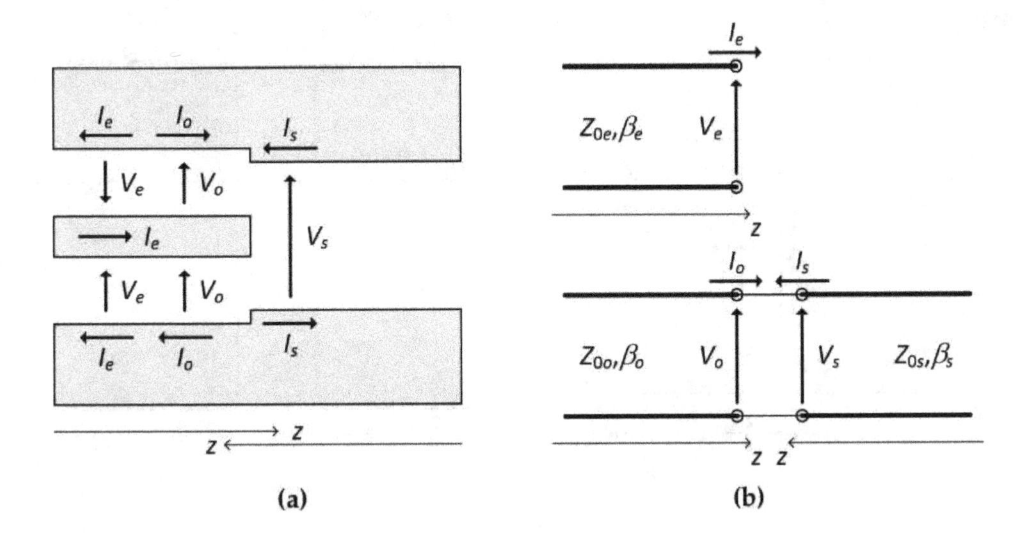

Figure 4. (a) Symmetric CPW-to-slotline transition. (b) Multimodal circuit model.

for the symmetric CPW-to-slotline transition is that of **Figure 4(b)**. As can be seen, a multimodal circuit model confines the contributions of each mode present in a transition into a different port.

2.2.2. Impedances connecting the two outer CPW strips and air bridges

Suppose an impedance connects the two outer CPW strips as shown in **Figure 5(a)**. This circuit can model an air bridge (a conducting wire connecting the two outer CPW strips, with an impedance $Z = 0$ ideally), but also more complicated situations. The even mode does not interact with the impedance since the two outer CPW conductors have the same even-mode electric potential. Therefore, the impedance behaves as a shunt impedance for the odd mode, and it is transparent to the even mode. Thus, its circuit model is that of **Figure 5(b)**. As can be seen, an air bridge ($Z = 0$) blocks the propagation of the odd mode by short-circuiting it. By controlling the value of Z, for instance, by means of MEMS switches, the amount of odd mode that propagates from the left side of the CPW to the right one can be controlled without affecting the propagation of the even mode.

2.2.3. Asymmetric shunt impedances in a CPW

In the two previous examples, the even and odd modes behaved in a different way at the analyzed transitions but did not interact between them due to the symmetry of the transitions. When the transitions are asymmetric, as it is the case for the asymmetric shunt impedances connecting the strips of the CPW shown in **Figure 6(a)**, the modes interact between them. The behavior of this transition is not obvious, but it can be rigorously modeled by the circuit shown in **Figure 6(b)** [37]. As can be seen, in this case, there is an energy balance between even and odd modes (there is a circuit connection between the even- and odd-mode ports), provided that the impedances Z_A and Z_B are different. Again, by controlling the values of Z_A and Z_B by means of MEMS switches, the amount of energy transfer among modes can be controlled. This transition and other described in [38–40] are the base for building multimodal uniplanar reconfigurable circuits [6, 9–11, 58] using MEMS switches.

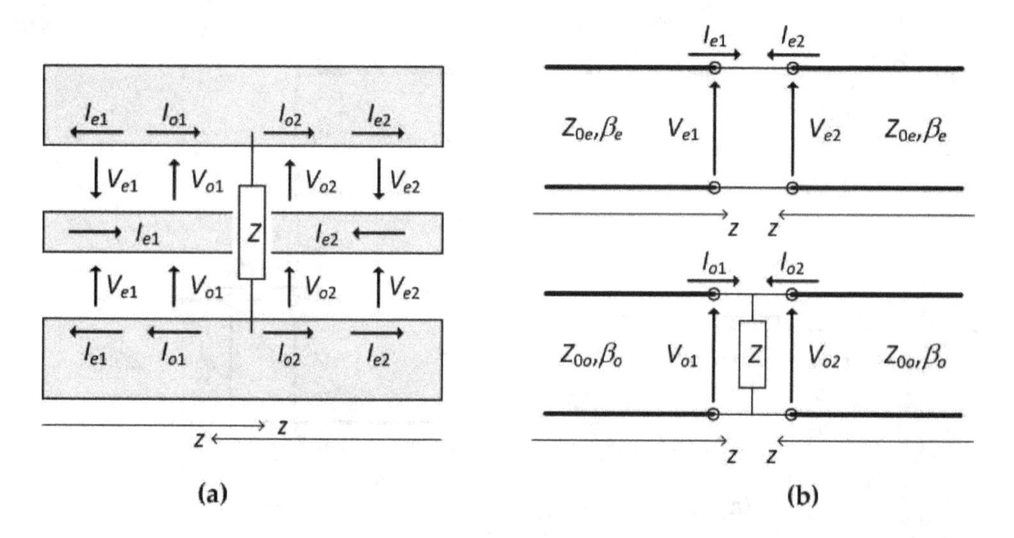

Figure 5. (a) Impedances connecting the two outer CPW strips. (b) Multimodal circuit model.

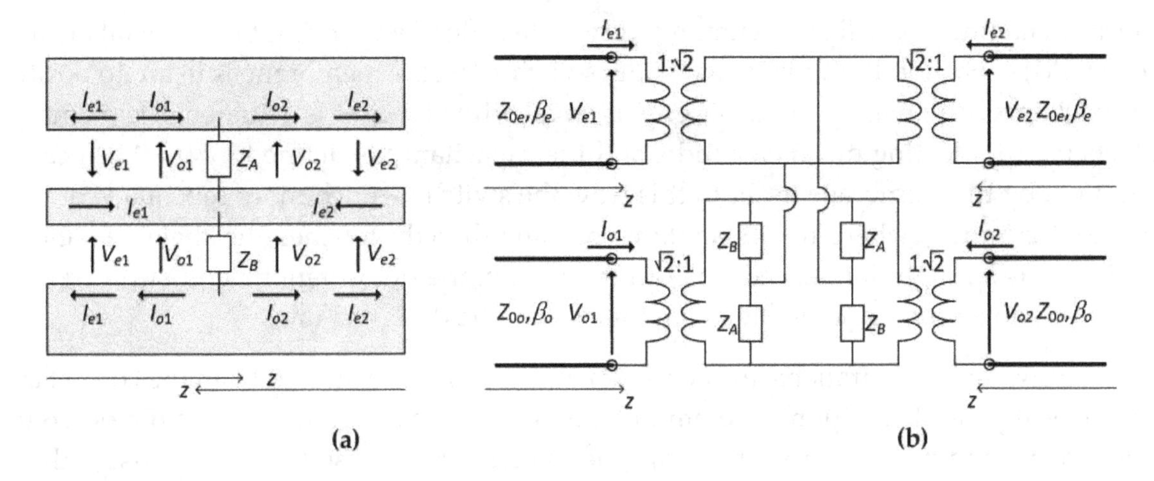

Figure 6. (a) Asymmetric shunt impedances connecting the strips of a CPW. (b) Multimodal circuit model.

3. Fabrication technology for electrostatically actuated RF-MEMS switches

3.1. RF-MEMS technology platform

A flexible technology platform has been developed and optimized at the FBK Institute (Trento, Italy) for the fabrication of RF-MEMS. Basic components (like low-loss CPW, microstrip line and slotline, ohmic [41, 42] and capacitive [43, 44] switches, variable capacitors and inductors) can be integrated in complex reconfigurable RF circuits. Many kinds of devices were produced, mainly for space and communication application, like switching matrices [45, 46], tunable and switchable phase shifters [47], reconfigurable antennas, impedance matching networks [48], VCOs [49, 50], and tunable filters. Depending on the used substrate, high-resistivity silicon (<40 GHz), or fused quartz (>40 GHz), the working frequency range spans from sub-GHz up to more than 100 GHz.

The base process requires eight lithography masks but, depending on the requirements, it can easily be expanded to deposit and pattern metal on the wafer backside to realize microstrip lines or antennas and to obtain devices suspended over thin membranes by locally removing the substrate. A wafer-to-wafer or a cap-to-die-bonding module is also available to encapsulate the delicate MEMS moving parts [51].

RF signal lines and ground area are made of thick electroplated gold to reduce insertion losses while actuation electrodes and DC-bias signal lines are made of a high-resistivity polysilicon to minimize coupling with adjacent RF lines. The movable and suspended structures of the electrostatically actuated switches, which can be either cantilevers or clamped-clamped beams, are made by gold deposited over a sacrificial photoresist layer having the thickness of the required air gap, while switch underpass lines and other conductors are made of a thin Al film. On ohmic-contact switches, the gold-to-gold contact area is defined by underneath polysilicon protruding dimples to ensure a repeatable contact force and a uniform and reproducible low contact resistance. On capacitive-contact switches, the contact capacitor is made by depositing a thin silicon oxide dielectric and an upper floating metal (FLOMET) electrode over the metal underpass line, obtaining a very well-defined and reproducible metal insulator metal (MIM) capacitor. In this way, when the switch-movable membrane is in an up position, the capacitance, due mainly to the air gap, is small while when it is actuated, the membrane contacts the top floating metal electrode, and the capacitance is defined by the MIM capacitor and not by the membrane itself. In this way, the switch is much more repeatable than the usual configuration, where the movable membrane directly touches the dielectric and the capacitance is strongly influenced by both the membrane deformations and surface roughness leading to a capacitance value much lower than the designed one.

For all the switch configurations, the actuation electrodes are separated from the contact area. This makes it possible to optimize them independently to sustain the high actuation voltage (up to 100 V) and reducing the charging phenomena. It is possible either to use a thicker dielectric over polysilicon to limit the electric field or better to use a dielectric-free configuration removing all the dielectric and using a matrix of mechanical stoppers to prevent short circuits. The height of the stoppers has to be designed in order to obtain an air gap between movable bridges and electrodes which is thick enough for isolation at the bias voltage used.

3.2. Fabrication process

The basic fabrication process for silicon substrate is reported in [20, 52] and illustrated in **Figure 7**, where a schematic cross section of an ohmic switch is represented. For high-frequency devices, the losses of the silicon are too high and quartz (fused silica) is preferred. Only minor adjustments are required to process transparent substrates.

The fabrication process starts with the oxidation of the high-resistivity (>5000 Ω·cm) 150-mm diameter silicon wafers in order to obtain a 1-μm-thick silicon oxide isolation layer. A 630-nm thick layer of polysilicon is then deposited by low-pressure chemical vapor deposition (LPCVD) and doped by ion implantation to obtain a sheet resistance of about 1600 Ω/sq. The polysilicon structures are defined by lithography and dry etching using chlorine-based gas plasma, and the residual photoresist is removed by an oxygen plasma (**Figure 7(a)**). An annealing at 925°C for 1 h in nitrogen atmosphere is required to diffuse and to electrically activate the B ions. To electrically isolate the polysilicon, 300 nm of silicon dioxide is deposited by LPCVD at 718°C (TEOS). When a backside conductive layer is required for microstrip lines or devices like phased array antennas, an aluminum film is sputtered and patterned on the wafer backside. The process continues on the front side with a lithography and dry etching to open holes in the TEOS layer for contacting the underneath polysilicon (**Figure 7(b)**). To realize connection lines, a conductive metal consisting of Ti/TiN diffusion barrier and Al1%Si is sputtered and patterned by dry etching (**Figure 7(c)**). The total thickness is the same as that of polysilicon to minimize distortion of the switch bridges crossing over both metal underpass and polysilicon actuation electrodes. A 100-nm thick SiO_2 deposited by PECVD is used as dielectric for capacitive contacts as well as for metal isolation. Holes in the oxide (vias) are realized by lithography and dry etching to contact the underneath metal and for the dielectric-free actuation electrodes (**Figure 7(d)**).

To realize the bottom part of the gold-to-gold contacts of ohmic switches as well as an electrically floating metal layer for capacitive-contact switches, a 5-nm Cr adhesion layer and 150-nm Au are deposited by an electron beam gun, patterned and wet etched (**Figure 7(d)**).

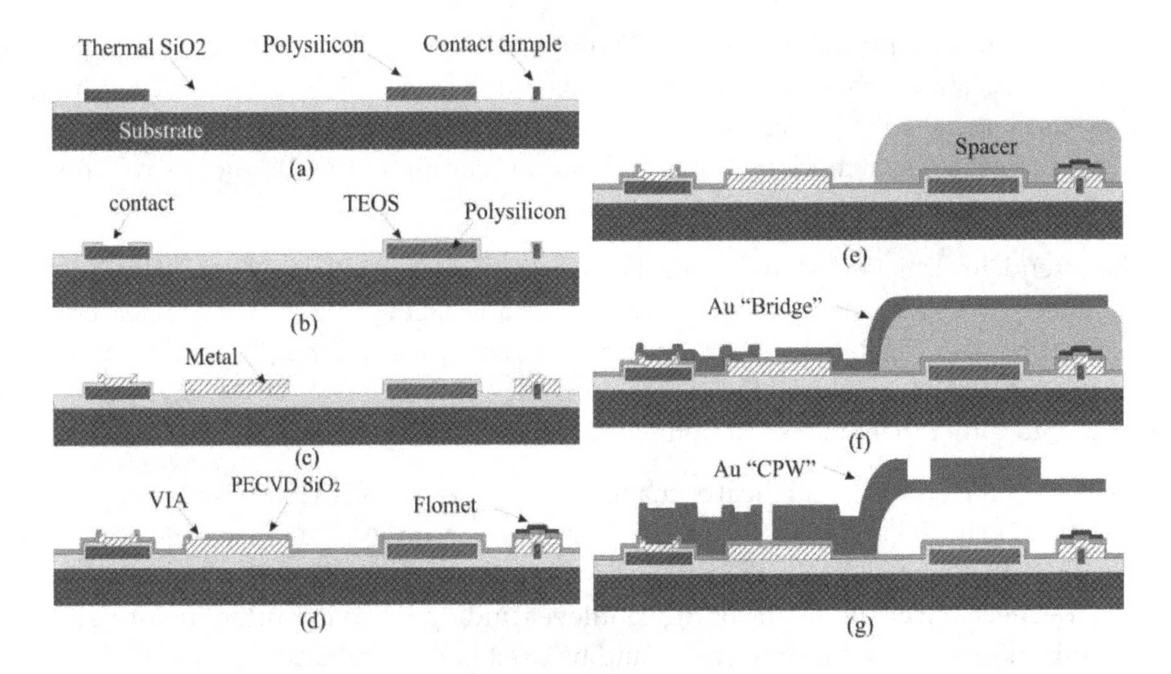

Figure 7. Depiction of the fabrication process flow on a schematic ohmic-switch cross section.

A photoresist sacrificial layer (spacer) is lithographically defined under movable structures and suspended air bridges because later it can be easily removed by oxygen plasma to form an air gap (**Figure 7(e)**). To make the RF structures, a conductive seed layer of 2.5 nm of Cr and 25 nm of Au is deposited by e-beam, patterned using thick AZ 4562 positive resist, and a 1.8-μm-thick first gold layer (bridge) having a slightly tensile residual stress is selectively grown by electroplating (**Figure 7(f)**). A second 3.5-μm-thick gold layer (CPW) is then defined by AZ 4562 and electroplated. The thinner Au bridge layer is used to make the suspended and movable structure while both layers are superimposed to obtain thicker low resistance signal lines and ground areas. To better control the deformation of the movable parts of the switch, it is possible to use the bridge layer for deformable suspension legs and bridge plus CPW layer for a stiffer main body that moves rigidly, almost without deformations. This concept is applied in the fabricated devices, described in Section 5. To complete the fabrication, the seed layer is removed by wet etching, and the suspended structures are released by removing the spacer underneath by an oxygen plasma (**Figure 7(g)**).

4. Electromechanical models for electrostatically actuated RF-MEMS switches: energy considerations

Mechanical design plays an important role in the RF behavior of MEMS switches because it couples important parameters such as the required actuation voltage (also called pull-in voltage, $V_{pull\text{-}in}$), actuation time, release time, and the appearance of a bouncing phenomenon after release, which delays a complete release of the switch. The pull-in voltage is commonly calculated assuming a static mechanical behavior in RF-MEMS switches. For compatibility of RF-MEMS with low-voltage CMOS and BiCMOS technologies [14], charge-pump techniques are often used [53]. Nonetheless, a current trend is to decrease the MEMS high $V_{pull\text{-}in}$, because charge pump has some limitations given the ever-reducing voltages in CMOS and BiCMOS [54].

One strategy is to dynamically drive the RF-MEMS switches with an input step voltage waveform, which has shown that can decrease the actuation voltage [55]. However, the induced dynamic mechanical behavior can cause an important bouncing phenomenon after release and deeply affect the switch RF/microwave isolation relevant to the RF/microwave behavior, disabling fast-switching applications.

Another trend to achieve fast-switching is to use actuation voltages beyond pull-in. The accumulated electrostatic energy will generate mechanical energy that will be released in the form of mechanical oscillations (bouncing) of the switch membrane [30–35]. To accurately approach RF-MEMS mechanical design, the dynamic behavior of RF-MEMS switches should be considered rather than only static behavior.

The analysis of the electromechanical exchange of energy in the RF-MEMS is an analytical tool that can provide inside knowledge on the required minimum $V_{pull\text{-}in}$. It also takes into account the rebound after release due to the increased actuation voltages [26]. During the switching process, the mechanical membrane or the cantilever undergoes an important deformation. To capture this influence, nonlinear terms should be used in the mechanical model [35].

Figure 8 shows the schematic of the one-dimensional (1D) lumped-mass model that, combined with classical Newtonian mechanics, can be used to predict the behavior under applied electrostatic forces of an RF-MEMS switch (either a membrane or a cantilever). If air damping is considered the only non-conservative force, then the equation of the motion for the 1D model shown in **Figure 8** is

$$m_{ef}\ddot{y} + b_{ef}\dot{y} + k_{ef}y = F_o \tag{1}$$

where m_{eff} is the effective mass of the mechanical structure, b_{eff} is the damping coefficient ($=\sqrt{m_{eff}k_{eff}}/Q$, where Q is the mechanical quality factor), $k_{eff} = k_1 + k_3\,y^2$, where k_1 and k_3 are the spring constants in the direction of the motion of the mechanical structure, and F_o is the summation of all external forces applied (i.e., the electrostatic force).

The study of the energy exchange of the mechanical system not only provides the position of the contact point of the MEMS switch but also provides deep insight to the required $V_{pull\text{-}in}$, the maximum rebound height, and the actual actuation/release times. For the lumped model shown in **Figure 8**, the total energy of the system E is expressed as

$$E = \left(\tfrac{1}{2}m_{eff}\dot{y}^2\right) + \left(\tfrac{1}{2}k_1 y^2 + \tfrac{1}{2}k_3 y^4\right) + \left(\frac{\varepsilon_o A}{2\,(g_i - y)^2}V^2\right) \tag{2}$$

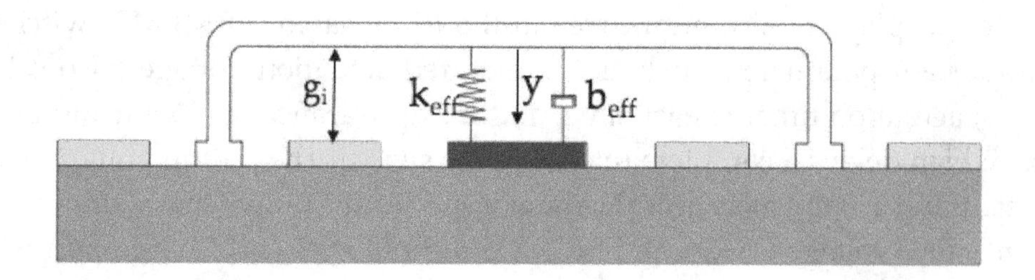

Figure 8. 1D mechanical lumped-mass model of an RF-MEMS switch.

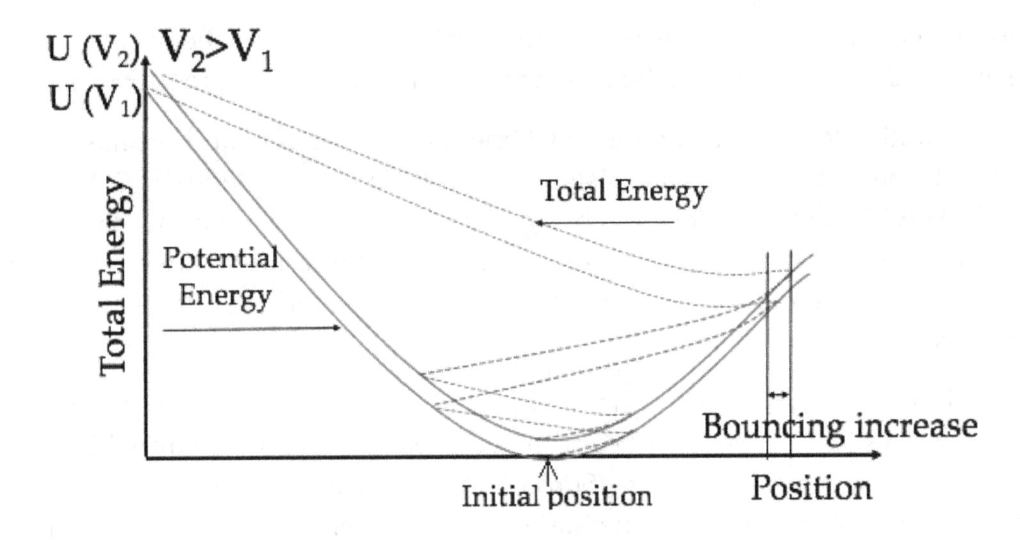

Figure 9. Simulated evolution of the switch total energy after release from the actuated position.

where V is the applied actuation voltage, ε_0 is the dielectric constant, g_i is the initial gap between the electrodes, and A is the area of the electrodes. In Eq. (2), the first term in brackets is the kinetic energy, the second is the potential energy U_m stored as the mechanical deformation of the structure, and the third is the electrical potential energy U_e. The different thermal processes in the RF switch fabrication (explained in Section 3) can induce an intrinsic residual stress. This effect produces a nonlinear mechanical behavior which can be modeled as a nonlinear spring with spring constants k_1 and k_3.

Figure 9 shows the evolution of the total energy E with respect to the position of the contact point of the mechanical structure along the gap when the switch is released from the actuated position. It can be observed that the total energy E evolves due to inertia and damping forces but is bounded by the potential energy curve ($U = U_m + U_e$), and the change on the potential energy accumulated can modify both the rebounds and the isolation time.

From Eq. (2), the actuation voltage can be obtained since at the point of instability (or switching) $\partial E/\partial y = 0$. In case of nonlinear mechanical behavior, the resulting expression of $V_{pull\text{-}in}$ is as follows:

$$V_{pull\_in} \approx \sqrt{\frac{g_i^3}{4\,\varepsilon_o A}\left(\frac{k_3 g_i^2}{8} + k_1\right)} \tag{3}$$

Eq. (3) is used in the simulations of the fabricated switches, discussed in the next Section.

5. Electrostatically actuated RF-MEMS switch configurations for reconfigurable uniplanar circuits

This section presents electrostatically actuated switch configurations which can easily be integrated in reconfigurable uniplanar circuits. All the considered devices were fabricated using the eight-mask surface micromachining process from FBK explained in Section 3. The structures are composed of a 1.8-μm-thick gold layer and reinforced with a 3.5-μm-thick superimposed gold frame to increase the rigidity of the cantilever or the bridge. The switches were designed taking into account the mechanical analysis described in Section 4.

To reduce the initial deformation of the switch membrane, different authors have reported on the effective stiffness of common suspensions types [26–28]. A coupled-field 3D finite element analysis (FEA) with ANSYS® Workbench™ can be used to model the mechanical structure and tune the measured initial deformation according to the residual stress produced by the fabrication process. The robustness of the design to manufacturing stresses can also be studied with this software.

The switch designs presented use either a clamped-clamped suspension or alternative suspension techniques such as straight-beam, curved-beam, or folded-beam which reduce the initial stress and the actuation voltage. Some FEA results are shown to assess the ability of the proposed suspensions to absorb the initial stress. Hysteresis measurements are also presented to show the featured pull-in and pull-out voltages. At the end of the section, **Table 1** summarizes the switch dimensions and the main mechanical parameters for the different switches.

The RF behavior of the switches, including equivalent circuits in ON and OFF states, is discussed in detail in Section 6.

5.1. Cantilever-type switches for switchable asymmetric shunt impedances

Figure 10 shows a cantilever-type ohmic-contact switch, able to synthesize asymmetric shunt impedances in a CPW to control the CPW even mode as explained in Section 2.2.3. The cantilever is placed above a rectangular notch in the upper lateral ground plane to which is anchored using a suspension composed of two 16-μm-wide beams, providing a low spring constant value and a low pull-in voltage. The beams may be either straight-shaped (**Figure 10(a)**) or semi-circular-shaped (**Figure 10(b)**). The latter is used to reduce the initial deformation of the switch due to the residual stress [27] produced by the fabrication process [20, 52]. To compute the initial deflection, a 3D FEA using ANSYS® was performed. The initial stress values used for the simulation were $\sigma_2 = 58$ MPa (gold layer with a thickness $t_2 = 1.8$ μm) and $\sigma_1 = 62$ MPa (gold layer with a thickness $t_1 = 3.5$ μm). As shown in **Figure 10(c)**, the simulated initial deformation was −0.19 μm. The devices were fabricated on a quartz substrate ($\varepsilon_r = 3.8$) with a thickness $h = 300$ μm and an air gap $g_i = 1.6$ μm. An isolated, high-resistivity polysilicon lower electrode is placed in the notch underneath the cantilever. Three ohmic contacts are defined on the bottom edge, and small dimples (12 × 10 μm^2) of polysilicon are placed underneath, creating contact bumps to reduce the contact resistance and enhance the RF behavior.

5.2. Bridge-type switches

Bridge-type switches can be used to perform both ohmic contacts and capacitive contacts in uniplanar circuits for multiple applications. In contrast to the cantilever-type switches, the bridge-types are symmetric structures and therefore (as discussed in Section 2.2), when actuated, they are able to control one of the fundamental CPW modes (either even or odd), leaving the other mode ideally unaffected. Some minor effects such as a small even-mode parasitic

Parameter	Ohmic-cantilever straight/semicircle	Capacitive, clamped/ folded	Ohmic-series	Ohmic-parallel (SAB)
Figure	10(a)/10(b)	11(a)/11(b)	13(a)	14(a)
Supporting beam radius (μm)	–/16	—	—	—
Meander length, *a* (μm)	—	–/30	—	30
Supporting beam width (μm)	16	–/10	—	10
Supporting beam length, *b* (μm)	40	–/75	—	75
Supporting beam length, *e* (μm)	—	–/45	—	77.5
Membrane width, *c* (μm)	90	90	100	90
Window width, *d* (μm)	—	—	—	60
Membrane length, *f* (μm)	170	580/230	580	165
Bottom electrodes area (μm^2)	13,500	2 × 22,800/2 × 7650	2 × 22,800	2 × 7650
Contact area (μm^2)	3 × 10 × 30	100 × 90*/50 × 90*	2 × 10 × 90	2 × 10 × 12
Spring constant, *keff* (N/m)‡	2.3/1.6	104.5/32.5	104.5	15.4
Pull-in voltage, *Vpull_in* (V)†	10.6/8.8	60/14	49	14
Resonant frequency (KHz)	8‡/6.6‡	16.25 ‡/25.3‡	16.25†,‡	17.1‡

†Measured.

‡Calculated using the mechanical analysis of Section 4.

*Floating metal area.

Table 1. Dimensions and mechanical parameters of the switches.

capacitance (in case of the ohmic-contact parallel switch discussed in Section 5.2.3) occur but with a very limited impact on the circuit behavior.

5.2.1. Capacitive-contact parallel switch

Figure 11 shows two fabricated capacitive-contact parallel switches. The change between ON and OFF states is performed by moving the suspended membrane, which can be actuated through bias pads connected to two symmetrical polysilicon electrodes placed under the membrane. A floating metal (FLOMET) strip is placed on top of the dielectric under the membrane. The overlapping area between FLOMET and a multi-metal layer under the bridge (CPW center conductor) defines a MIM capacitor, as described in Section 3. When the actuation voltage is equal or higher than the pull-in voltage V_{pull_in}, the bridge collapses on the FLOMET, producing a reproducible, constant-value down-state capacitance. The resistance of the ohmic contact between the membrane and the FLOMET strip fixes the amount of switch

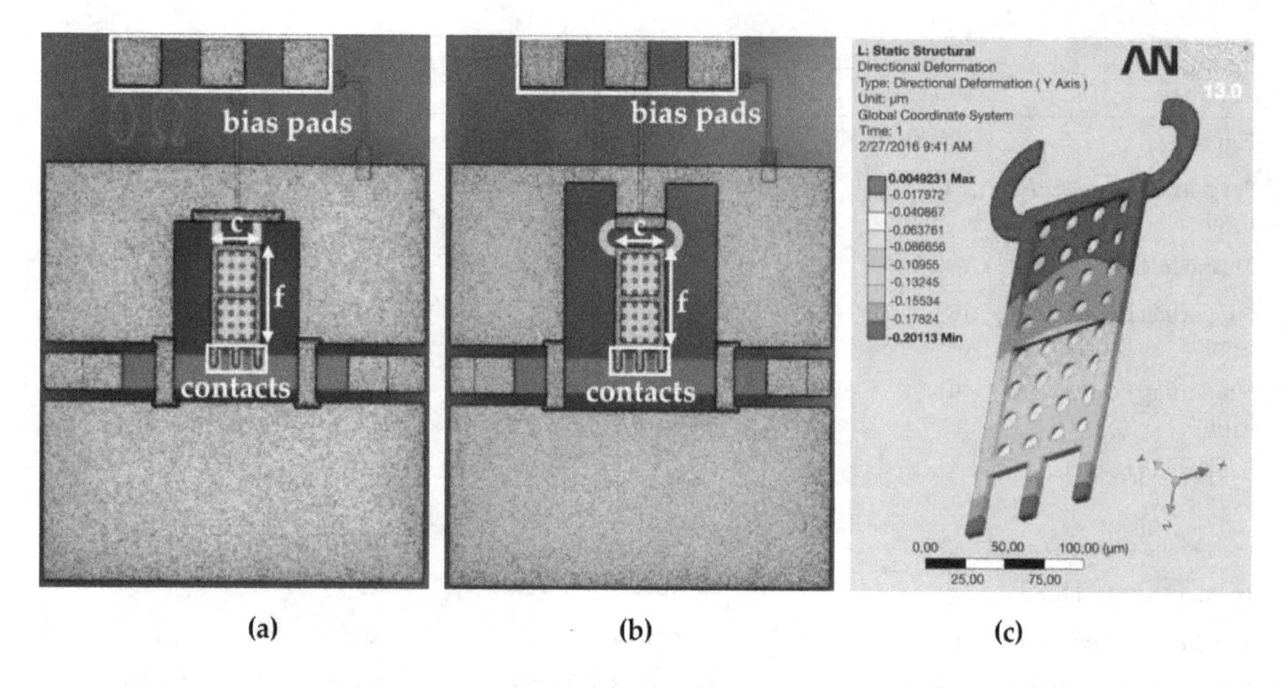

Figure 10. Cantilever-type ohmic-contact switch. (a) Straight-shaped suspension. (b) Semi-circular-shaped suspension. (c) Simulation of initial deformation due to residual stress on the semi-circular suspension device.

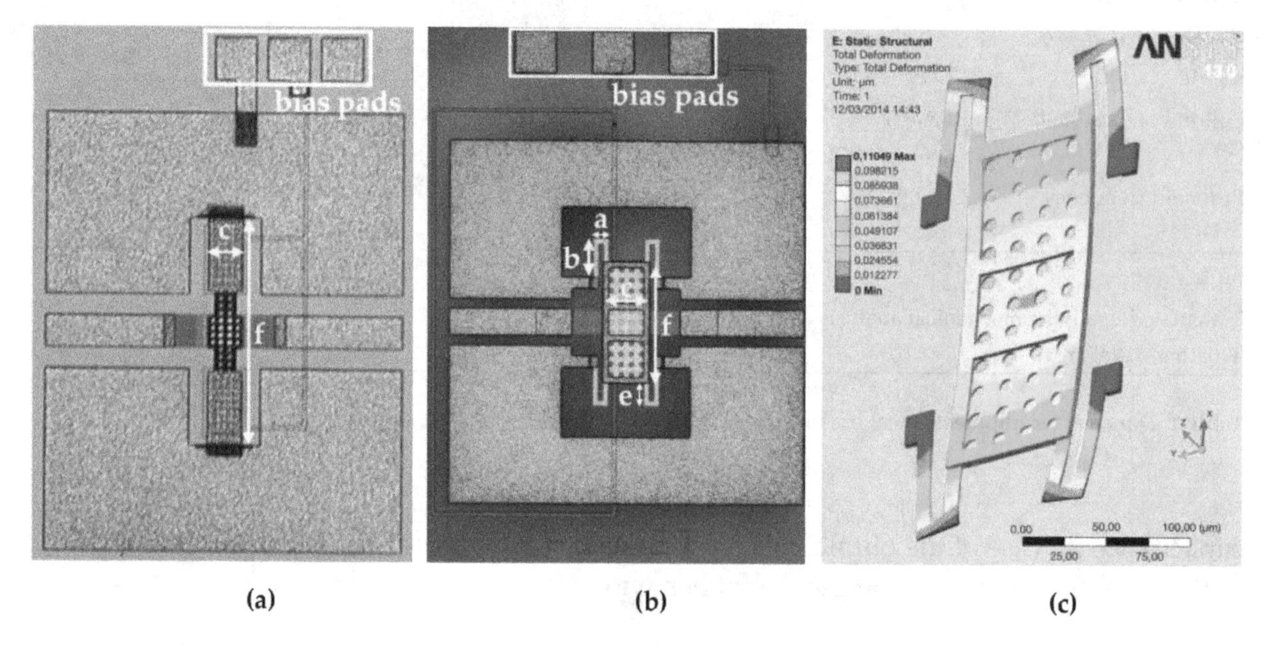

Figure 11. Capacitive-contact parallel switch. (a) Fabricated switch using a clamped-clamped membrane. (b) Fabricated switch using a folded-beam suspension. (c) for a device with a folded-beam suspension: Simulation of initial deformation due to residual stress.

RF insertion loss. The MIM capacitor combined with the membrane inductance defines an RLC circuit in the down state. The RF equivalent circuit details are given in Section 6.

The capacitive switch shown in **Figure 11(a)** uses a clamped-clamped membrane suspension. The device was fabricated on a quartz substrate with a thickness h = 500 µm and an air gap g_i = 2.7 µm. The area of each actuation electrode is 120 × 190 µm². The MIM capacitor area is 90 × 100 µm², which gives a capacitance of 3.8 pF. The capacitive switch of **Figure 11(b)** uses a folded-beam suspension. It was fabricated on a quartz substrate with a thickness h = 300 µm and an air gap $g_i = 1.6$ µm. Each actuation electrode has an area of 90 × 85 µm². The MIM capacitor area is 90 × 50 µm², which gives a capacitance of 1.5 pF (without taking into account the fringing fields) calculated, as with the previous switch, using 2.5-D planar-simulation software. As with the cantilever switch shown in **Figure 10(b)**, the deformation of the bridge due to stress gradients was simulated using ANSYS® 3D FEA. As shown in **Figure 11(c)**, the maximum initial deflection is 0.11 µm. The simulated stiffness constant is $k_{eff} = 32.5$ N/m, and the calculated pull-in voltage is $V_{pull\_in} = 15.6$ V. **Figure 12** shows the measured hysteresis of the switch. The measured pull-in voltage when the isolation is higher than 12 dB at 10 GHz is $V_{pull\_in} = 14$ V.

5.2.2. *Ohmic-contact series switch*

A photograph of an ohmic-contact series switch is shown in **Figure 13(a)**. It was fabricated on a quartz substrate with a thickness $h = 500$ µm and an air gap $g_i = 2.7$ µm The CPW line does not have continuity under the membrane. When the membrane is in its up state, the switch is OFF, while when the membrane is in a down state, the switch is ON, since the metallic membrane puts into contact the two sides of the CPW line. Two electrodes are placed under the membrane at both sides to generate the actuation force. The area of each actuation electrode is 120 × 190 µm². As with the previous switches, the bias pads are isolated from the membrane using thin high-resistivity silicon bias lines.

For this switch, the switching and release times were key parameters in radiometric applications, as discussed in Section 7. The measured switching and release times are 100 µs and 15 µs, respectively. To more accurately assess the switch behavior after the membrane release (evolving from ON state to OFF state), the energy model discussed in Section 4 was applied as follows. The evolution in time of the switch transmission coefficient magnitude $|S_{21}(t)|$ (isolation) at a given RF frequency ($f = 3$ GHz) was measured after removing the applied actuation voltage. Then, from the energy equation (Eq. (2)), which can be solved numerically, the evolution of the capacitance $C(y_r)$ was calculated. Next, using the equivalent circuit for this particular switch design (**Figure 18(a)**), the switch transmission coefficient magnitude as a function

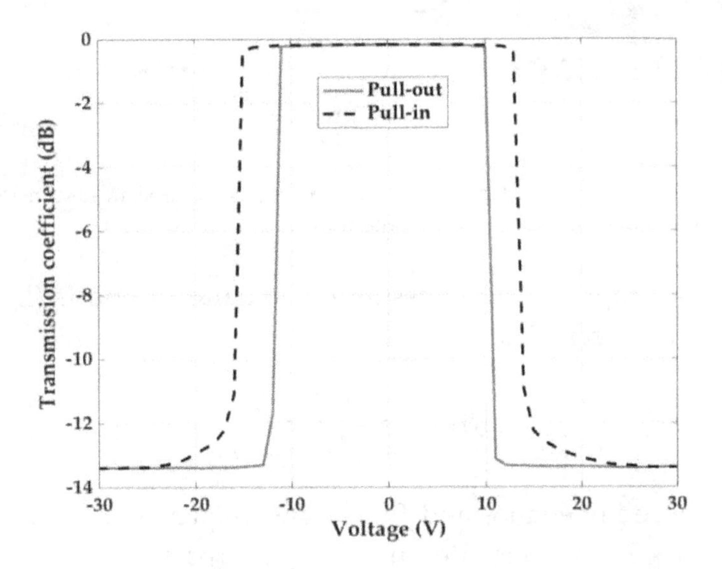

Figure 12. Measurement of hysteresis of the switch with a folded suspension (**Figure 11(b)**) showing pull-in and pull-out traces.

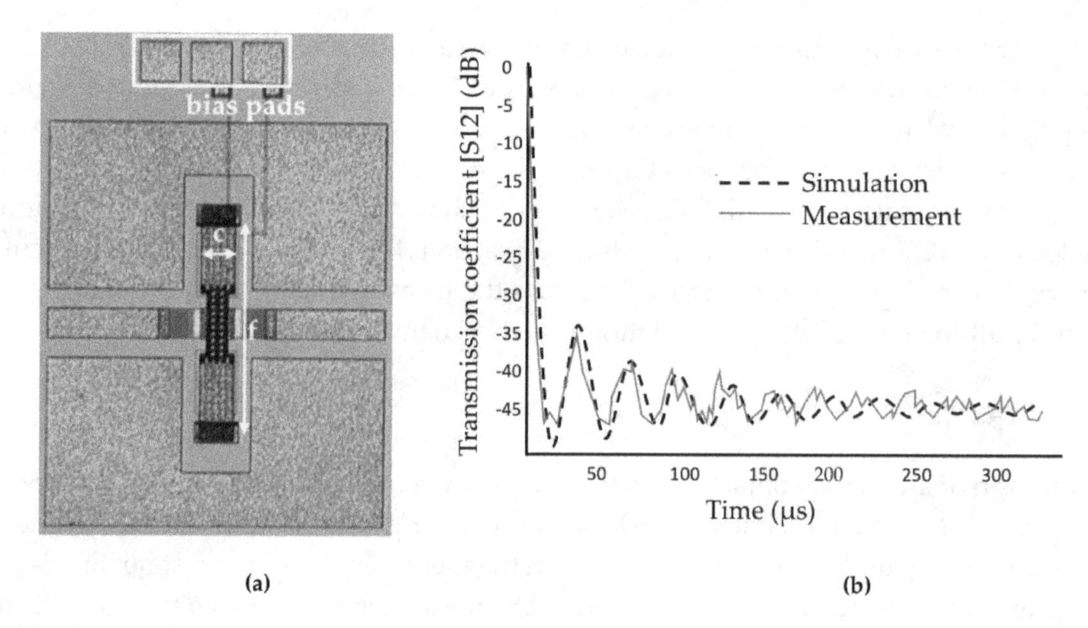

Figure 13. Ohmic-contact series switch. (a) Fabricated switch. (b) Measured and simulated time evolution of the microwave isolation after release (switch going from ON state to OFF state).

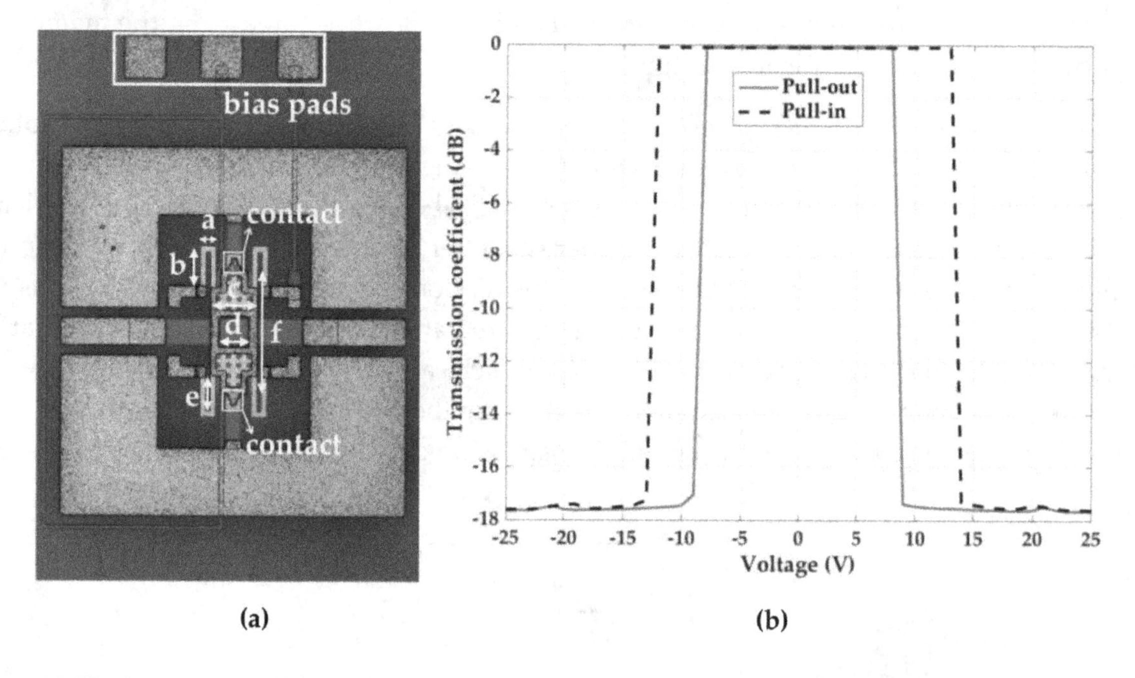

Figure 14. Ohmic-contact parallel switch (switchable air bridge) using a folded-beam suspension. (a) Fabricated switch. (b) Measurement of hysteresis showing pull-in and pull-out traces.

of frequency $|S_{21}(f)|$ was computed which, assuming no parasitic effects ($L_p = 0$ and $C_p = 0$) and no inner line sections, is expressed as

$$|S_{21}(f)| = \frac{4\pi Z_0 fC(y_r)}{\sqrt{1 + (4\pi Z_0 fC(y_r))^2}} \tag{4}$$

where Z_0 is the reference impedance and f is the RF frequency. **Figure 13(b)** compares the simulated evolution of $|S_{21}(t)|$ calculated using Eq. (4) (particularized at $f = 3$ GHz) to the measured $|S_{21}(t)|$ for the series ohmic-contact switch shown in **Figure 13(a)**, showing a good agreement.

5.2.3. Ohmic-contact parallel switch (switchable air bridge)

Figure 14(a) shows a switchable air bridge (SAB) that can be used in CPW reconfigurable multimodal circuits for a selective use of the CPW odd mode. The device was fabricated on a quartz substrate with a thickness h = 300 μm and an air gap g_i = 1.6 μm. The structure features a reinforced gold membrane with two ohmic contacts at the edges. The bridge membrane is anchored to isolated islands by folded-beam suspensions, made of a 1.8-μm-thick, 10-μm-wide single gold layer. When an actuation voltage equal or higher than the pull-in voltage V_{pull_in} is applied, the bridge collapses over an underpass metal layer connected to the ground planes of the CPW. The SAB has an air gap of 1.6 μm, and the distance between the membrane and the bottom ohmic contacts over the underpass metal layer is 1.3 μm. The measured hysteresis of the ohmic switch is shown in **Figure 14(b)**. The pull-in voltage was measured when the isolation is higher than 17 dB at 10 GHz. The measured pull-in voltage is V_{pull_in} = 14 V, and the simulated stiffness constant is k_{eff} = 15.4 N/m. The first-mode mechanical resonant frequency f_{0m} = 17.1 kHz was obtained from the electromechanical analysis (ANSYS®).

The air gap of ohmic contacts placed on the top and bottom electrodes could be affected by stress gradients during fabrication. As with the previous switches, the deformation of the bridge was simulated using ANSYS® 3D FEA. The structure can handle positive- and negative-stress gradients without compromising on the function of the switch. **Figure 15(a)** shows the deformation of the bilayer membrane. For this case, the simulated maximum initial deflection is smaller than 0.16 μm.

The measured topography of the device just after fabrication (**Figure 15(b)**) shows a very good agreement with the 3D FEA results, thus validating this analysis. This model can then be used to extract the nonlinear stiffness values that in turn can modify the potential energy curve shown in **Figure 9**.

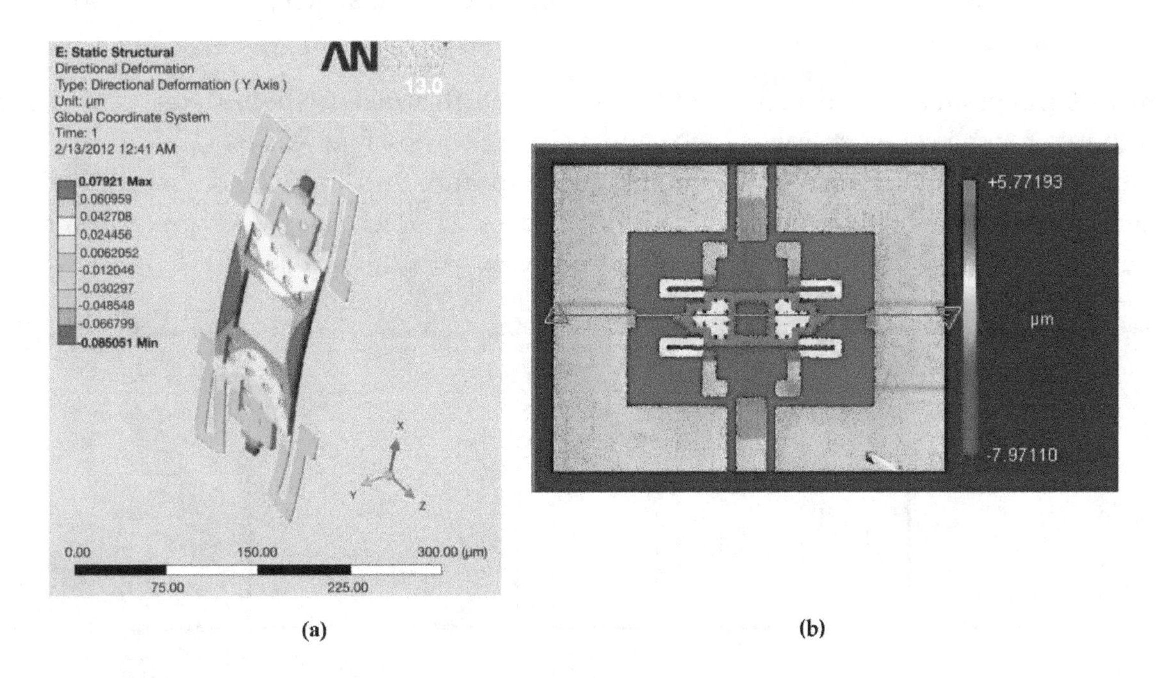

Figure 15. (a) Initial deformation due to residual stress. (b) Measured topography of the device just after fabrication.

Table 1 shows the dimensions and the main mechanical parameters of the switches presented in this section. The membrane/cantilever height is 1.6 μm in all cases except for the capacitive-contact parallel switch with the clamped-clamped membrane and the ohmic-contact series switch, both featuring a membrane height of 2.7 μm. While the pull-in voltages are measured, the spring constant and the mechanical resonant frequency are computed using the 3D FEA method presented in Section 4. For the ohmic-contact series switch, the mechanical resonant frequency was also measured using the method reported in [56].

6. RF-MEMS switch electrical modeling and characterization

6.1. Parallel and series switch models

6.1.1. Capacitive-contact parallel switch

The configuration of a capacitive-contact parallel switch is shown in **Figure 11**. It is able to control the propagation of the CPW even mode and simultaneously suppress the CPW odd mode because the lateral metal planes are permanently connected through the switch membrane. When the switch is in its "up" (OFF) state, the capacitance between the elevated membrane and the RF line underneath (C_{OFF}) is very small. When a bias voltage greater than the pull-in voltage is applied between the membrane and the lower electrodes, the membrane moves down and gets in contact with the floating metal deposited on top of the RF line, implementing the switch "down" (ON) state. In this case, the switch capacitance (C_{ON}) is that of the parallel-plate capacitor defined between the floating metal pad and the underneath RF line section, across a thin oxide layer. The thinner the oxide layer and the larger the overlapping area between the floating metal and the RF line, the larger is the ON capacitance C_{ON}. Thus, the device acts as a switched capacitance. For a highly efficient switch, it is desired to have a high capacitance ratio C_{ON}/C_{OFF}.

Figure 16 shows an equivalent circuit of the switch, which applies to both states. Capacitance C accounts for the switched capacitance and adopts two possible values, C_{ON} and C_{OFF}. The resistance R_C and inductance L_C are membrane resistance and inductance, respectively, at either side of the capacitive contact. The RF frequency at which the switch presents a maximum insertion loss in its ON state is given by the approximate expression $f_0 = (1/\pi) \cdot \sqrt{2 \cdot L_C \cdot C_{ON}}$.

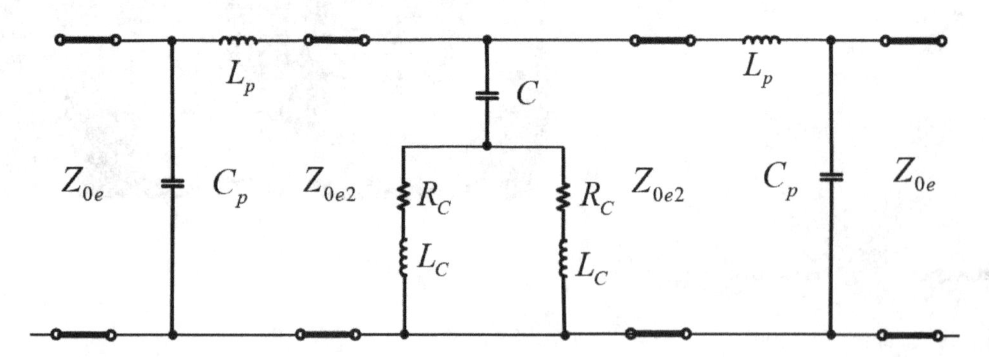

Figure 16. Capacitive-contact parallel-switch equivalent circuit.

A proper combination of the switch ON capacitance C_{ON} and inductance L_C is selected to yield the desired operating frequency f_0 and membrane dimensions for a given fabrication technology. Short even-mode CPW transmission line sections are considered at each side of the membrane, with characteristic impedances Z_{0e} and Z_{0e2} which depend on the CPW line and slot dimensions. The inductance L_p and capacitance C_p are modeling the small parasitic effects due to changes in the width of the CPW central strip and lateral ground planes and can be obtained by adjusting their values to fit a 2.5 D electromagnetic simulation of the CPW structure. The CPW propagation constant and characteristic impedance are also obtained from the electromagnetic simulation of a straight CPW line section of the same dimensions excited with a CPW even mode.

6.1.2. Parallel and series ohmic-contact switches

The configuration of an ohmic-contact parallel switch (switchable air bridge or SAB) is shown in **Figure 14**. This switch is used to efficiently control the CPW odd-mode propagation. When the switch is in its "up" (OFF) state, the capacitance between the elevated membrane and the two RF ohmic contacts at either lateral metal plane is negligible (C=C_{OFF} < 1 fF) because the overlap area for contacts is extremely small. When a bias voltage greater than the pull-in voltage is applied between the membrane and the lower electrodes, the membrane moves down and the two edges get in contact with the lateral metal planes, implementing a "bridge" which performs a double ohmic contact (switch "down"—ON—state). The membrane has a window in the center part to reduce the capacitance between the membrane and the central CPW strip, in such a way to minimize the impact on the propagation of the CPW even mode. This capacitance, obtained from measurement, is 47 fF.

Figure 17 shows an equivalent circuit of the switch for the OFF state (**Figure 17(a)**) and ON state (**Figure 17(b)**). The same parasitic inductance L_p and capacitance C_p as in the circuit of **Figure 16** can be observed, but will certainly have different values because they are now modeling the CPW odd mode. The CPW line sections now refer to the CPW odd mode with

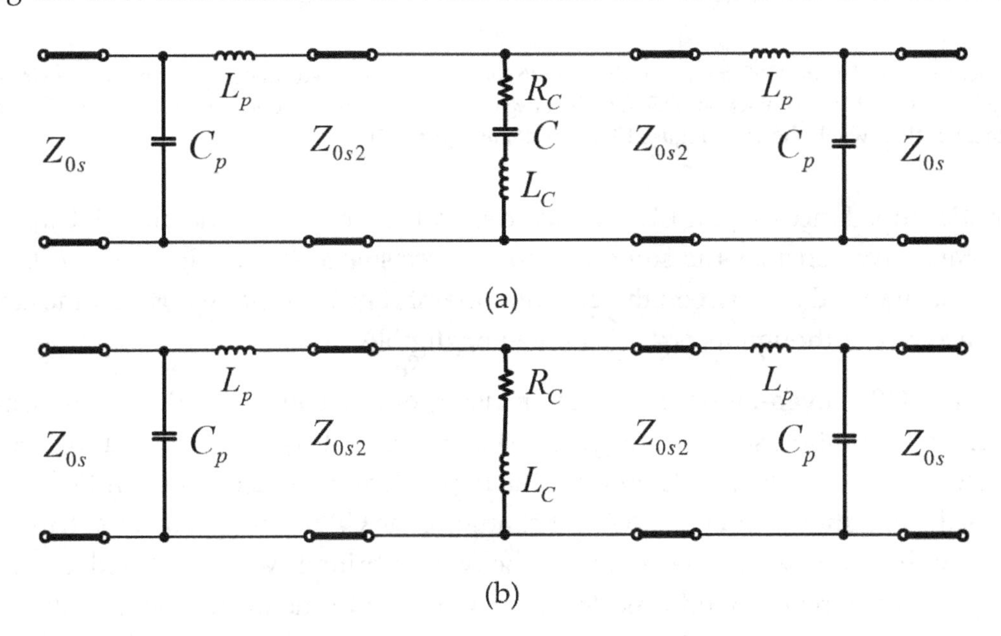

Figure 17. Ohmic-contact parallel-switch equivalent circuit. (a) OFF. (b) ON.

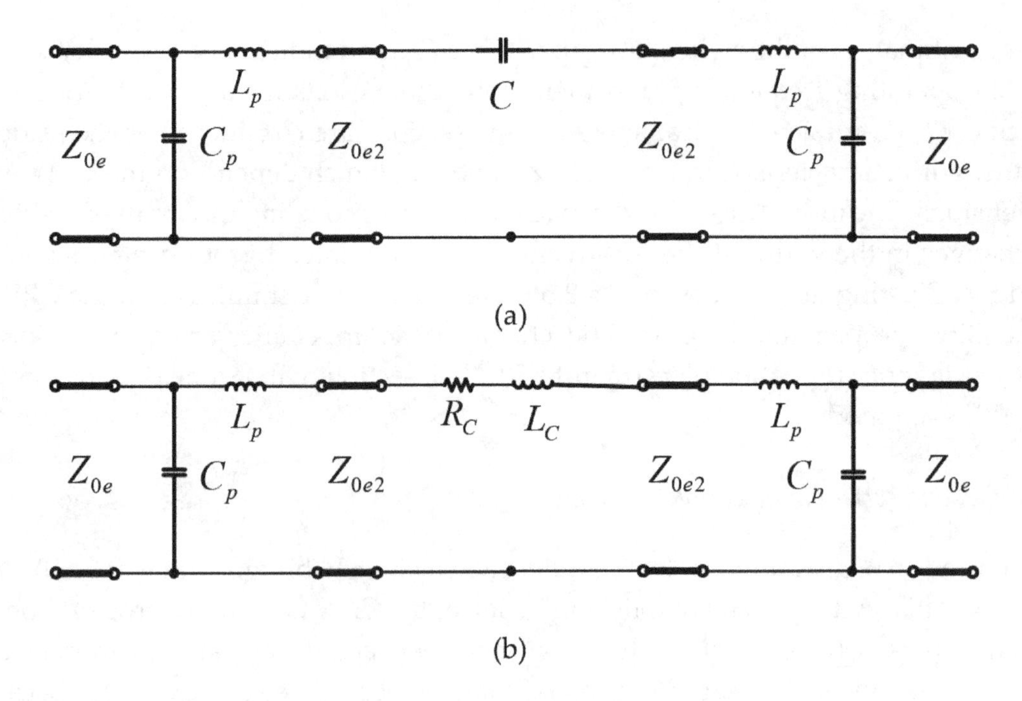

Figure 18. Ohmic-contact series-switch equivalent circuit. (a) OFF. (b) ON.

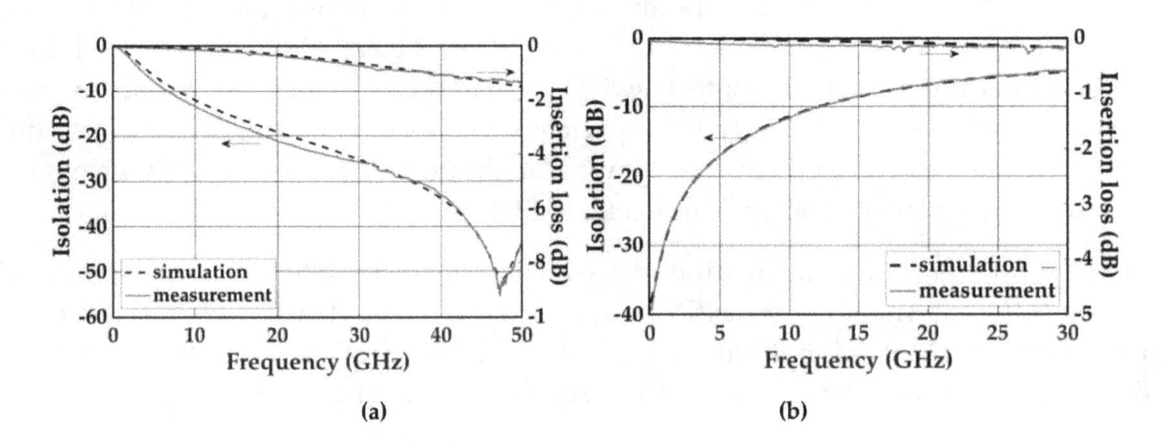

Figure 19. Measured isolation (OFF) and insertion loss (ON) of the fabricated switches compared to simulations of circuit models. (a) Capacitive-contact parallel switch shown in **Figure 11(b)** (model in **Figure 16**). (b) Cantilever-type ohmic-contact parallel switch shown in **Figure 10(a)** (model in **Figure 17**).

characteristic impedances Z_{0s} and Z_{0s2} which depend on the CPW line and slot dimensions. For both states, we consider the same membrane resistance R_C and inductance L_C. In OFF state, the capacitance C_{OFF} between the elevated membrane and the two RF ohmic contacts is taken into account, although its effect is almost negligible.

To control the CPW even-mode propagation using ohmic-contact switches, the cantilever-type switch configuration shown in **Figure 10** can be used. It is observed that the odd mode is suppressed using air bridges. Therefore, the equivalent circuit for this kind of switches is the same as the one shown in **Figure 17**, but changing the CPW line characteristic impedances Z_{0s} and Z_{0s2} with Z_{0e} and Z_{0e2}, respectively. If the two air bridges were removed, the CPW line is able to propagate the CPW odd mode simultaneously to the even mode. In this case, the cantilever-type switch configuration can be used to generate the CPW odd mode whenever the switch is in its ON state. This structure is favorably used in reconfigurable multimodal filters, as explained in Section 7.2.

The CPW even mode can be controlled without suppressing the odd mode by using the bridge-type ohmic-contact series switch configuration shown in **Figure 13**. In this case, the equivalent circuit for the even mode is shown in **Figure 18**. In the OFF state, the switch behaves as a small series capacitance ($C = C_{OFF} \approx 6$ fF), and it can be described as a series *R-L* circuit in the ON state.

6.2. Experimental characterization

Figures 19 and **20** compare the measured to simulated results of the switches presented in Section 5 and modeled in Section 6.1. The switch insertion loss (ON state) and isolation (OFF state) are plotted as a function of frequency.

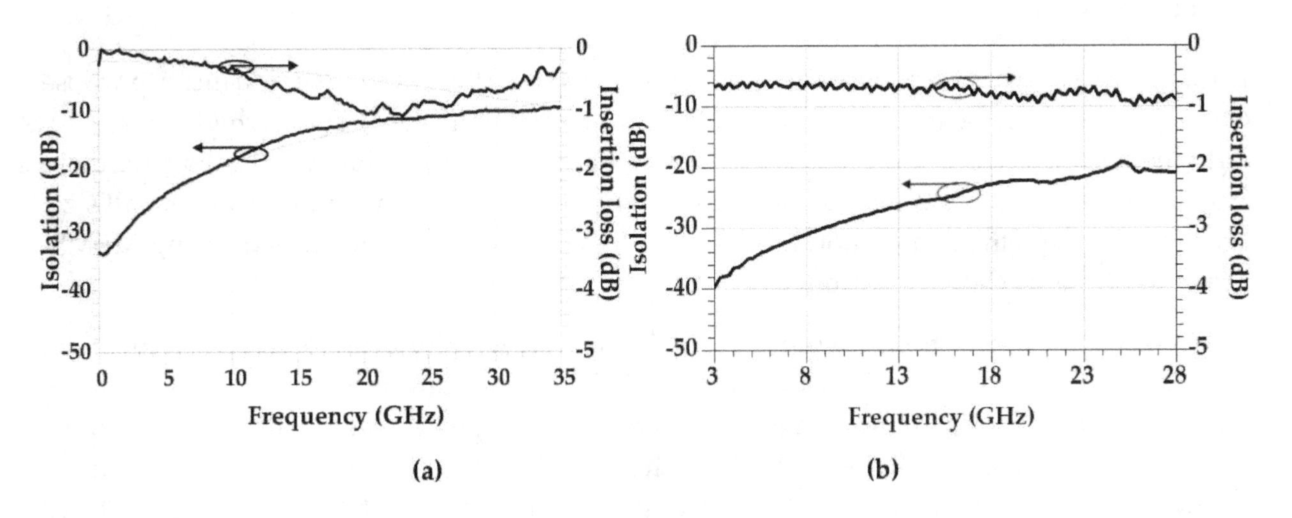

Figure 20. (a) Measured isolation (OFF) and insertion loss (ON) of the fabricated ohmic-contact parallel switch (SAB) shown in **Figure 14(a)** compared to simulations using the circuit model shown in **Figure 17**. (b) Measured isolation (OFF) and insertion loss (ON) of the fabricated bridge-type ohmic-contact series-switch shown in **Figure 13(a)**.

Parameter	Capacitive: direct clamp/folded		Ohmic: parallel		Ohmic: cantilever		Ohmic: series	
	OFF	ON	OFF	ON	OFF	ON	OFF	ON
L_p (nH)	0	0	0	0	0.02	0.02	0	0
C_p (fF)	8.5	8.5	3	3	0	0	6	6
L_C(pH)	28 / 7	28 / 7	180	180	100	100	0	230
C (fF)	1 / 30.5	3800 / 1530	0.9	–	5.1	–	3.35	–
R_C (Ω)	0.27 / 0.1	0.27 / 0.1	0.3	3.2	0	1	0	2.4
Z_{0x} (Ω)*	76.1 / 78.6	76.1 / 78.6	100	100	45.3	45.3	97	97
Z_{0x2}(Ω)*	118.7 / 148.5	118.7 / 148.5	195	195	63.5	63.5	152	152

*x = e (even) or o (odd) according to **Figures 16–18**.

Table 2. Equivalent circuit elements of **Figures 16–18** (switches **Figures 10, 11, 13,** and **14**).

In **Table 2**, the values of the equivalent circuit elements obtained to fit measurement are listed. The capacitive switch features a high capacitance ratio (C_{ON}/C_{OFF} = 50.2). The ohmic switches feature low insertion loss (<1 dB) and high isolation (>20 dB) for f < 10 GHz (parallel switch) and for f < 25 GHz (series switch).

7. Design of multimodal reconfigurable microwave and mm-wave circuits

This section presents two applications of the RF-MEMS presented in Section 5 when combined with the multimodal circuits presented in Section 2: 180° phase switches and bandpass filters. All of them were fabricated using the process described in Section 3.

7.1. Phase switches

A 180° phase switch is a circuit that shifts the phase between 0 and 180°. This is a multi-purpose element required in the microwave and millimeter-wave applications, such as high-sensitivity radiometers or electronic beam steering in phased-array antennas. The specifications for these systems are, in most cases, a broadband operation, a low power consumption, and small size. They can be implemented monolithically using HBT/HEMT-based MMICs or using MEMS-based solutions as shown subsequently.

Figure 21(a) shows a compact, uniplanar 180° phase switch fabricated on a quartz substrate [9]. It is based on two different back-to-back (BTB) CPW-to-slotline transitions [39] (symmetric and antisymmetric transitions, respectively), creating two phase paths with a relative transmission phase shift between them of 180°. Each path is selected using two single-pole-double-throw (SPDT) switches. The SPDT consists of two ohmic-contact MEMS series switches

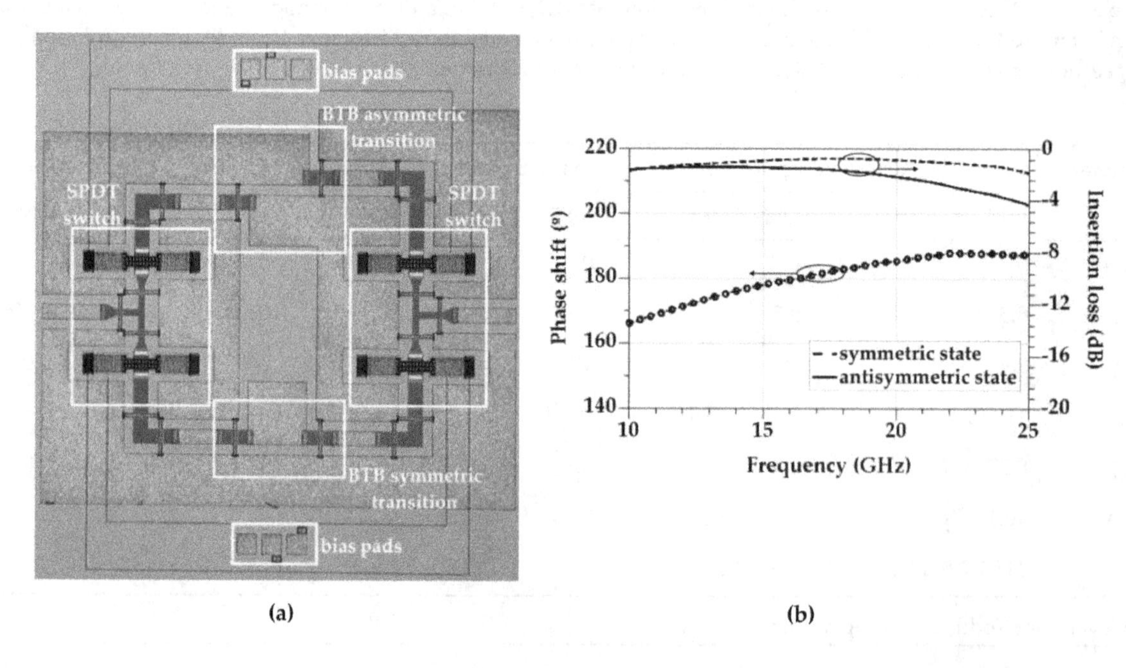

Figure 21. (a) Manufactured 180° phase switch. (b) Measured insertion loss (symmetric and antisymmetric paths) and measured phase shift between the two paths.

(described in Section 5.2.2) and a power divider. **Figure 21(b)** shows the measured performance of the circuit, featuring a 180° ± 5° phase shift between both states in a bandwidth of 35% (14–20 GHz), with an insertion loss smaller than 2 dB in both paths.

Figure 22(a) shows the photograph of a second compact, uniplanar 180° phase switch fabricated on a quartz substrate [10]. In this case, it is based on an air-bridged CPW cross [57]. The two CPW arms of the cross are loaded with capacitive-contact MEMS switches (described in Section 5.2.1). The two phase-switch states (0°/180°) are obtained by actuating the MEMS switches in opposite states (ON/OFF and OFF/ON), resulting in a multimodal interaction between the two CPW modes (even and odd) at the air-bridged cross. The CPW-to-slotline transition [39] at the input port and the CPW taper at the output port are included in order to enable the measurement of the circuit with a probe station. **Figure 22(b)** shows the measured results of phase shift between both states and insertion loss, featuring 180° +1.8°/−1° and ± 0.1 dB insertion-loss unbalance, respectively, in a very wide bandwidth (5–25 GHz). The measured insertion loss is better than 2 dB in 10–20-GHz frequency band.

7.2. Uniplanar bandpass filters

Figure 23(a) shows a second-order bandwidth-reconfigurable bandpass filter, which was fabricated on a quartz substrate (h = 500 μm) [11]. The $\lambda_0/2$ slotline resonators are coupled by a slotline short-circuit (K_{12}). The filter features multimodal immittance inverters (MIIs) based on CPW-to-slotline transitions [37, 58] which are embedded in the input and output slotline resonators. As shown in **Figure 23(b)**, two cantilever-type ohmic-contact MEMS switches (described in Section 5.1) are used to enable reconfigurable MIIs. When actuated, the switches modify the input and output coupling of the filter (K_{01} and K_{23}), resulting in a change in the filter's fractional bandwidth (FBW). To keep the central frequency constant, another cantilever MEMS switch (also shown in **Figure 23(b)**) is integrated in K_{12} and actuated simultaneously. **Figure 23(c)** shows the filter measured results. It features two FBW states of 0.082 (when the inner switches are actuated) and 0.043 (when the outer switches and the impedance inverter switch are actuated), while maintaining a constant center frequency (18.9 GHz).

Figure 24(a) shows a second-order bandpass filter fabricated on a 5-KΩ-cm high-resistivity silicon substrate (ε_r = 11.9, h = 200 μm), which uses switchable air bridges or SABs [6] (shown in **Figure 24(b)**) similar to the SAB of **Figure 14(a)**. The SABs are described in Section 5.2.3. Like

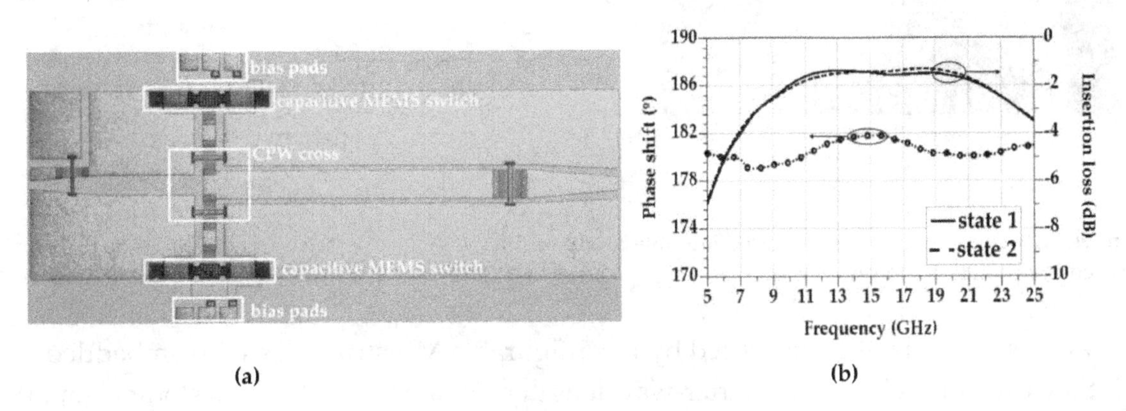

Figure 22. (a) Manufactured 180° phase switch. (b) Measured insertion loss for the two states and phase shift between the two states.

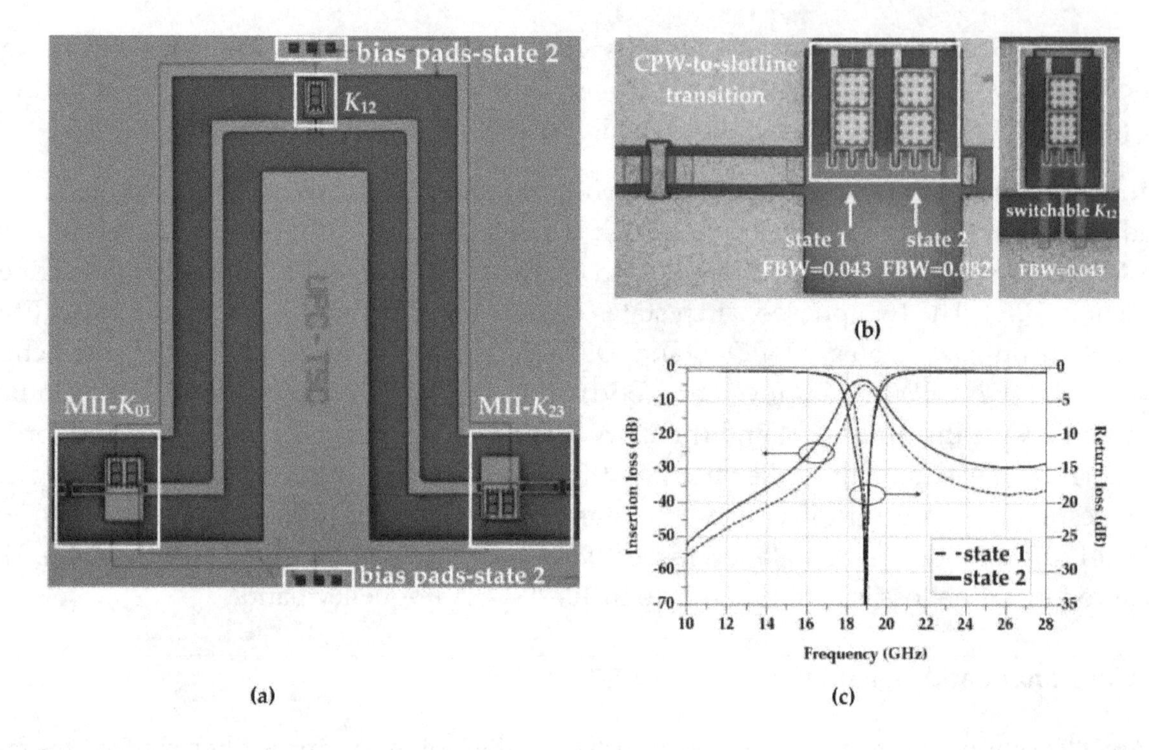

Figure 23. (a) Second-order bandwidth-reconfigurable bandpass filter. (b) Detail of reconfigurable MIIs and reconfigurable inductive coupling K_{12}. (c) Filter measured insertion loss and return loss.

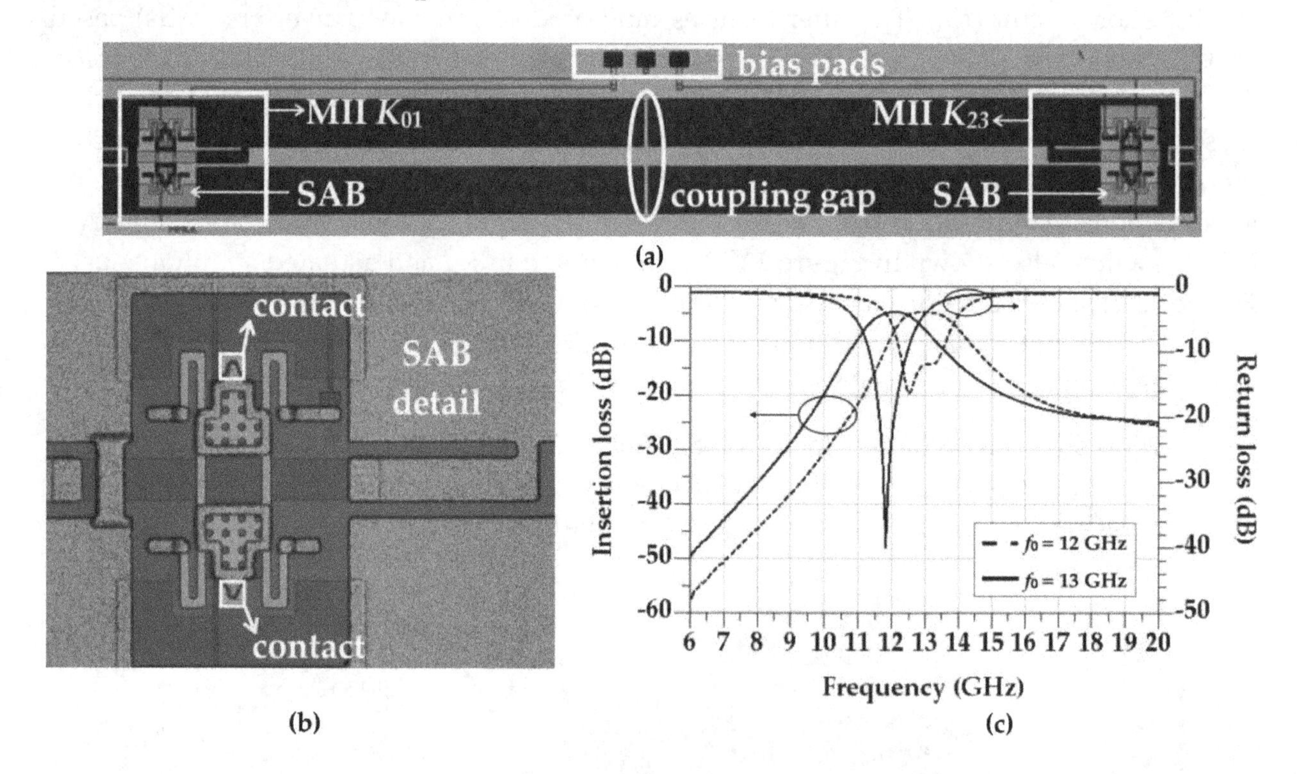

Figure 24. (a) Second-order frequency-reconfigurable bandpass filter. (b) Detail of reconfigurable MIIs and SAB. (c) Filter measured insertion loss and return loss.

the previous example, the filter is fed by reconfigurable MII structures with embedded CPW-to-slotline transitions but uses quarter-wavelength resonators coupled by a slotline gap. When actuated, the SABs reduce the resonators' effective length, which in turn increases the resonance frequency. As a result, the filter center frequency is shifted up. **Figure 24(c)** shows the filter measured results. It features two operating frequencies, f_0 = 12 GHz and f_0 = 13 GHz, with a constant fractional bandwidth (FBW) of 14%. The measured filter IL is 4.6 dB in both states.

8. Conclusion

In this chapter, the design considerations of RF-MEMS switches aimed to be integrated in uniplanar multimodal reconfigurable circuits for operation in the microwave and mm-wave bands have been presented. Different configurations of series and parallel switches featuring ohmic and capacitive contact have been discussed in detail, including the analysis of mechanical topologies which minimize the initial deformation due to residual stress. The fabrication process has been described, which provides the required flexibility to integrate the MEMS switches into higher-level RF communication systems. The switch RF behavior has been modeled using suitable equivalent circuits for both (ON/OFF) switch states. The models can easily be embedded into complex multimodal environments which enable multipurpose, compact designs. The fabricated switches have been experimentally characterized in terms of hysteresis, RF isolation (OFF state), and RF insertion loss (ON state), demonstrating an excellent behavior. As practical examples of the application into communication systems, some of the proposed switches have successfully been integrated into 0°/180° phase switches and into reconfigurable filters with either center frequency or fractional-bandwidth reconfiguration, showing excellent performances.

Acknowledgements

This work was supported by the Spanish MEC under Projects TEC2013-48102-C2-1/2-P, TEC2016-78028-C3-1-P and grant BES-2011-051305, the Mexican CONACYT under fellowship 410742, and the Unidad de Excelencia Maria de Maeztu MDM-2016-0600 financed by the AEI, Spain.

Author details

Lluis Pradell[1*], David Girbau[2], Miquel Ribó[3], Jasmina Casals-Terré[4], Antonio Lázaro[2], Adrián Contreras[1], Marco Antonio Llamas[1], Julio Heredia[1], Flavio Giacomozzi[5] and Benno Margesin[5]

*Address all correspondence to: pradell@tsc.upc.edu

1 Department of TSC, Technical University of Catalonia (UPC), Barcelona, Catalonia, Spain

2 Department of EEEA, Rovira i Virgili University, Tarragona, Catalonia, Spain

3 Department of ET, La Salle-Ramon Llull University, Barcelona, Catalonia, Spain

4 Department of EM, UPC, Barcelona, Catalonia, Spain

5 Fondazione Bruno Kessler, Trento, Italy

References

[1] Bakri-Kassem M, Mansour RR. High power latching RF MEMS switches. IEEE Transaction on Microwave Theory and Techniques. 2015:222-232. DOI: 10.1109/TMTT.2014.2376932

[2] Girbau D, Pradell L, Lázaro A, Nebot A. Electrothermally-actuated RF-MEMS switches suspended on a low-resistivity substrate. IEEE/ASME Journal of Microelectromechanical Systems. 2007:1061-1070. DOI: 10.1109/JMEMS.2007.904744

[3] Mennai A, Bessaudou A, Cosset F, Guines C, Passerieux D, Blondy P, Crunteanu A. High Cut-off Frequency RF Switches integrating a Metal-Insulator Transition Material. In: IEEE MTT-S Int. Microwave Symposium; May 2015; Phoenix. New York: IEEE. pp. 1-3

[4] DiNardo S, Farinelli P, Giacomozzi F, Mannocchi G, Marcelli R, Margesin B, Mezzanotte P, Mulloni V, Russer P, Sorrentino R, Vitulli F, Vietzorreck L. Broadband RF-MEMS based SPDT. In: 36th European Microwave Conference; September 2006; Manchester. New York: IEEE. pp. 1727-1730

[5] Ocera A, Farinelli P, Cherubini F, Mezzanotte P, Sorrentino R, Margesin B, Giacomozzi F. A MEMS-Reconfigurable Power Divider on High Resistivity Silicon Substrate. In: IEEE/MTT-S International Microwave Symposium; May 2007; Honolulu. pp. 501-504

[6] Contreras A, Casals-Terré J, Pradell L, Giacomozzi F, Iannacci J, Ribó M. A Ku-band RF-MEMS frequency-reconfigurable multimodal bandpass filter. International Journal of Microwave and Wireless Technologies. 2014:277-285. DOI: 10.1017/S1759078714000567

[7] Cazzorla A, Sorrentino R, Farinelli P. Double-actuation extended tuning range RF MEMS Varactor. In: 45th European Microwave Conference; September 2015; Paris. pp. 937-940

[8] Vähä-Heikkilä T, Varis J, Tuovinen J, Rebeiz GM. A reconfigurable 6-20 GHz RF MEMS impedance tuner. In: IEEE MTT-S International Microwave Symposium; June 2014; Fort Worth. New York: IEEE. pp. 729-732

[9] Llamas MA, Girbau D, Ribó M, Pradell L, Lázaro A, Giacomozzi F, Margesin B. MEMS-based 180° phase switch for differential radiometers. IEEE Transaction on Microwave Theory and Techniques. 2010:1264-1272. DOI: 10.1109/TMTT.2010.2045558

[10] Llamas MA, Girbau D, Ribó M, Pradell L, Giacomozzi F, Colpo S. RF-MEMS uniplanar 180° phase switch based on a multimodal air-bridged CPW cross. IEEE Transaction on Microwave Theory and Techniques. 2011:1769-1777. DOI: 10.1109/tmtt.2011.2140125

[11] Contreras A, Ribó M, Pradell L, Casals-Terré J, Giacomozzi F, Iannacci J. K-band RF-MEMS uniplanar reconfigurable-bandwidth bandpass filter using multimodal immittance inverters. Electronics Letters. 2013:704-706. DOI: 10.1049/el.2013.0681

[12] Mansour RR. RF MEMS-CMOS device integration: An overview of the potential for RF researchers. IEEE Microwave Magazine. 2013:39-56. DOI: 10.1109/MMM.2012.2226539

[13] Bakri-Kassem M, Fouladi S, Mansour RR. Novel high-Q MEMS curled-plate variable capacitors fabricated in 0.35-μm CMOS technology. IEEE Transaction on Microwave Theory and Techniques. 2008:530-541. DOI: 10.1109/TMTT.2007.914657

[14] Fouladi S, Mansour RR. Capacitive RF MEMS switches fabricated in standard 0.35-mm CMOS technology. IEEE Transaction on Microwave Theory and Techniques. 2010:478-486. DOI: 10.1109/TMTT.2009.2038446

[15] Riverola M, Uranga A, Torres F, Barniol N, Marigó E, Soundara-Pandian M. A reliable fast miniaturized RF MEMS-on-CMOS switched capacitor with zero-level vacuum package. In: 2017 IEEE MTT-S International Microwave Workshop Series on Advanced Materials and Processes for RF and THz Applications (IMWS-AMP); September 2017. pp. 1-3

[16] Kaynak M, Ehwald KE, Drews J, Scholz R, Korndörfer F, Knoll D, Tillack B, Barth R, Birkholz M, Schulz K, Sun YM, Wolansky D, Leidich S, Kurth S, Gurbuz Y. BEOL embedded RF-MEMS switch for mm-wave applications. In: IEEE MTT-S International Electron Devices Meeting; December 2009; Baltimore/New York: IEEE; p. 1-4

[17] Kaynak M, Wietstruck M, Zhang W, Drews J, Barth R, Knoll D, Korndörfer F, Scholz R, Schulz K, Wipf C, Tillack B, Kaletta K, Suchodoletz MV, Zoschke K, Wilke M, Ehrmann O, Ulusoy AC, Purtova T, Liu G, Schumacher H. Packaged BiCMOS embedded RF-MEMS switches with integrated inductive loads. In: IEEE MTT-S International Microwave Symposium; June 2012; Montreal. New York: IEEE; p. 1-3

[18] Rynkiewicz P, Franc, A-L, Coccetti, F, Tolunay-Wipf, S, Wietstruck M, Kaynak M, Prigent G. Tunable dual-mode ring filter based on BiCMOS embedded MEMS in V-band. In: Asia Pacific Microwave Conference; November 2017; Kuala Lumpur/New York: IEEE; p. 124-127

[19] Rebeiz GM. RF MEMS, Theory, Design and Technology. Hoboken: Wiley; 2003. 483p. DOI: 10.1002/0471225282

[20] Giacomozzi F, Mulloni V, Colpo S, Iannacci J, Margesin B, Faes A. A flexible fabrication process for the fabrication of RF MEMS devices. In: International Semiconductor Conference; Octuber 2011; Sinaia. pp. 155-158

[21] Benoit RR, Barker NS. Superconducting tunable microstrip gap resonators using low stress RF MEMS fabrication process. Journal of the Electron Devices Society. 2017:239-243. DOI: 10.1109/JEDS.2017.2706676

[22] Pelliccia L, Farinelli P. Sorrentino R. High- tunable waveguide filters using Ohmic RF MEMS switches. IEEE Transaction. on Microwave Theory and Techniques. 2015:3381-3390. DOI: 10.1109/TMTT.2015.2459689

[23] Shah U, Reck T, Frid H, Jung-Kubiak C, Chattopadhyay G, Mehdi I, Oberhammer J. A 500-750 GHz RF MEMS waveguide switch. IEEE Transactions on Terahertz Science and Technology. 2017:326-334. DOI: 10.1109/TTHZ.2017.2670259

[24] Tolunay-Wipf S, Göritz A, Wipf C, Wietstruck M, Burak A, Türkmen E, Gürbüz Y, Kaynak M. 240 GHz RF-MEMS Switch in a 0.13 μm SiGe BiCMOS Technology. In: IEEE Bipolar/BiCMOS Circuits and Technology Meeting; October 2017; Miami. pp. 54-57

[25] Nadaud K, Roubeau F, Pothier A, Blondy P, Zhang L-Y, Stefanini R. High Q zero level packaged RF-MEMS switched capacitor arrays. In: 11th European Microwave Integrated Circuits Conference; Octuber 2016; London/New York: IEEE; p. 448-451

[26] Young WC, Budynas RG, Sadegh AM. Roak's Formulas for Stress and Strain. 8th ed. New York: McGraw-Hill; 2012

[27] Wong WC, Azid IA, Majlis BY. Theoretical analysis of stiffness constant and effective mass for a round-folded beam in MEMS accelerometer. Journal of Mechanical Engineering. 2011;**57**:517-525. DOI: 10.5545/sv-jme.2009.151

[28] Fedder GK. Simulation of microelectromechanical systems [thesis]. Berkeley: Department of EECS University of California; 1994

[29] Contreras A, Casals-Terre J, Pradell L, Ribó M, Heredia J, Giacomozzi F, Margesin B. RF-MEMS switches for a full control of the propagating modes in uniplanar microwave circuits and their application to reconfigurable multimodal microwave filters. Microsystem Technologies. 2017:5959-5975. DOI: 10.1007/s00542-017-3379-8

[30] Llamas, MA, Girbau D, Pradell L, Lázaro A, Giacomozzi F, Colpo S. Characterization of Dynamics in RF-MEMS Switches. In: 10th International Symposium. RF MEMS and RF Microsystems (MEMSWAVE); 6-8 July 2009; Trento. pp. 117-120

[31] Fargas-Marques A, Casals-Terre J, Shkel A. Resonant pull-in condition in parallel-plate electrostatic actuator. Journal of Microelectromechanical Systems. 2007:1044-1053. DOI: 10.1109/JMEMS.2007.900893

[32] Casals-Terre J, Fargas-Marques A, Shkel A. Snap-action Bistable micromechanisms actuated by nonlinear resonance. Journal of Microelectromechanical Systems. 2008:1082-1093. DOI: 10.1109/JMEMS.2008.2003054

[33] Elata D, Bamberger H. On the Dynamic Response of Electrostatic MEMS Switches. Journal of Microelectromechanical Systems. 2008:236-243. DOI: 10.1109/JMEMS.2007.908752

[34] Leus V, Elata D. A new efficient method for simulating the dynamic response of electrostatic switches. In: 22nd International Conference on Micro Electro Mechanical Systems (MEMS); January 2009; Sorrento. New York: IEEE; p. 1115-1118

[35] Casals-Terre J, Llamas MA, Girbau Pradell L, Lázaro A, Giacomozzi F, Colpo S. Analytical energy model for the dynamic behavior of RF MEMS switches under increased actuation voltage. Journal of Microelectromechanical Systems. 2014:1428-1439. DOI: 10.1109/JMEMS.2014.23147

[36] Contreras A, Casals-Terré J, Pradell L, Giacomozzi F, Colpo S, Iannacci J, Ribó M. A RF-MEMS switchable CPW air-bridge. In: 7th European Microwave Integrated Circuits Conference; October 2012; Amsterdam/New York: IEEE; p. 441-444

[37] Ribó M, Pradell L. Circuit model for mode conversion in coplanar waveguide asymmetric shunt impedances. Electronics Letters. 1999:713-715. DOI: 10.1049/el:19990507

[38] Ribó M, de la Cruz J, Pradell L. Circuit model for slotline-to-coplanar waveguide asymmetrical transitions. Electronics Letters. 1999;1153-1155. DOI: 10.1049/el:19990779

[39] Ribó M, Pradell L. Circuit model for coplanar-slotline tees. IEEE Microwave Guided Wave Letters. 2000:177-179. DOI: 10.1109/75.850369

[40] Ribó M, Pradell L. Circuit model for a coplanar –Slotline cross. IEEE Microwave Guided Wave Letters. 2000:511-513. DOI: 10.1109/75.895085

[41] Gaddi R, Bellei M. Gnudi A, Margesin B, Giacomozzi F. Low-Loss Ohmic RF-MEMS Switches with Interdigitated Electrode Topology. In: Symposium on design, test, Integration and Packaging of MEMS/MOEMS (DTIP 2004); 2004. pp. 161-166

[42] Farinelli P, Margesin B, Giacomozzi F, Mannocchi G, Catoni S, Marcelli R, Mezzanotte P, Vietzorreck L, Vitulli F, Sorrentino R, Deborgies F. A low contact-resistance winged-bridge RF-MEMS series switch for wide-band applications. Journal of the European Microwave Association. 2007;**3**:268-278

[43] Calaza C, Margesin B, Giacomozzi F, Rangra K, Mulloni V. Electromechanical characterization of low actuation voltage RF MEMS capacitive switches based on DC CV measurements. Microelectronic Engineering. 2007:1358-1362. DOI: 10.1016/j.mee.2007.01.196

[44] Mulloni V, Solazzi F, Resta G, Giacomozzi F, Margesin B. RF-MEMS switch design optimization for long-term reliability. Analog Integrated Circuits and Signal Processing. 2014:323-332. DOI: 10.1007/s10470-013-0220-x

[45] De Angelis G, Lucibello A, Proietti E, Marcelli R, Bartolucci G, Casini F, Farinelli P, Mannocchi G, Di Nardo S, Pochesci D, Margesin B, Giacomozzi F, Vendier O, Kim T, Vietzorreck L. RF MEMS ohmic switches for matrix configurations. International Journal of Microwave and Wireless Technologies. 2012:421-433. DOI: 10.1017/S1759078712000074

[46] Diaferia F, Deborgies F, Di Nardo S, Espana B, Farinelli P, Lucibello A, Marcelli R, Margesin B, Giacomozzi F, Vietzorreck L, Vitulli F. Compact 12×12 switch matrix integrating RF MEMS switches in LTCC hermetic packages. In: 44th IEEE European Microwave Conference; 6-9 October 2014. Rome/New York: IEEE; p. 199-202

[47] Bastioli S, Di Maggio F, Farinelli P, Giacomozzi F, Margesin B, Ocera A, Pomona I, Russo M, Sorrentino R. Design and Manufacturing of a 5-bit MEMS Phase Shifter at K-band. In: European microwave In: Integrated Circuit Conference (EuMIC 2008); 27-30 October 2008. Amsterdam/New York: IEEE; p. 338-341

[48] Bedani M, Carozza F, Gaddi R, Gnudi A, Margesin B, Giacomozzi F. A reconfigurable impedance matching network employing RF-MEMS switches. In: Symposium on Design, Test, Integration and Packaging of MEMS and MOEMS (DTIP); 25-27 April 2007; Stressa

[49] Gaddi R, Gnudi A, Franchi E, Guermandi D, Tortori P, Margesin B, Giacomozzi F. Reconfigurable MEMS-enabled LC-tank for multi-band CMOS oscillator. In: IEEE MTT-S International Microwave Symposium; 17 June 2005; Long Beach. New York: IEEE. pp. 1353-1356

[50] Cazzorla A, Farinelli P, Urbani L, Cacciamani F, Pelliccia L, Sorrentino R, Giacomozzi F, Margesin B. MEMS-based LC tank with extended tuning range for low phase-noise VCO. International Journal of Microwave Wireless Technologies. 2017:249-258. DOI: 10.1017/S1759078715001579

[51] Giacomozzi F, Mulloni V, Colpo S, Faes A, Sordo G, Girardi S. RF-MEMS packaging by using quartz caps and epoxy polymers. Microsystem Technologies. 2014;**12**:1941-1948. DOI: 10.1007/s00542-014-2256-y

[52] Giacomozzi F, Iannacci J. RF MEMS technology for next-generation wireless communications. In: Handbook of Mems for Wireless and Mobile Applications. Woodhead Publishing; 2013. pp. 225-257. DOI: 10.1533/9780857098610.1.225

[53] Alameh AA, Nabki F. A 0.13-μm CMOS dynamically reconfigurable charge pump for electrostatic MEMS actuation. IEEE Transaction on Very Large Scale Integration Systems (VLSI). 2017:1261-1270. DOI: 10.1109/ICECS.2014.7050074

[54] Ismail Y, Lee H, Pamarti S, Yang C-KK. A 34V charge pump in 65nm bulk CMOS technology. In: IEEE International Solid-State Circuits Conference; February 2014; San Francisco. New York: IEEE. pp. 408-409

[55] Sattler R, Plötz F, Fattinger G, Wachutka G. Modeling of an electrostatic torsional actuator: Demonstrated with an RF MEMS switch. Sensors Actuators A: Physical. 2002:337-346. DOI: 10.1016/S0924-4247(01)00852-4

[56] Mercier D, Blondy P, Pothier A. Monitoring mechanical characteristics of MEMS switches with a microwave test bench. In: ESA-ESTEC 4th Round Table on Micro/Nano Technologies for Space; May 2003

[57] Llamas MA, Ribó M, Girbau D, Pradell L. A rigorous multimodal analysis and design procedure of a uniplanar 180° hybrid. IEEE Transaction on Microwave Theory Techniques. June 2009;**16**:1832-1839. DOI: 10.1109/TMTT.2009.2022881

[58] Contreras A, Ribó M, Pradell L, Blondy P. Uniplanar bandpass filters based on multimodal immitance inverters and end-coupled slotline resonators. IEEE Transaction on Microwave Theory Techniques. 2013:77-88. DOI: 10.1109/TMTT.2012.2226743

2

Milliwatt-Level Electromagnetic Induction-Type MEMS Air Turbine Generator

Minami Kaneko, Ken Saito and Fumio Uchikoba

Abstract

In this chapter, an electromagnetic induction-type MEMS air turbine generator that combined with the MEMS technology and the multilayer ceramic technology is proposed. Three types of MEMS air turbine generators that included the different bearing systems, shape of the rotor and shape of the magnetic circuits are discussed to achieve the high output power. In the MEMS air turbine, the purpose is to achieve high-speed rotational motion. As a result of the comparison between the different structures, a rim-type rotor and a miniature ball bearing system showed the high rotational speed than a flat-type rotor and a fluid dynamic bearing system. The maximum rotational speed of the fabricated air turbine was 290,135 rpm. Moreover, it is important to introduce the magnetic flux to the magnetic circuit. By the multilayer ceramic technology, the three-dimensional coil in miniature monolithic structure was fabricated. The magnetic core that was designed to introduce the magnetic flux showed the low magnetic flux loss. The fabricated MEMS air turbine and the multilayer ceramic magnetic circuit were combined, and the miniature electromagnetic induction-type generator was achieved. The output power was 2.41 mVA, when the load resistance and the output voltage were 8 Ω and 139 mV, respectively.

Keywords: electromagnetic induction type, multilayer ceramic technology, air turbine generator, ferrite, milliwatt level

1. Introduction

Micro elector mechanical systems (MEMS) are used for various fields. The most conventional field is a miniature electronics field as sensors. These are made from single-crystal silicon. Characteristics of the MEMS are a fine pattern, a high accuracy, and a high-aspect ratio. MEMS process is based on the integrated circuit (IC) production process, and it can form a fine pitch pattern on a planer silicon structure. Moreover, the high accuracy and the high aspect ratio pattern are realized by a Bosch process [1]. By this process, it is possible to form miniature mechanical parts and fabricate an acceleration sensor, a gyroscope sensor, and so on. The miniature sensors progress a miniaturization of an electronic device, and it supports an information society. To realize the Internet of Things (IoT) society, the miniature, an enormous amount of the sensor is required, and these are usually produced by the MEMS.

One of a field of the MEMS process, the mechanical field is researched because it can form a miniature high aspect ratio pattern. Miniature MEMS actuators are researched for miniature mechanical systems [2–5]. Moreover, attention is paid to microrobots that have miniature silicon mechanical components. These robots have miniature body structure and miniature actuator [6–9]. The miniature actuators and the miniature robot can be used for the medical field. The miniature structure can work in the narrow space such as inside of a human body. For example, an endoscope microrobot has been researched [10].

The MEMS can realize the development of the information society and the medical field. However, a miniature power source is required for the electronic device such as the sensor, the actuator and the microrobot. Conventionally, a lithium-ion-secondary battery is used as the small- and high-power density source, but it is too large for these electronic devices. Moreover, the power density of the lithium-ion-secondary battery is approaching to the theoretical limit. Therefore, the MEMS power generators have been studied for miniature power supply system. To keep a micro scale, many researchers use a piezoelectric vibration power generator [11, 12]. It uses only a material characteristic without complex mechanical structure. However, these generators harvest the force to move the device by the environmental vibration, and then, its power is too small to be used for main power source.

On the other hand, ultra-micro gas turbine (UMGT) that used an electrostatic type was reported by the MIT group [13]. It was a remarkable generator because of its extremely high energy density in small size. The advantage of the electrostatic-type power generator is that the components are based on planar structure, and it is easy to fabricate by the MEMS process. A lot of studies on the electrostatic-type MEMS generator have been reported [14]. However, it shows charge saturation and high internal impedance; as a result, an output current of the electrostatic-type generator becomes small. For this reason, an electromagnetic induction type that is usually used a commercial size generator has been studied in the MEMS generator. The electromagnetic induction type shows low output impedance and high output power.

The conventional electromagnetic induction-type generator has a magnet, moving part such as the turbine structure, and a magnetic circuit including a magnetic core. The winding wire magnetic circuit forms a three-dimensional coil structure. In the MEMS process, the miniature moving part can realize, but it is difficult to form the three-dimensional structure coil by using the MEMS process. Therefore, the planer structures such as a spiral, a meandering or equivalent shape are employed for the electromagnetic induction-type MEMS generator. The complex three-phase coil pattern made from a copper conductor that was arranged on a plane substrate has reached milliwatt to watt class [15, 16]. However, the planar structure coil requires the large area if the turn number increases because the conductor patterns extend parallel on the surface. Also, it is difficult to use the magnetic material. Therefore, to catch the magnetic flux, long length coil is required. As a result, it shows high internal resistance and small output power. In order to realize the miniature electromagnetic induction-type generator, the miniature three-dimensional structure coil that has the magnetic core is required.

The multilayer ceramic technology is suitable for forming the miniature three-dimensional structure coil. This technology is the fabrication technology of a miniature electronic component like a multilayer ceramic inductor. It can form a three-dimensional circuit pattern inside a ceramic material. A ferrite ceramic material shows excellent magnetic property, and the multilayer ceramic technology will realize the miniature monolithic structure that has the three-dimensional structure coil and the magnetic core.

In this chapter, the miniature electromagnetic induction-type air turbine generator is proposed and developed. The proposed generator is combined with the miniature MEMS air turbine structure and the miniature multilayer ceramic magnetic circuit. To achieve milliwatt-level output power, it is required that high rotational speed of the rotor. Therefore, the air turbine is discussed optimization of a rotating structure. The various complex structure multilayer ceramic magnetic circuits are proposed. The proposed magnetic circuits are discussed and optimized the shape to catch the magnetic flux. Moreover, the fabricated air turbine generator demonstrates the rotational motion and power generation.

2. Design and concept of MEMS air turbine generator

2.1. Electromagnetic induction-type MEMS air turbine generator

The electromagnetic induction type is employed for developed MEMS air turbine generators. Proposed MEMS air turbine generators are combined with the MEMS mechanical parts and the ceramic electronic part. In the MEMS air turbine structure, a magnet is connected to the rotor. The rotor shows a rotational motion by the inletting fluid. This motion occurs the electromagnetic induction revolving-field.

The design concept of the air turbine is the high-speed rotational motion. The rotational structure at a bearing system and a rotor blade form are compared. In the bearing system, the fluid dynamic bearing system and a miniature ball bearing structure are discussed. The rotor blade forms influence the rotational motion. In this chapter, a flat-type rotor blade and a rim-type rotor blade are compared. Therefore, three types of air turbine structures are designed and fabricated. The design concepts of the ceramic magnetic circuit are a miniature three-dimensional structure coil and the introduction of the magnetic flux. The magnetic material designs of the circuit are analyzed, and these are compared. Moreover, the arrangement of the magnetic circuit is an important factor to the miniature electromagnetic induction-type generator. The arrangement of the magnetic circuit is discussed as the generator that combined with the air turbine and the magnetic circuit.

2.2. Mechanical parts of generator

The mechanical parts of the generator are the air turbine structure. These are made from a miniature silicon parts, and the miniature structure parts are fabricated by the MEMS process. To achieve the high-speed rotational motion, two types of bearing system such as a fluid-dynamic bearing system and a miniature ball bearing system are designed. The one of the bearing system, a fluid-dynamic bearing system is employed. In the miniature structure, the friction force influence to the rotational motion. Therefore, a contactless-type miniature bearing system is advantaged for the miniature silicon air turbine. However, it is difficult to rotate with stability by this method. Moreover, design of the air turbine will be complex because it requires flow passages to rotating and floating.

Another one is the miniature ball bearing. It made from a mechanical process, and it can suppress an eccentric motion of the rotor because the ball bearing holds the rotor directly. The rotor and the magnet are held a shaft through the ball bearing structure. It requires the initial torque to achieve the rotational motion, but it is desired a stable rotational motion and the simple structural design. By using the miniature ball bearing, the rotational part and the magnet part can separate.

The MEMS air turbine designs are shown in **Figures 1** and **2**. **Figure 1** shows the flat-type rotor blade and the fluid dynamic bearing system air turbine. Image (a) and (b) of the **Figure 2** employs the miniature ball bearing structure for the bearing system. Image (a) uses the flat-type rotor blade air turbine and (b) uses the rim-type rotor blade air turbine, respectively.

The air turbine parts of **Figure 1** are seven silicon structural layers and the rotor that has the magnet. The upper layers form the air passage structure. Through this passage, air is passed to the stator and it generates the rotational motion of the rotor. The ring-shaped magnet is

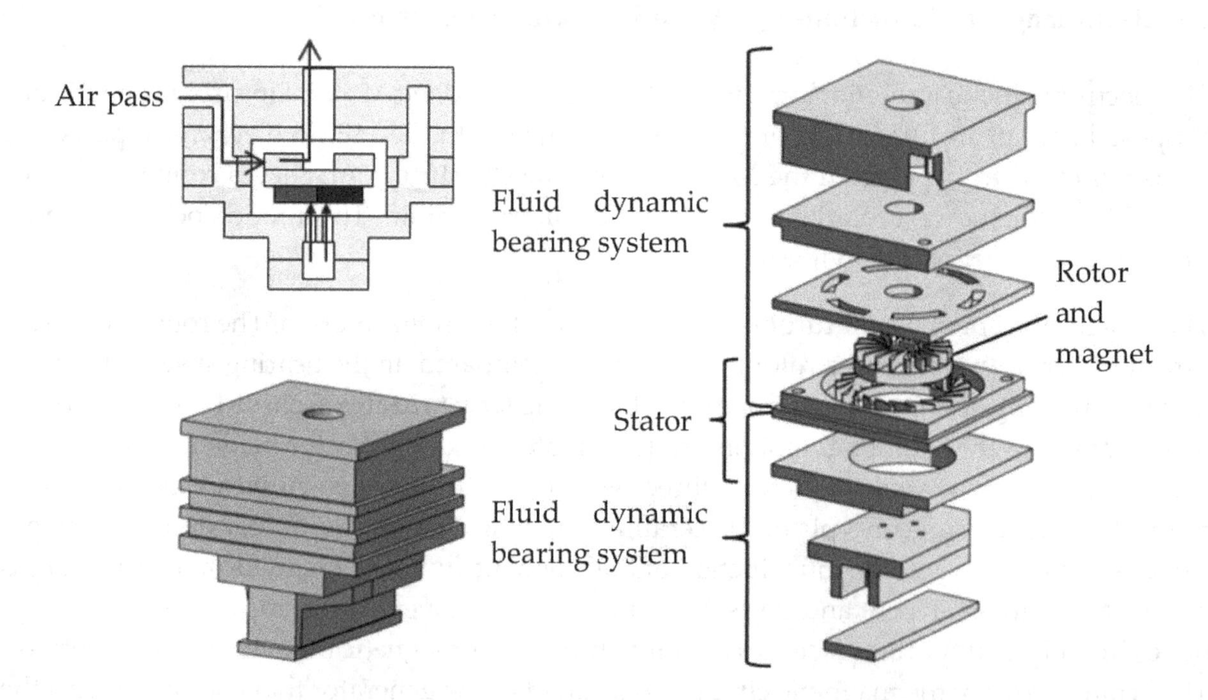

Figure 1. Design of MEMS air turbine structure that has flat-type rotor blade and fluid dynamic bearing system.

attached to the rotor and placed in a center of the stator. The lower layers form the passage for the fluid dynamic bearing system. The shape of the lower layers that form the fluid dynamic bearing system is bump structure. This structure realizes a short distance between the magnet and the magnetic circuit. However, the gap between the magnet and the magnetic circuit is more than 800 μm. The designed flat-type rotor has the plate structure under the rotor blade. The plate receives the air through the levitation passage, and then, the air is released from an outlet. The rotational air through the stator is introduced to the center of the rotor, and then, the air passed the same outlet. The designed dimensions of the air turbine are 3.0, 3.0 and 3.0 mm, length, width and height, respectively.

Designed turbines in **Figure 2** have a difference at the arrangement of the miniature ball bearing. The flat-type rotor blade (a) has one miniature ball bearing inside the air turbine structure. The air turbine parts are made from 11 silicon layers and 3 silicon rotor parts. The compressed air flows from side inlet, and the down flow air passes to the stator and the rotor. The released air passes a top outlet. The dimensions of the miniature ball bearing are 2.0 mm (outer diameter), 0.6 mm (inner diameter) and 0.8 mm (height), respectively. It is made by a martensitic stainless steel. The rotor is putted on the ball bearing. The magnet is arranged under the bearing structure. Then, these are connected to the ball bearing through the shaft. The diameter of the shaft is 0.593 mm, the material is cemented carbide. The flat-type rotor shape is the same design with the fluid dynamic bearing system air turbine. The magnet and a magnetic yoke are combined to suppress a leaked magnetic flux. The ring-shaped magnet is neodymium magnet 2-pole radial direction. Its dimensions are outer diameter 3.0 mm, inner diameter 1.0 mm and height 0.5 mm, respectively. The ring-shaped magnet yoke is formed by a silicon steel sheet. The dimensions of the magnet yoke are outer diameter 3.0 mm, inner diameter 1.0 and height 0.38 mm, respectively. The size of the designed air turbine is 5.20 mm, 5.20 and 4.50 mm, length, width and height, respectively.

In the image (b) of **Figure 2**, two ball bearings are used for holding the rotor and the magnet. The bearings are placed above and below the rim-type rotor. The air turbine structure is constructed by seven silicon layers, rotor and magnet supporting part. The air passage for rotational motion of the rotor is formed around the rim-type rotor, and the compressed air flows from side inlet to side outlet. Thickness of the rotor blade is 750 μm, and the rotor blades are formed on the side wall of the rotor. The dimensions of the rim type are 5.20, 5.20, 4.60 mm length, width and height, respectively. The ball bearing-type air turbine can separate the rotor and the magnet because these are held by the shaft. Therefore, the gap between the magnet and the magnetic circuit is shorter than the fluid dynamic bearing-type air turbine. The gap dimension of the design (a) is 220 μm and (b) is gapless design.

The fine pattern of the rotor and each layer were fabricated by the photolithography process. In this process, the miniature components were fabricated from single crystal silicon wafer. Each silicon wafer for the parts was washed, deposited with an aluminum layer by physical vapor deposition, and coated with a photoresist. The designed pattern was exposed to the resist layer and developed by soaking in the developer. The aluminum layer on the specimen was then chemically etched, leaving an imprint of the designed pattern. The patterned wafer was dry etched by high aspect ratio inductively coupled plasma etching combined with

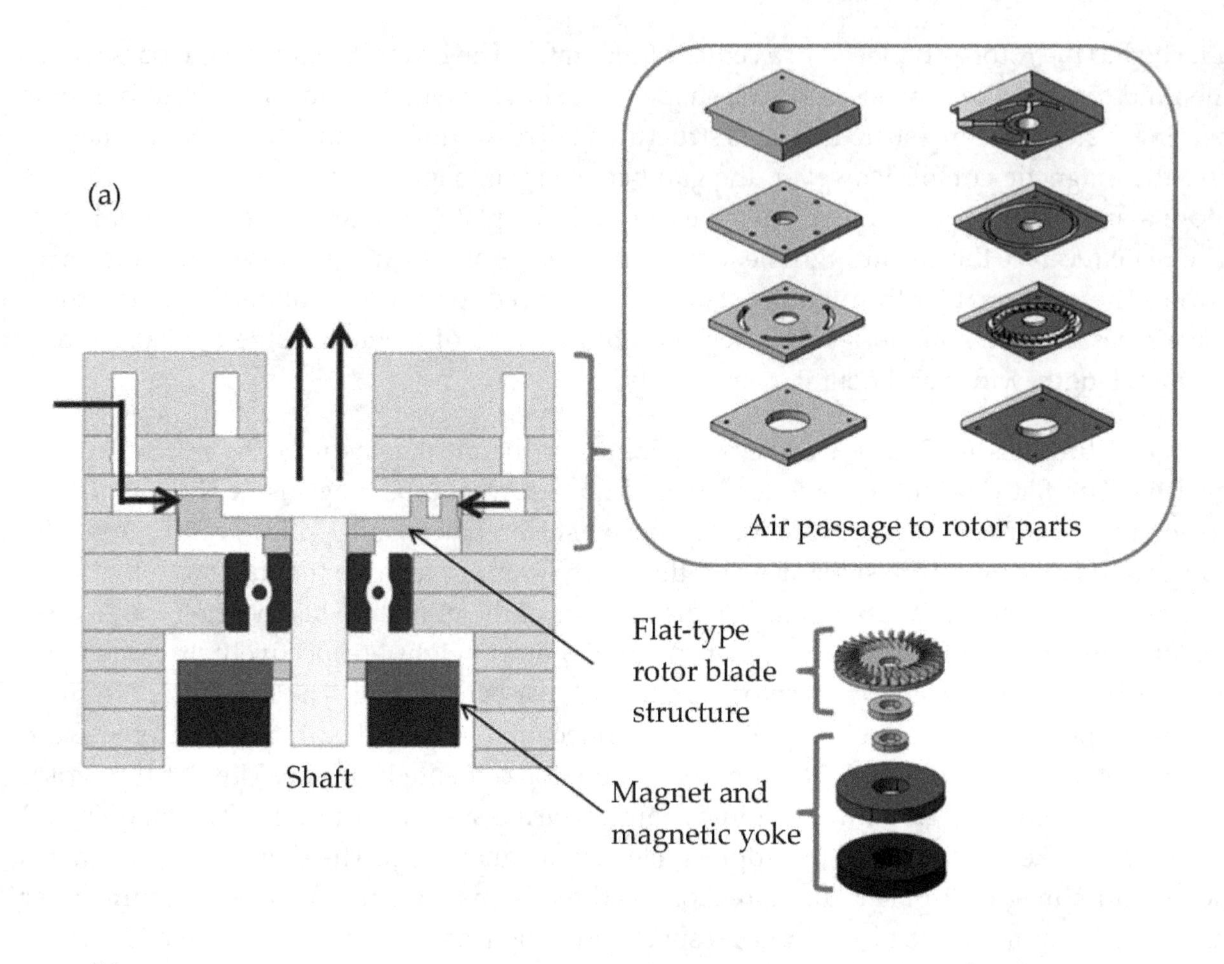

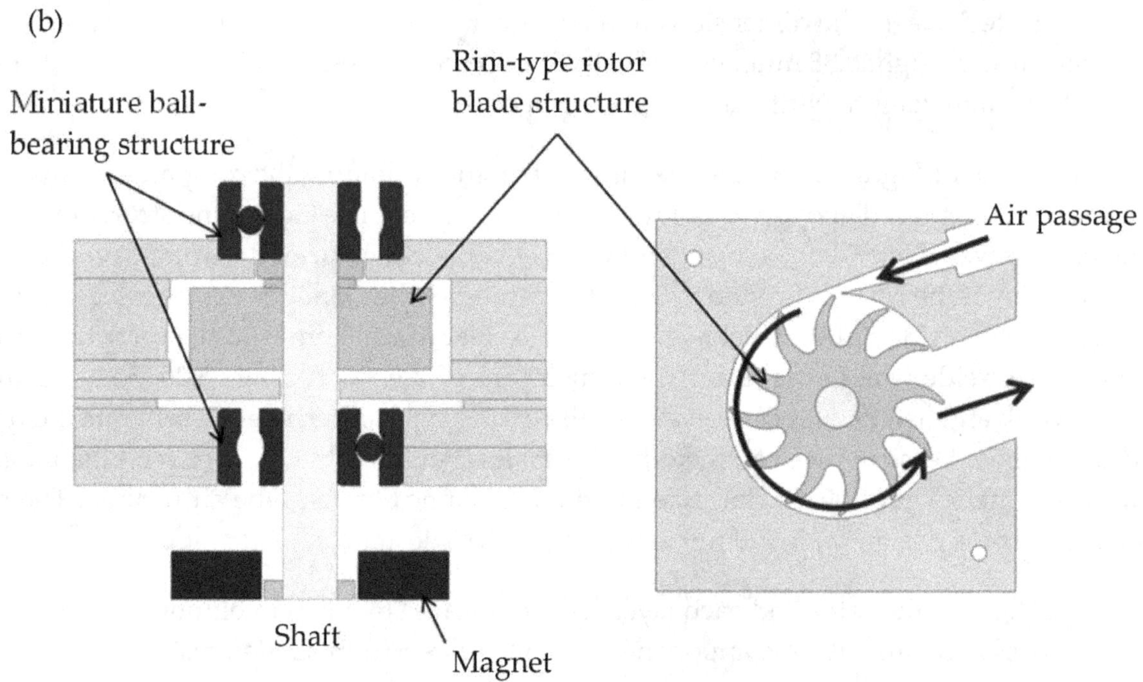

Figure 2. Design of MEMS air turbine structures (a) flat-type rotor blade and miniature ball bearing and (b) rim-type rotor blade and miniature ball bearing.

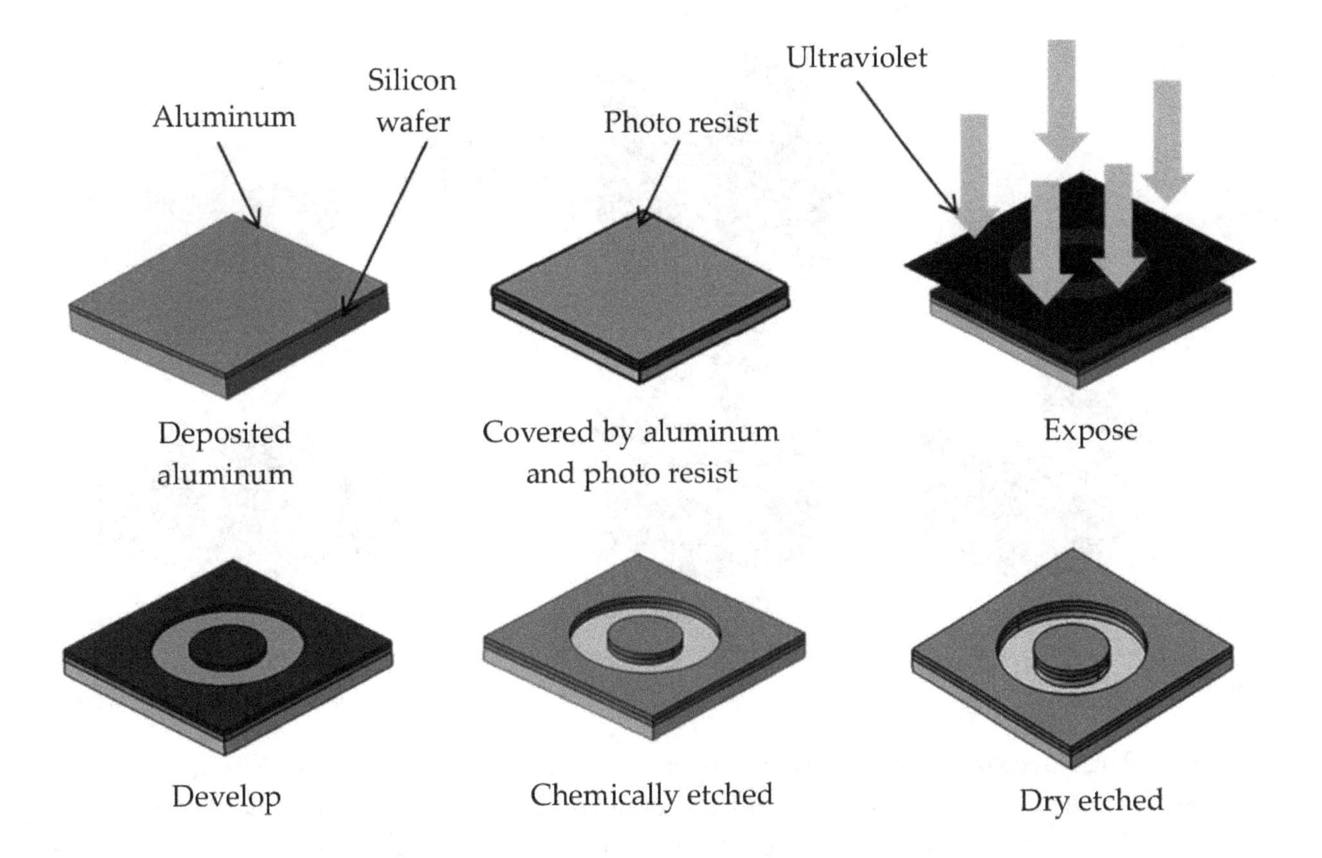

Figure 3. Fabrication process for miniature components of designed MEMS air turbine.

a Bosch process [1]. The parts were achieved after removing the aluminum and washing. Through these processes, the silicon structural components were fabricated. The obtained components were assembled by hand and alignment pin. **Figure 3** shows the schematic illustration of the fabrication process.

2.3. Electronic parts of generator

The electronic part for the electromagnetic induction-type MEMS air turbine generator is the magnetic circuit. To achieve the miniature generator, the miniature magnetic circuit that has a three-dimensional structure coil and a magnetic core for introducing a magnetic flux is required. Moreover, low internal resistance is an important factor to achieve high output power. Therefore, the three-dimensional coil pattern is required. The multilayer ceramic technology is used for the magnetic circuit. Optimization of the magnetic circuit design and the fabrication process are explained.

Designs and analyzed results of the magnetic circuit for the miniature generators are shown in **Figures 4** and **5**. **Figure 4** shows designs for the fluid dynamic bearing system-type air turbine generator. The shape of (a) is a step-wise shape, and (b) is a horseshoe-shaped structure. These magnetic circuits have two-pole coil structures. Each pole has 12 turn coil structure, and they are connected at the connecting layer. Therefore, the designed magnetic circuits have 24

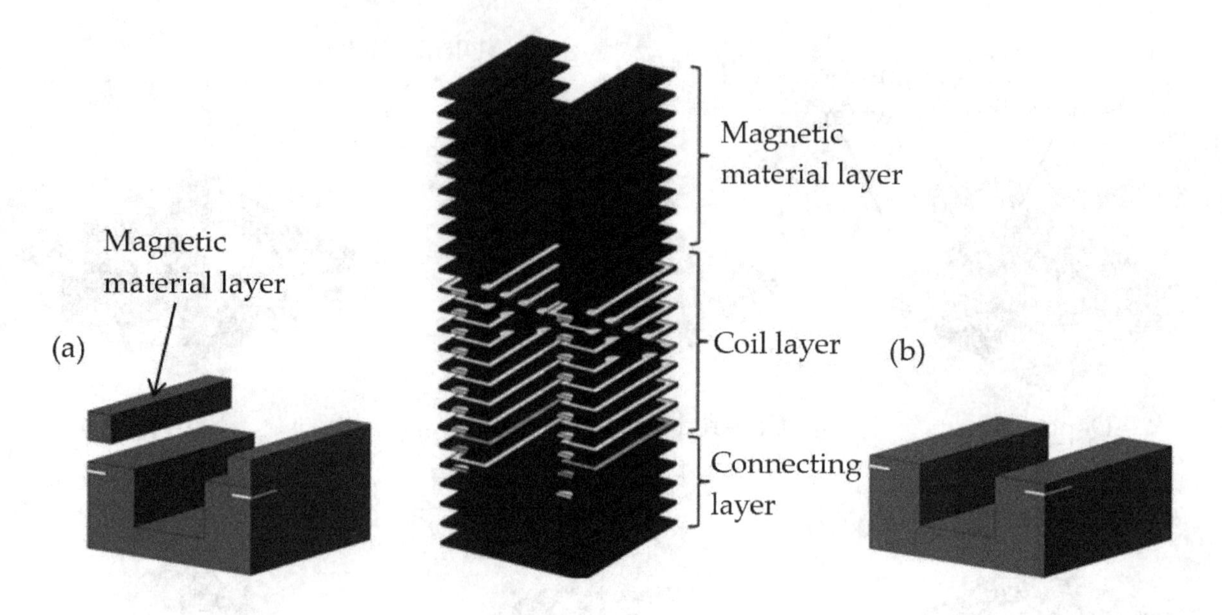

Figure 4. Design of the magnetic circuits for fluid dynamic bearing system generator: (a) step-wise shape and (b) horseshoe shape.

turn coil. The horseshoe-shaped circuit has only the coil layer and the connecting layer, and the step-wise shape circuit has the magnetic material layer. When the magnetic circuit and the air turbine are combined, the magnet inside the air turbine is placed between the magnetic material layers. The magnetic flux loss is compared between the magnetic circuits. The results are shown in **Figure 5**. The magnetic flux is introducing from the magnet to magnetic circuit in (a). On the other hand, (b) shows the magnetic flux loss more than (a). In the both structures, the dimensions are 3.5 and 3.5 mm, length and width, respectively. The heights are 2.0 mm in the step-wise shape and 1.2 mm in the horseshoe shape.

The output powers of these designs magnetic circuit are compared. In this evaluation, a spindle machine is used for power evaluations.

Figure 6 shows the around-type magnetic circuit. It is used for the miniature ball bearing-type air turbine generator. Designed magnetic circuit is constructed by four pieces. The coil layers

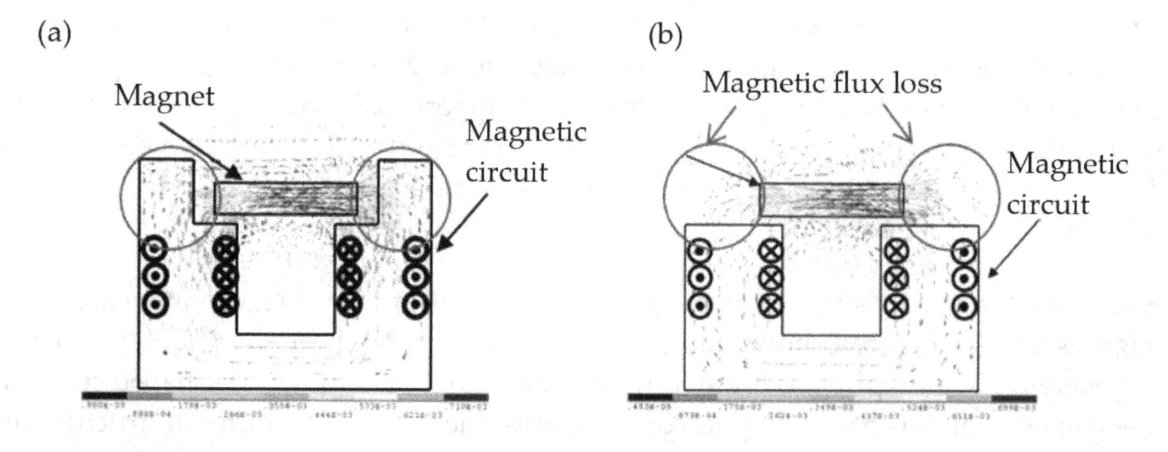

Figure 5. Analyzed results of the magnetic circuits: (a) step-wise shape and (b) horseshoe shape.

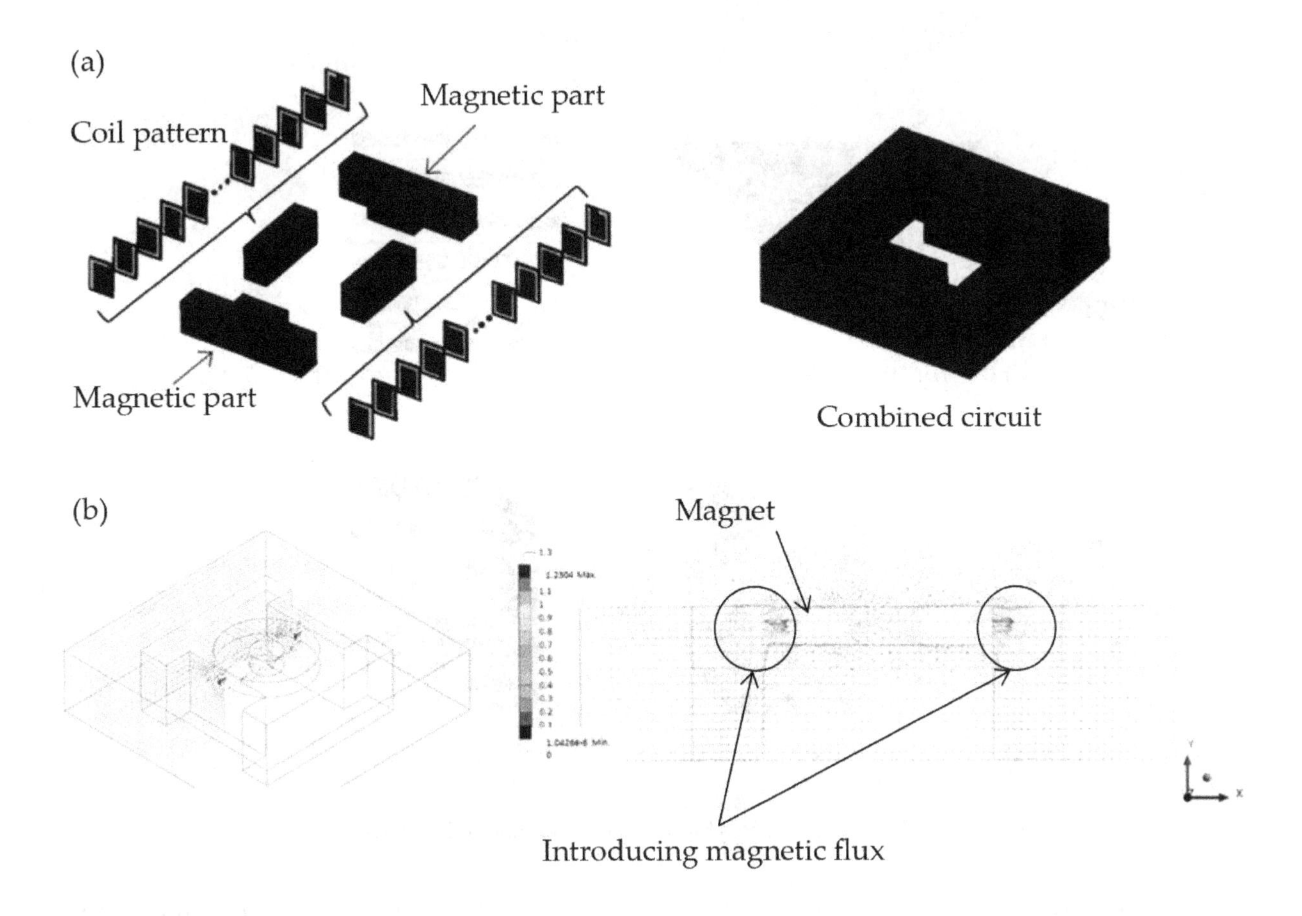

Figure 6. Design of around-type magnetic circuit for ball bearing-type air turbine (a) schematic illustration of circuit design, (b) analyzed result.

are arranged in side parts, and these have 50 turns each other. The total number of coil turns is 100 turns; these are bonded with the magnetic parts for introducing the magnetic flux. By this process, the closed magnetic circuit is obtained. The magnetic parts are introducing the magnetic flux from the magnet. In **Figure 6**, the analyzed result is shown in (b). By this result, the magnetic flux through the magnetic parts were observed. Dimensions of the designed circuit are 7.40, 8.50, 2.40 mm, length, width and height, respectively.

The magnetic ceramic material is low-temperature co-fired Ni-Cu-Zn ferrite with the permeability of 900. The compositional ratio is 49.2 Fe_2O_3–8.8 NiO-10 CuO-32ZnO. The Ni-Cu-Zn ferrite can be fired at around 900°C. Therefore, it is possible to use the low-resistance silver conductive material.

The fabrication process for the miniature magnetic circuits is the green sheet process, that is, the multilayer ceramic technology. In this process, the ceramic slurry was made for forming a sheet structure. This sheet is called the green sheet. The slurry in our fabrication was made of the mixture of the ferrite ceramic powder, binder, dispersing agent, plasticizer and organic materials. The through hole for connecting the under layer was machined, and then the coil pattern was printed on the ferrite green sheet by the screen printing technology. The material of the coil pattern was silver paste as a conductor paste. The multiple sheets were stacked, and the specimen was diced into the designed part. By using the magnetic ceramic for the green

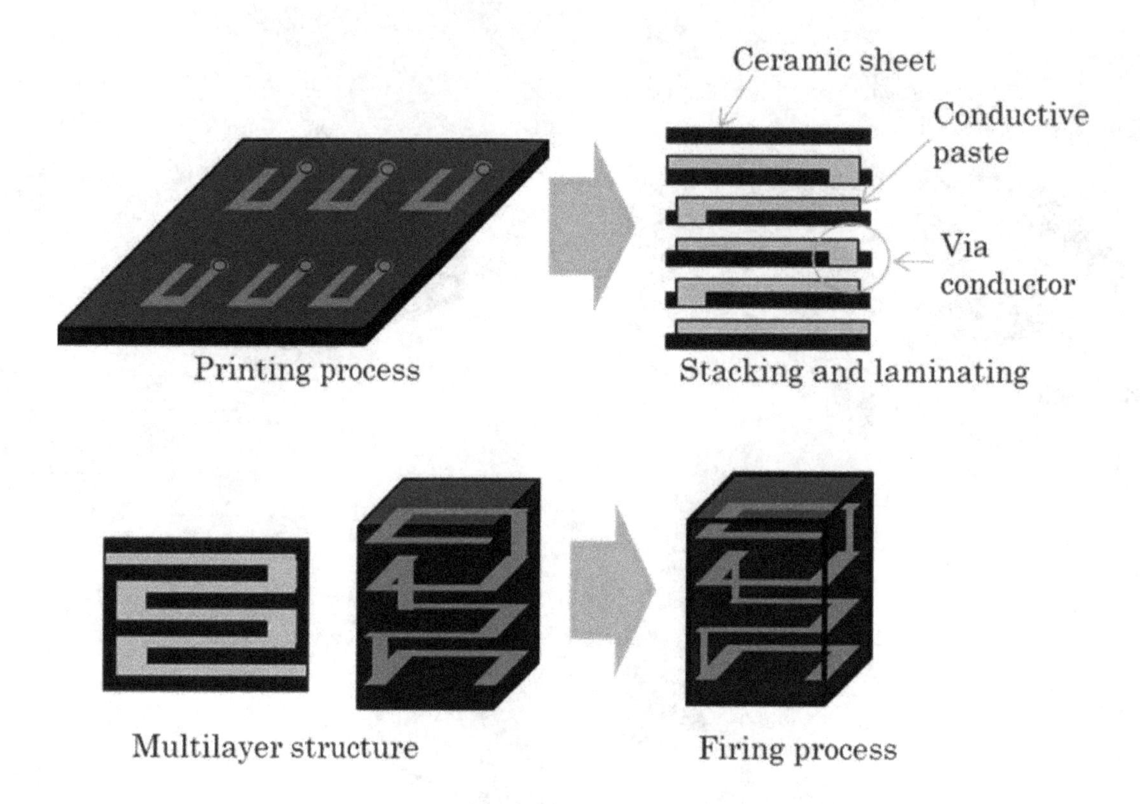

Figure 7. Schematic illustration of fabrication process for multilayer ceramic coil pattern.

sheet, the magnetic core is formed simultaneously. Through this process, the obtained specimen was a planar structure that had the miniature coil pattern inside the magnetic ceramic. **Figure 7** shows the schematic illustration of the fabrication process for the multilayer ceramic coil pattern.

In order to combine the MEMS air turbine, more complicated structure is required. Each part was combined for the design structure. After that, the specimen was fired in the electric furnace. In the around-type coil, the pieces were firing. After that, the pieces were combined because a shrinkage process deforms the ceramic coil. Through these processes, the objective structure was completed. A schematic illustration of the combined process for the complex structure coil is shown in **Figure 8**.

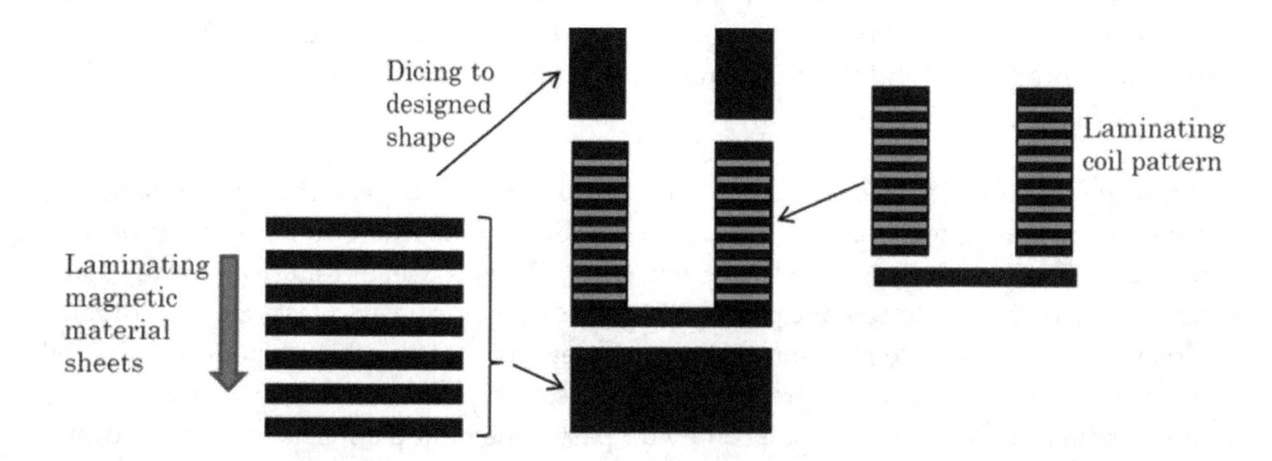

Figure 8. Schematic illustration of the combined process for the complex structure coil.

2.4. Experimental procedure

The combined MEMS generator was evaluated on the output power. The compressed nitrogen gas was injected to the MEMS generator. The rotational speed and the output voltage were measured. Load resistances were connected to the ceramic magnetic circuit. The output waveform was measured by an oscilloscope.

3. Results and discussion

The fabricated MEMS air turbines and the multilayer ceramic magnetic circuits were evaluated. The combined electromagnetic induction-type MEMS air turbines were evaluated on the power generation.

3.1. Fluid dynamic bearing system air turbine generator

The fabricated components of the fluid dynamic bearing system air turbine and the assembled MEMS air turbine are shown in **Figure 9**. Designed dimensions and the measured dimensions are shown in **Table 1**. As a result, it is found that the error was less than 5 μm.

Figure 10 shows the fabricated multilayer ceramic magnetic circuits. The dimensions of the step-wise shape multilayer ceramic circuit were 3.40, 3.47 and 1.88 mm, length, width and height, respectively. Inductance and DC resistance were 5.35 μH and 0.53 Ω. The dimensions of the horseshoe shape circuit were 3.25, 3.49 and 1.34 mm, length, width and height, respectively. Inductance and DC resistance were 5.85 μH and 0.94 Ω.

The result of the power generation experiment by the spindle machine is shown in **Table 2**. The load resistance of 1 Ω was connected to both magnetic circuits. The rotational speed of the spindle machine was 300,000 rpm. By the results, the maximum output power of the

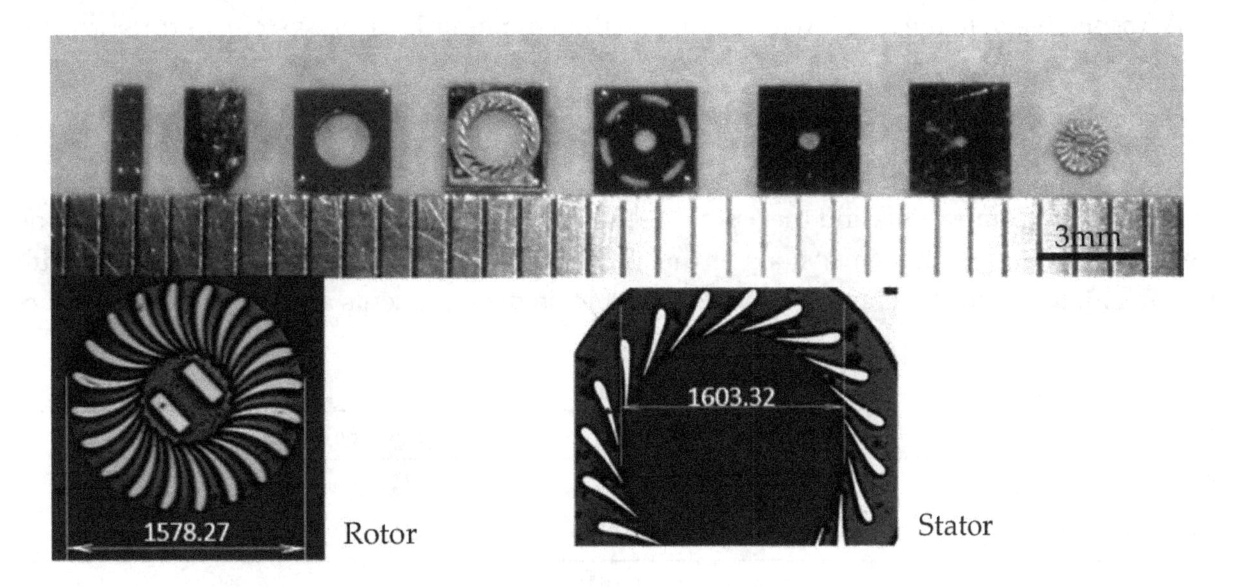

Figure 9. Fabricated fluid dynamic bearing system air turbine components.

	Design dimension (μm)	Measured dimension (μm)
Rotor diameter	1580	1578.27
Stator diameter	1600	1603.32

Table 1. Dimensions of the MEMS air turbine components.

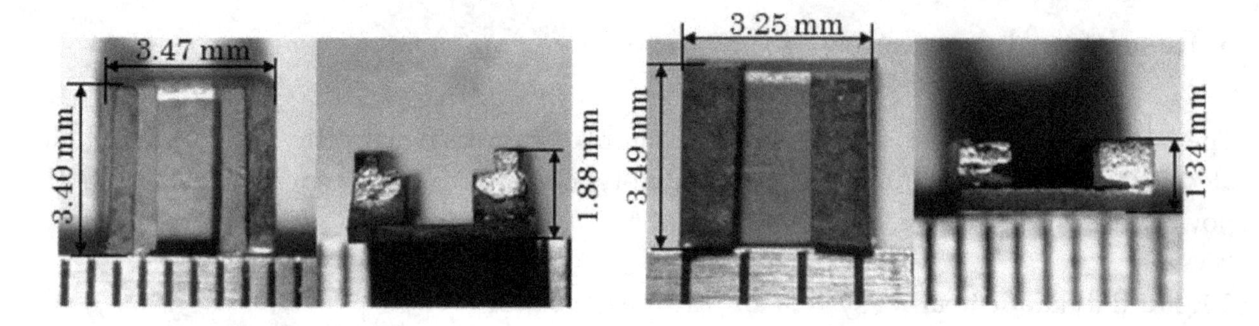

Figure 10. Fabricated multilayer ceramic magnetic circuit: (a) step-wise shape and (b) horseshoe shape.

step-wise shape was 1.47 mVA, and the output power of the horseshoe shape was 0.72 mVA. The step-wise shape magnetic circuit showed the larger output power than the horseshoe shape magnetic circuit. It is agreed from the result of the analyses. Therefore, it is found that the magnetic material surrounding the magnet improved the output power.

The fabricated MEMS air turbine and the multilayer ceramic magnetic circuit were combined. **Figure 11** shows the MEMS air turbine generator and its output voltage waveform. The step-wise shape magnetic circuit that shows the internal resistance of 1.05 Ω was used. The dimensions of the MEMS air turbine generator were 3.50, 3.47 and 3.86 mm, length, width and height, respectively. The maximum rotational speed was 30,000 rpm on the condition of 0.28 MPa. The load resistance of 1 Ω was connected to the output of the magnetic circuit. The maximum output voltage and the output power of the generator were 1.32 mV and 1.74 μVA, respectively. In the rotational motion, the fluid dynamic bearing system air turbine rotor showed the eccentric motion.

3.2. Miniature ball bearing air turbine generator

The fabricated components and the assembled structure of the flat-type rotor blade air turbine are shown in **Figure 12**. Dimensions of the structure were 5.23, 5.20, 4.51 mm length, width and height, respectively. **Figure 13** shows the fabricated components and the assembled struc-

	Output voltage [mV]	Output power [mVA]
Step-wise shape	26.8	1.47
Horseshoe shape	38.4	0.72

Table 2. Power generation results of fabricated magnetic circuit using spindle machine.

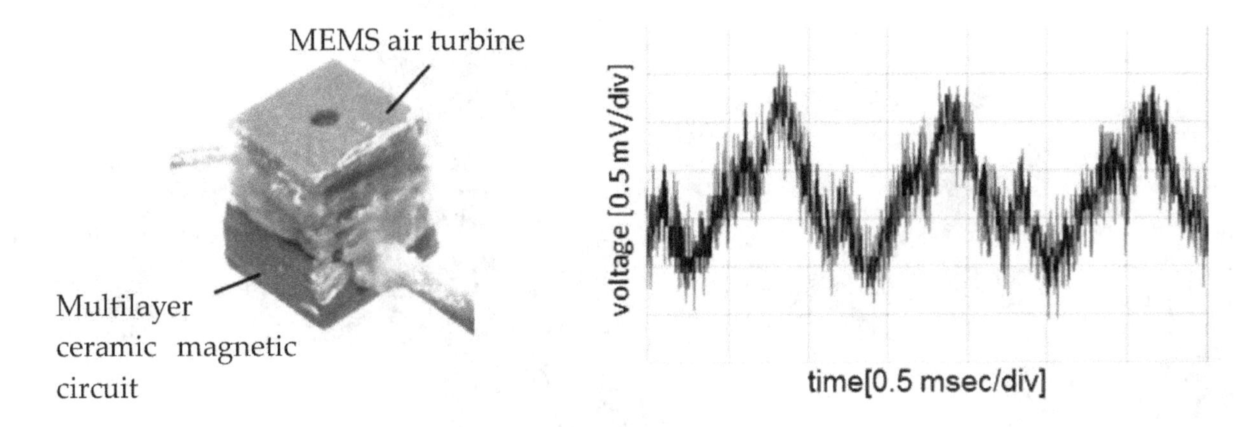

Figure 11. Fabricated MEMS air turbine generator and its output voltage waveform.

ture of the rim-type rotor blade air turbine. Dimensions were 5.24, 5.37, 4.64 mm length, width and height, respectively. These fabricated MEMS air turbine were evaluated on the rotational speed. The rotational speed was measured by a hall sensor. The comparison of the rotational speed is shown in **Figure 14**. The flow rate changed from 0 to 1.0 l/min. at pressure of 0.3 MPa. Experimental result shows that the rim-type system had a superior potential to the planar type. The flux change dependents the rotational speed because the magnet attached to the rotor. The high rotational speed air turbine is advantage for high output power. Therefore, to achieve high output power, the rim-type air turbine was employed the generator.

The fabricated multilayer ceramic magnetic circuit was shown in **Figure 15**. Dimensions of the circuit were 7.40, 8.47, 2.36 mm, length, width and height, respectively. The measured DC resistance was 2 Ω.

The fabricated rim-type air turbine and the magnetic circuit were combined. **Figure 16** shows the combined electromagnetic induction-type MEMS air turbine generator. Dimensions of the generator were 7.40, 8.47, 5.82 mm, length, width and height, respectively. When the maximum

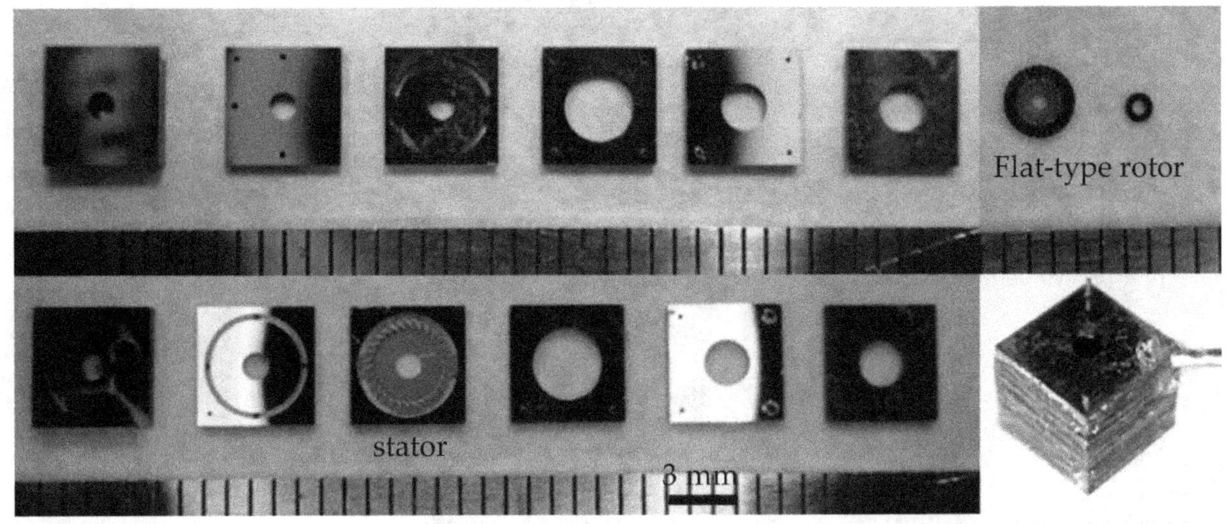

Figure 12. Fabricated components and assembled structure of flat-type rotor blade air turbine.

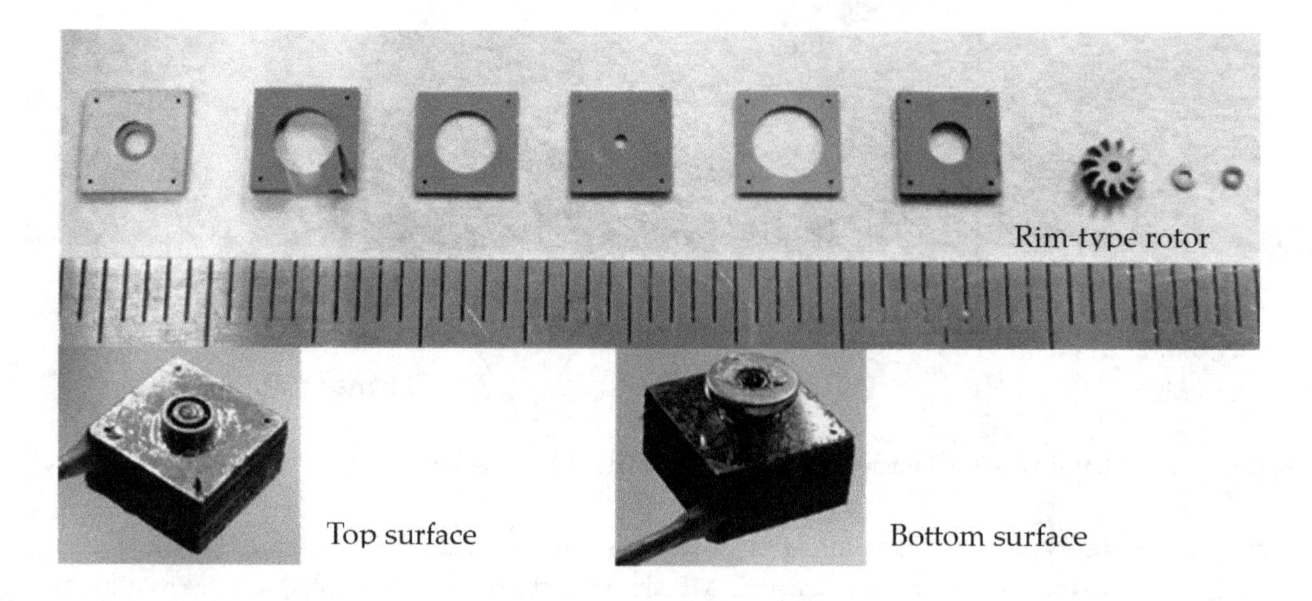

Figure 13. Fabricated components and assembled structure of rim-type rotor blade air turbine.

rotational speed was 290,135 rpm, the inlet flow was 2.4 l/min and the pressure was 0.3 MPa. The output voltage and the output power at each load resistance are shown in **Figure 16**. The maximum output power was 2.41 mVA when the load resistance was 8 Ω, and the output voltage of 139 mV was shown. The output waveform at the load resistances was 8 and 1 kΩ, as shown in **Figure 17**.

The output voltage $V = R_L I$ is given by the following equation: (1) when the load resistance R_L is connected to the generator and the current I flows through the circuit. In this equation, it is necessary to consider the influence of the voltage drops by the self-inductance L and the internal resistance r of the connected magnetic circuit.

$$V = R_L I = N\, d\varphi/dt - i\omega LI - rI \tag{1}$$

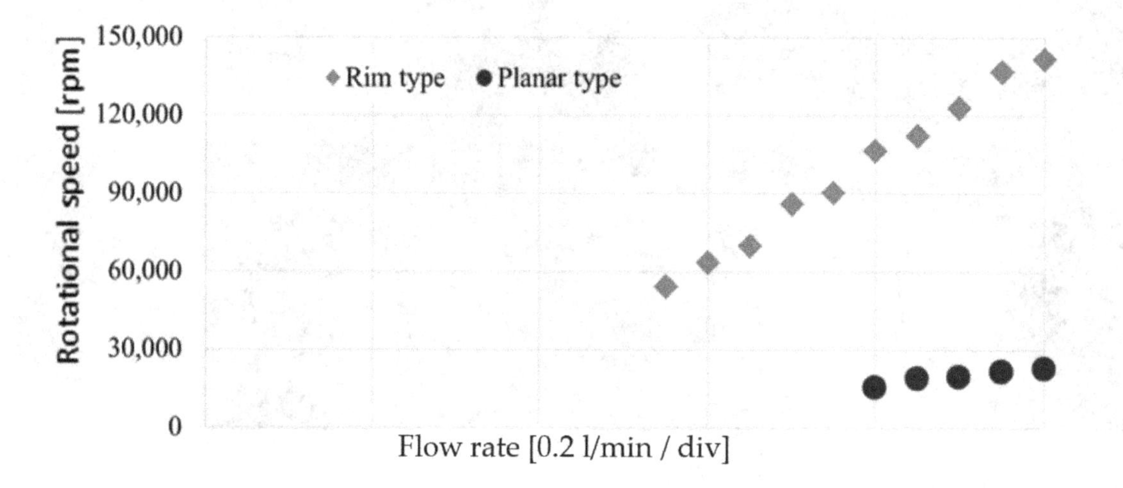

Figure 14. Comparison of rotational speed.

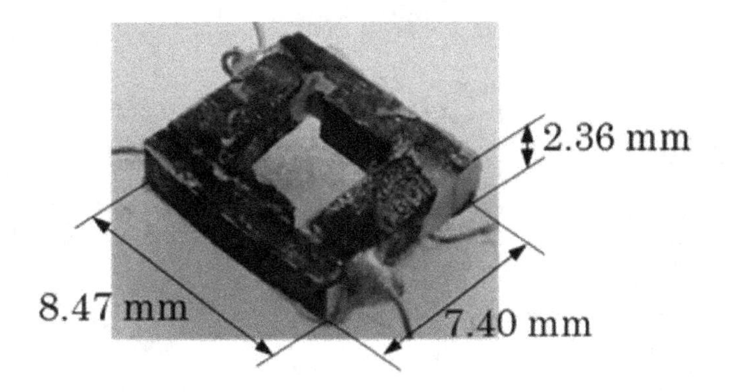

Figure 15. Fabricated multilayer ceramic magnetic circuit.

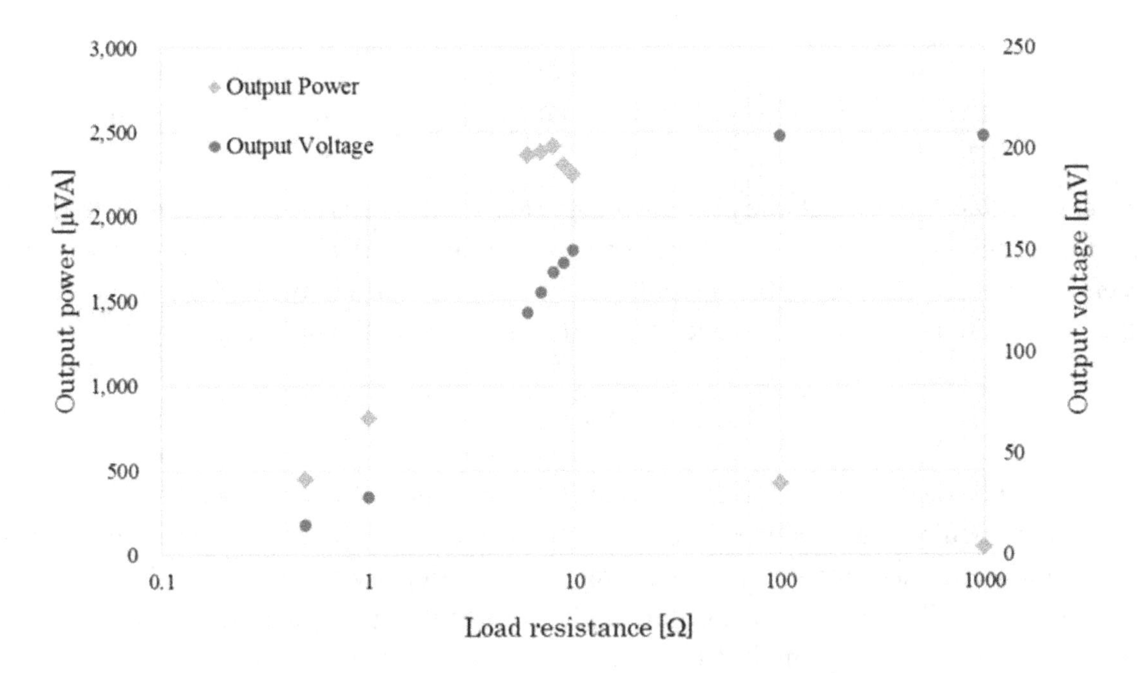

Figure 16. Output voltage and output power at each load resistance.

The magnetic flux passing the magnetic circuit is φ. N is the turn number of coil. L is the measured value at the equivalent frequency of 290,135 rpm (impedance analyzer: Agilent 4294A), and it was 241 μH. When the R_L was 1 kΩ, the voltage drop due to self-inductance and the internal resistance become small because the current carrying the circuit is sufficiently small. Therefore, the output voltage is approximated by $d\varphi/dt$. The surface magnetic flux density of the permanent magnet and the magnitude of the magnetic flux on the magnet surface are 0.159 T and 0.375 μWb, respectively. Theoretical value if the all flux enters in the circuit, the output voltage was calculated to be 570 mV. Compared with the output voltage that connected 1 kΩ, 36% of the maximum magnetic flux contributes to the actual power generation. On the other hand, when R_L of 8 Ω was connected, the reactance of ωL is calculated 7.3 Ω. As a result, values of ωL and r are close to R_L, and their influence appears. The absolute value of the output voltage is estimated as Eq. (2).

$$| V | = (N\, d\varphi/dt\, R_L) / \left[(R_L + r)^2 + \omega^2 L^2 \right]^{1/2} \quad (2)$$

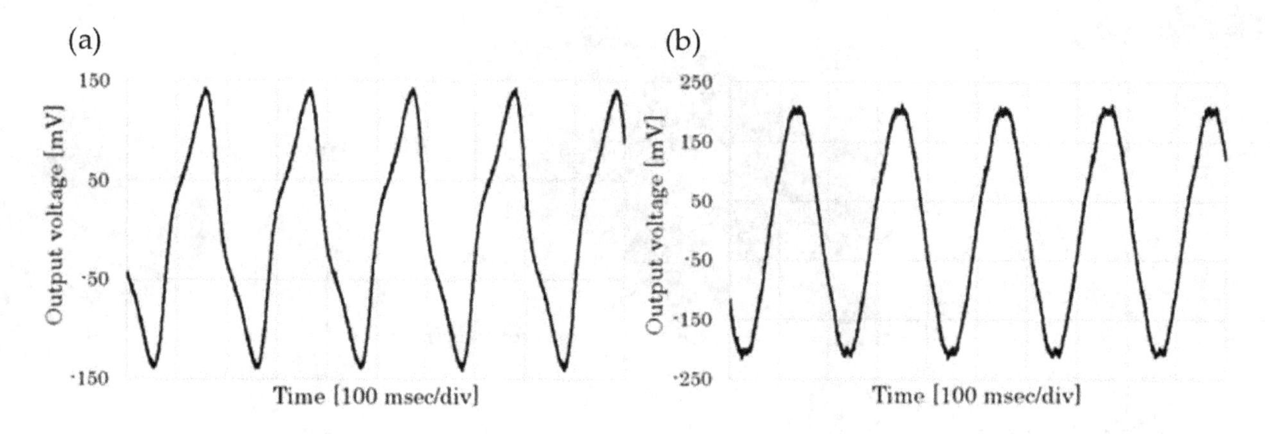

Figure 17. Output waveform at the load resistances were (a) 8 Ω and (b) 1 kΩ.

The calculated result of Eq. (2), |V| becomes 367 mV. If the contribution rate 36 % of the magnetic flux to this result, |V| become 132.12 mV. It almost coincides with output voltage at 8 ohm (**Figure 17(a)**).

In the electromagnetic induction type, the magnetic circuit occurs a braking torque to the permanent magnet. However, the rotational speed of **Figure 17(a)** and **(b)** shows almost equal. Therefore, the braking torque by current is sufficiently small in this turbine structure. The magnetic flux density from the magnetic circuit can be calculated at Eq. (3).

$$B = \mu_0 \mu_r NI \tag{3}$$

The vacuum permeability and the ferrite relative permeability are expressed in μ_0 and μ_r. As a result, B is calculated as 1.96 mT. The maximum braking torque occurs when the magnetization of the permanent magnet and the magnetic flux density of the magnetic circuit cross perpendicular. If all the magnetic flux contributes, the braking torque is 42.9 pNm. This value is considered as sufficiently small value.

4. Conclusions

The electromagnetic induction-type MEMS air turbine generator was proposed. In this chapter, three types of MEMS air turbine generators that included the different bearing systems, shape of the rotor blades and shape of the magnetic circuits were discussed to achieve the high output power. In the MEMS air turbine, the purpose was achieving high-speed rotational motion. The magnetic circuit required the miniature structure that had the three-dimensional coil, magnetic core and magnetic flux introduction design. Therefore, the multilayer ceramic technology and the ferrite ceramic were used. One of the developed air turbines employed the fluid dynamic bearing system and flat-type rotor. In the miniature structure, the contactless-type miniature bearing system is advantaged because the friction force is impact issue. Moreover, two types of magnetic circuits for the fluid dynamic bearing turbine generator were compared with the magnetic flux loss. By the power generation experiment, the stepwise shape circuit that had the magnetic material introducing the magnetic flux from the magnet was suitable to the generator. The fabricated MEMS air turbine generator showed the output power of 1.74 μVA, when the rotational speed was 30,000 rpm, the output voltage was 1.32 mV and the load resistance of 1 Ω was connected. However, it showed eccentric motion because it was not supported by structurally. Therefore, another one of the air turbines used the miniature ball bearing system. The developed ball bearing air turbines were compared with the rotational speed between the different rotor blades. As a result, the rim-type rotor blade showed high rotational speed than the flat-type rotor. Moreover, the ball bearing-type air turbine could separate the magnet from the rotor. Therefore, the short distance between the magnet and the magnetic circuit was realized. The shape of the magnetic circuit was around type that had the magnetic flux induction parts. To evaluate the power generation, the rim-type air turbine and the around-type multilayer ceramic magnetic circuit were combined. The maximum rotational speed was 290,135 rpm. The output power of fabricated MEMS air turbine generator was 2.41 mVA when the load resistance was 8 Ω and the output voltage of 139 mV was shown. By these results, the milliwatt-level MEMS air turbine generator was realized by the high-speed rotational motion structure that had the rim-type rotor blade and the miniature ball bearing system, and by introduction of the magnetic flux.

Acknowledgements

The sample of this study was fabricated by the facility at the Research Center for Micro Functional Devices, Nihon University. Part of this study was supported by the CST research project of Nihon University and by JSPS KAKENHI (16 K18055).

Author details

Minami Kaneko*, Ken Saito and Fumio Uchikoba

*Address all correspondence to: takato@eme.cst.nihon-u.ac.jp

Department of Precision Machinery Engineering, College of Science and Technology, Nihon University, Chiba, Japan

References

[1] Bhardwaj Jy K, Ashraf H. Advanced silicon etching using high-density plasmas. In: Proceedings of the SPIE Micromachining and Microfabrication Process Technology; 19 September 1995; Austin, TX, United States. pp. 224-233

[2] Long-Sheng F, Yu-Chong T, Muller RS. IC-processed electrostatic micro-motors. In: Proceedings of the Int. Electron Devices Meeting (IEDM '88). Technical Digest; 11-14 December 1988; San Francisco, CA, United States. pp. 666-669

[3] Zhang W, Zou Y, Lin T, Chau FS, Zhou G. Development of miniature camera module integrated with solid tunable lens driven by MEMS-thermal actuator. Journal of Microelectromechanical Systems. 2017;**26**:84-94. DOI: 10.1109/JMEMS.2016.2602382

[4] Berka MJ, Yadid-Pecht O, Mintchev MP, Wang GMEMS. Actuator for splinter-like skin penetration in glucose-sensing applications: Design and demonstration. In: Proceedings of 2016 IEEE SENSORS; 30 October-3 November 2016. Orlando, FL, USA: IEEE; 2017. DOI: 10.1109/ICSENS.2016.7808549

[5] Junagal K, Meena RS. Design and simulation of microstage having PZT MEMS actuator for 3D movement. In: Proceedings of International Conference on Advances in Computing, Communications and Informatics (ICACCI); 21-24 September; Jaipur, India. DOI: 10.1109/ICACCI.2016.7732375

[6] Donald BR, Levey CG, McGray CG, Paprotny I, Rus D. An untethered, electrostatic, globally controllable MEMS micro-robot. Journal of Microelectromechanical Systems. 2006;**15**(1):15. DOI: 10.1109/JMEMS.2005.863697

[7] Vogtmann D, Pierre St. R, Bergbreiter S. A 25 mg magnetically actuated microrobot walking at > 5 body lengths/sec. In: Proceedings of 2017 IEEE 30th International Conference on Micro Electro Mechanical Systems (MEMS); 22-26 January 2017; Las Vegas, NV, USA. pp. 179-182

[8] Murthy R, Stephanou HE, Popa DO. AFAM: An articulated four axes microrobot for nanoscale applications. IEEE Transactions on Automation Science and Engineering. 2013;**10**:276-284

[9] Elbuken C, Khamesee MB, Yavuz M. Design and implementation of a micromanipulation system using a magnetically levitated MEMS robot. IEEE/ASME Transactions on Mechatronics. 2009;**14**:434-445

[10] Yan G, Ye D, Zan P, Wang K, Ma G. Micro-robot for endoscope based on wireless power transfer. In: Proceedings of International Conference on Mechatronics and Automation (ICMA 2007); 5-8 August 2007; Harbin, China. DOI: 10.1109/ICMA.2007.4304140

[11] Beeby SP, Tudor MJ, White NM. Energy harvesting vibration sources for microsystems applications. Measurement Science and Technology. 2006;**17**:R175-R195

[12] Renaud M, Karakaya K, Sterken T, Fiorini P, Van Hoof C, Puers R. Fabrication, modelling and characterization of MEMS piezoelectric vibration harvesters. Sensors and Actuators A: Physical. 2008;**145-146**:380-386

[13] Epstein AH, Senturia SD. Macro power from micro machinery. Science. 1997;**276**:1211. DOI: 10.1126/science.276.5316.1211

[14] Janicek V, Husak M. Designing the 3D electrostatic microgenerator. Journal of Electrostatics. 2013;**71**:214-219

[15] Holmes AS, Hong G, Pullen KR. Axial-flux permanent magnet machines for micropower generation. Journal of Micro Electro Mechanical Systems. 2005;**14**:54-62

[16] Herrault F, Ji CH, Allen MG. Ultraminiaturized high-speed permanent-magnet generators for milliwatt-level power generation. Journal of Micro Electro Mechanical Systems. 2008;**17**:1376-1387

3

Study and Design of Reconfigurable Wireless and Radio-Frequency Components Based on RF MEMS for Low-Power Applications

Bassem Jmai, Adnen Rajhi, Paulo Mendes and Ali Gharsallah

Abstract

This chapter intends to deal with the challenging field of communication systems known as reconfigurable radio-frequency systems. Mainly, it will present and analyze the design of different reconfigurable components based on radio-frequency microelectromechanical systems (RF MEMS) for different applications. This chapter will start with the description of the attractive properties that RF MEMS structures offer, giving flexibility in the RF systems design, and how these properties may be used for the design of reconfigurable RF MEMS-based devices. Then, the chapter will discuss the design, modeling, and simulation of reconfigurable components based on both theoretical modeling and well-known electromagnetic computing tools such as ADS, CST-MWS, and HFSS to evaluate the performance of such devices. Finally, the chapter will deal with the design and perfor-mance assessment of RF MEMS-based devices. Non-radiating devices, such as phase shifter and resonators, which are very important components in the hardware RF boards, will be addressed. Also, three types of frequency reconfigurable antennas, for the three different applications (radar, satellite, and wireless communication), will be proposed and evaluated. From this study, based on theoretical design and electromagnetic computing evaluation, it has been shown that RF MEMS-based devices can be an enabling solution in the design of the multiband reconfigurable radio-frequency devices.

Keywords: RF MEMS capacitive switch, modeling, tunable, MMIC, phase shifter, RF resonator, frequency reconfigurable antennas

1. Introduction

The extraordinary evolution and the knowledge built-in the radio-frequency field were noticed in various applications such as militaries, medicine, and telecommunication. At the system level, one trend in the field of wireless telecommunications is the design of multiband and multimode devices, with an ever-increasing number of features, leading to the so pursued reconfigurable systems.

At present, reconfigurable systems have become very promising in a wide range of applications, including future services of wireless communication systems. However, the wide spread of wireless communication systems and the emergence of new wireless communication standards have introduced new challenges in the hardware design for transmitters and receivers. To tackle this problem, nowadays telecommunication systems need to use a significant number of tunable components, where the performance is degraded when compared to their equivalents at fixed frequencies.

A telecommunication system is said to be tunable or reconfigurable, when some of its characteristics (central frequency, bandwidth, polarization, etc.) can be modified by an external control signal (electrical, mechanical, thermal, etc.). Despite tunable components can be realized using many different designs, mainly, two approaches exist for tunability:

- The first way is achieved by the possibility to change the substrate permittivity (ferroelectric [1] and ferromagnetic [2]).
- The second way consists in a change of the capacitive or the inductive load by the addition of tunable radio-frequency integrated circuits (RFIC). This method relies on semiconductor devices (diode [3] and transistors [4]) or mechanical (RF MEMS [5]) components.

The RF MEMS devices feature low-power consumption, high linearity, wide bandwidth, and high dynamic range, which are among the most important requirements that each component must meet in order to achieve high-performance wireless systems.

This chapter will present the development of frequency and phase reconfigurable components, based on capacitive tunable RF MEMS.

2. RF MEMS technologies

RF MEMS has its origin in the MEMS systems, which are miniature electronic and/or mechanical systems designed to perform specific tasks. They consist of motors, gears, levers, electrical devices, or tiny sensors. These devices are used in many applications and their size range from a few micrometers to a few millimeters. By the late 1960s, MEMS systems were used as precise sensors of hydraulic pressure in aircraft. Today, these systems play an important and ever-increasing role in the fields of medicine (detection of organic cells), automotive (accelerometer in airbag triggering), entertainment (motion detection in a video game), and optics (micro mirrors), to mention just a few applications.

2.1. What is RF MEMS?

MEMS devices have found application in many different fields. As shown in **Figure 1**, the MEMS components can be classified into four main families [6]:

- **Sensors**: miniaturized systems, made from microtechnologies used for sensor applications in measurement and instrumentation fields, such as pressure sensors and capacitive accelerometer.
- **MOEMS**: this type of MEMS can be used in optical technologies, such as micro mirrors, optical switches, and optical cavities.
- **Bio-MEMS**: miniaturized systems, made from micro- and nanotechnologies derived from microelectronics (integrated circuits) which is intended to carry out experiments in biology/chemistry, such as DNA chip, microchemical reactor, and micro valves.
- **RF MEMS**: in the field of microwaves, they improve the performance of tunable devices to various functions, such as variable passive components, resonators, filters, and antennas.

In this way, RF MEMS is usually related with the application of MEMS technologies to develop systems that contribute to the RF system development. They can be used to increase the performance or to implement characteristics not achievable by other solutions, even if performance is slightly degraded. Despite a few drawbacks, the emergence of RF MEMS represents a revolution in the development of new radio-frequency systems. In fact, these elements should compete or replace certain semiconductor components in microwave applications. They are very compact (typically a few hundred square micrometers) and can be up to 50% smaller than semiconductor components performing the same function [7].

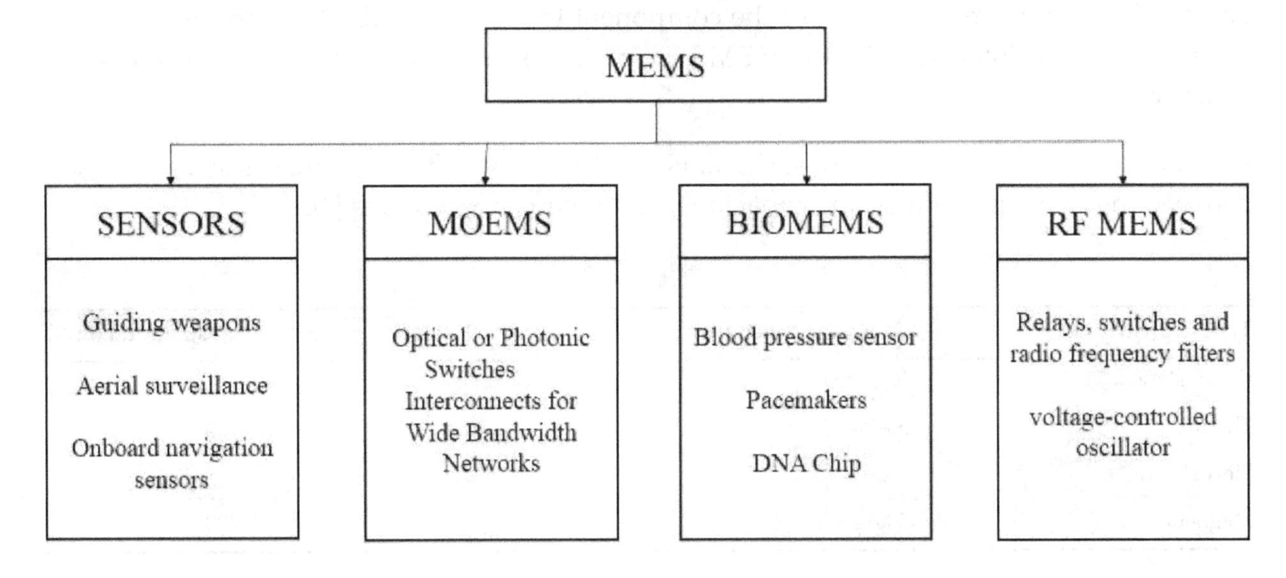

Figure 1. MEMS families.

2.2. RF MEMS switch as a building block

The microelectromechanical components enable the reconfiguration of electronic devices using mechanical movements. Using this feature, one building block widely used to enable a device's tunability is the MEMS switch, and we call it a MEMS microswitch. The microelectromechanical part of the microswitch, or varactor components, has the form of a mobile beam suspended and anchored to one of its ends. The beam can be built-in, or double-embedded. The main idea behind a tunable device is the fact that when a MEMS switch moves, besides switching from on and off states, it may exhibit a different RF load. And controlling such RF load, it is possible to tune different RF devices.

2.3. RF MEMS switch control mechanism

The mechanical movement of the beam is obtained by applying an actuating force. This actuating force is generally of an electrostatic nature [8, 9], but it can be thermal [10], piezoelectric [11], or magnetic [12]. **Table 1** is showing the comparative study of the different types of actuation [8, 13].

Electrostatic actuators are the most used components because they consume very little, occupy a very small volume, and has a short switching time. In this chapter, the electrostatically actuated RF MEMS will be explored as a solution for different applications.

2.4. RF MEMS switches topologies

RF MEMS microswitches are components intended to perform an electrical function through the control of a movable or mechanically deformable structure. There are two main types of RF MEMS components: the capacitive touch switch (contact: metal-dielectric-metal) and the resistive or ohmic contact switch (contact: metal-metal). In both cases, it is necessary to apply a force to the movable part of the component to move the MEMS beam. In this work, we will only be interested in RF MEMS capacitive shunt switch based on electrostatic actuators.

In order to summarize the previous points for RF MEMS, **Table 2** makes a comparison between the two types, ohmic and capacitive, and their configurations [14–17].

Type of switching	Switching speed (μs)	Switch size	Consumption (mW)
Electrostatic	0.05–200	Small	˜0
Thermal	50–200	Medium	<100
Piezoelectric	1–200	Medium	˜0
Magnetic	500–4000	Large	<200

Table 1. Comparison of the different types of actuation.

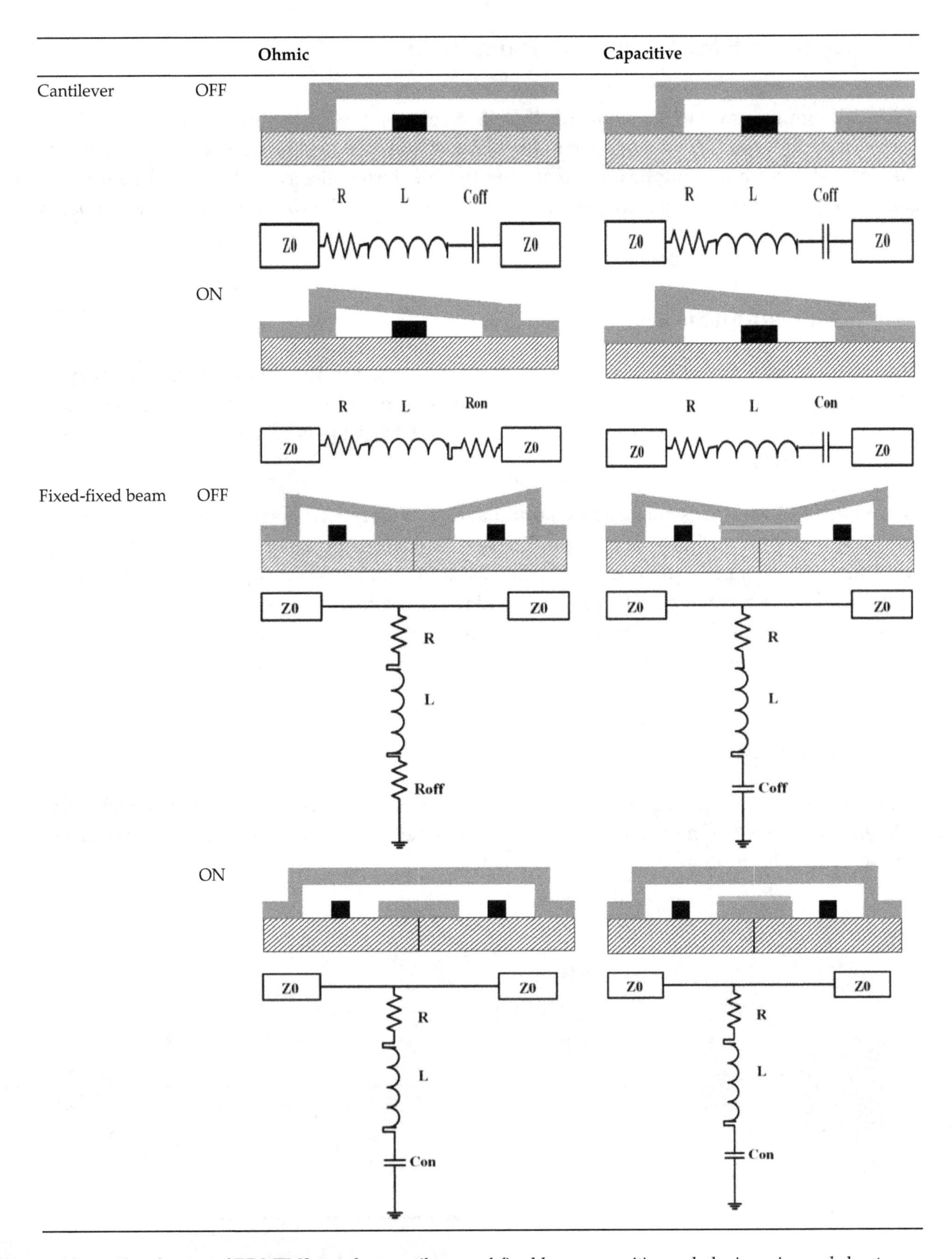

Table 2. Classification of RF MEMS switches: cantilever and fixed beam, capacitive and ohmic, series, and shunt electrical model of MEMS switches.

3. Capacitive RF MEMS mechanical modeling

Before designing any device using RF MEMS technology, it is required to understand the behavior of the main block, the switch. In this way, the first design step will be to compute the applied force a moving beam suffers due to an external electric field. It will be derived considering that the control signal is an electric potential applied between the movable beam and an activation electrode.

3.1. Capacitive RF MEMS device configuration

The performance, as well the analysis used to improve the RF MEMS device, is heavily dependent on the study of the bridge. **Figure 2** presents the proposed RF MEMS switch, with small dimensions (1200 × 900 × 681 μm^3). This RF MEMS is based on CPW technology (G/S/G) = (90/120/90).

This RF MEMS varactor structure has a multilayer configuration. The used substrate is based on silicon (Si) with thickness of 675 μm. The second layer is made of a silicon dioxide (SiO_2) with thickness equal to 2 μm and a CPW line circuit metal based on copper with thickness of 1 μm. The bridge has a depth of 1 μm, with ends attached to the groundline of the CPW by an epoxy (polymer based on negative-tone photoresist SU-8 2000.5 with 3 μm thickness). The dielectric is fabricated through a silicon nitride (Si_3N_4) with depth equal to 1 μm.

3.2. Mechanical model

The first step when modeling a RF MEMS device is to determine the electromechanical behavior of the switch, meaning that we want to understand how, and how much, the structure will move in response to an applied voltage.

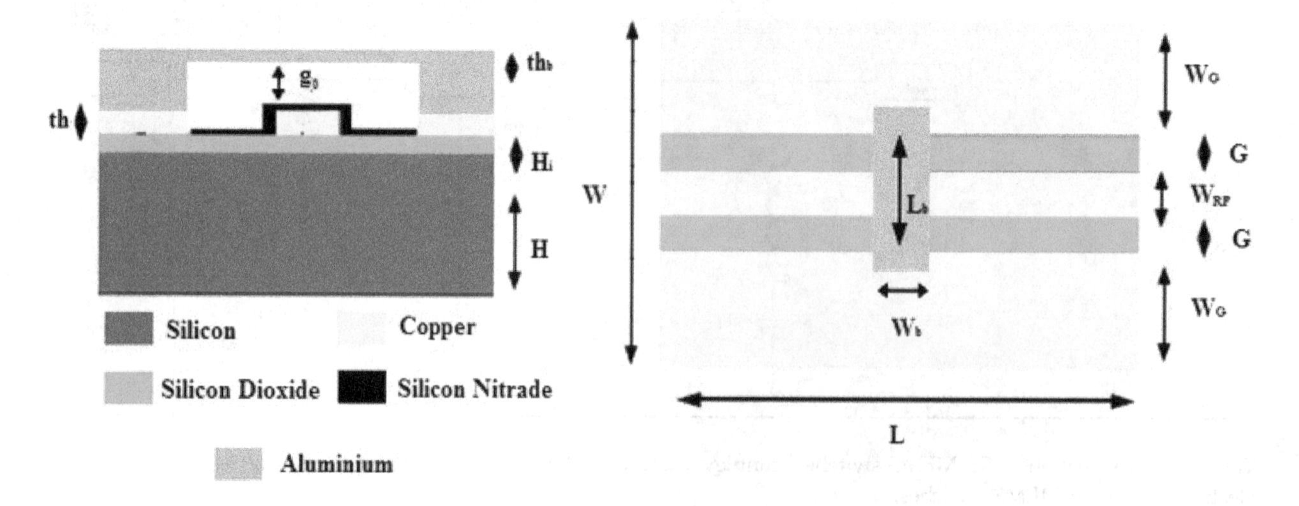

Figure 2. Design of the proposed capacitive RF MEMS device.

The uniform bridge of MEMS is geometrically simple since it is only a rectangular suspension located above a tape connected to contact pads by the sides of the same width, as shown in **Figure 3** [18].

Nominal capacity: the capacitance C between the two electrodes is given by:

The mechanical model of deformable flat capacity is given by Eq. (1), where g is the height between the low beam and the electrode. The bridge width is denoted by W_b and the length of the ground electrode by W:

$$C = \varepsilon_0 \frac{W_b W}{g} \tag{1}$$

The height depends on the voltage applied between the electrodes. In the absence of voltage (V = 0), the height is equal to g_0 and the capacity is named C_0 [8]:

$$dW_t = dW_e - dW_m = (V.dq) - (Fe.dg_e) \tag{2}$$

where q is the quantity of charge accumulated in the capacity and Fe is the electrostatic force.

The well-known equation of potential of electrostatic energy is given by Eq. (3):

$$U_E(j) = \frac{1}{2} C(V).V^2 \tag{3}$$

Then, the electrostatic force for a flat capacity can be expressed by Eq. (4):

$$F_e = \frac{\partial We}{\partial g} = -\frac{1}{2}\varepsilon_0 \frac{W_b W}{g^2} V^2 \tag{4}$$

The mechanical behavior of the beam can be modeled by a spring of constant kz. This induces a mechanical force (F_m) exerted by the bridge. This force is the opposite of Fe and it is defined by Eq. (5):

$$F_m = -F_e = k_z(g_0 - g) = -\frac{1}{2}\varepsilon_0 \frac{W_b W}{g^2} V^2 \tag{5}$$

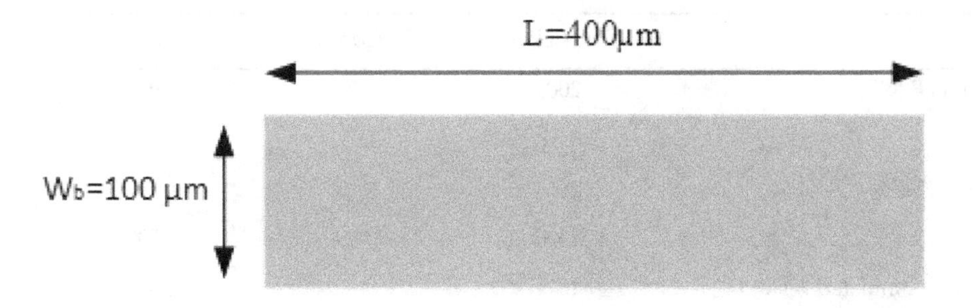

Figure 3. The beam of the MEMS under analysis.

From this equation, we can then conclude a relationship between the height of the air gap g and the applied voltage V:

$$0 = g^3 - g_0 g^2 + \frac{1}{2k_m} \varepsilon_0 W_b W V^2 \quad (6)$$

The relationship between the applied voltage and the spacing g parameter is given in Eq. (7):

$$V_p = \sqrt{\frac{2k_z}{\varepsilon_0 W_b W} g^2 (g_0 - g)} \quad (7)$$

$$k_z = \frac{1}{2}\left(32Ew\left(\frac{t}{l}\right)^3 + 8\sigma(1 - \vartheta)w\left(\frac{t}{l}\right)\right). \quad (8)$$

where E is the Young's modulus, σ is the residual stress of the beam, υ is the Poisson's coefficient, t is the thickness, and l is the length of the bridge.

In **Table 3**, the mechanical properties (Young's modulus, Poisson's ratio, residual stress, and density) of four different bridge materials (nickel, copper, aluminum, and gold) are presented.

In terms of control voltage, comparatively the nickel presents a bad choice, since the required voltage to obtain some deflection is near twice the voltage that is required for gold and aluminum. However, the gold price presents an obstruction, being the best choice, in this comparative study, the aluminum.

The bridge was simulated using COMSOL multiphysic and reaches a deflection equal to 2 μm. The obtained simulation results are given in **Figure 4** for applied voltage equal to 25 V. The relationship between the capacitance and the applying voltage is shown in **Figure 5**.

3.3. RF MEMS switch design parameters

We will present next the relevant parameters that define the variable MEMS capacity.

Tuning range: the "tuning range" or variation of the capacity is an important factor of the variance MEMS capacities. It is defined as

Material	Nickel	Copper	Aluminum	Gold
Young's modulus E (GPa)	200	120	69	79
Poisson's ratio	0.31	0.355	0.345	0.42
Residual stress σ (MPa)	20	20	20	20
Density (kg/m^3)	8900	8960	2700	19,300
Pull-down voltage simulated result [V]	46	34	25	22

Table 3. Mechanical parameters of different materials of the bridge.

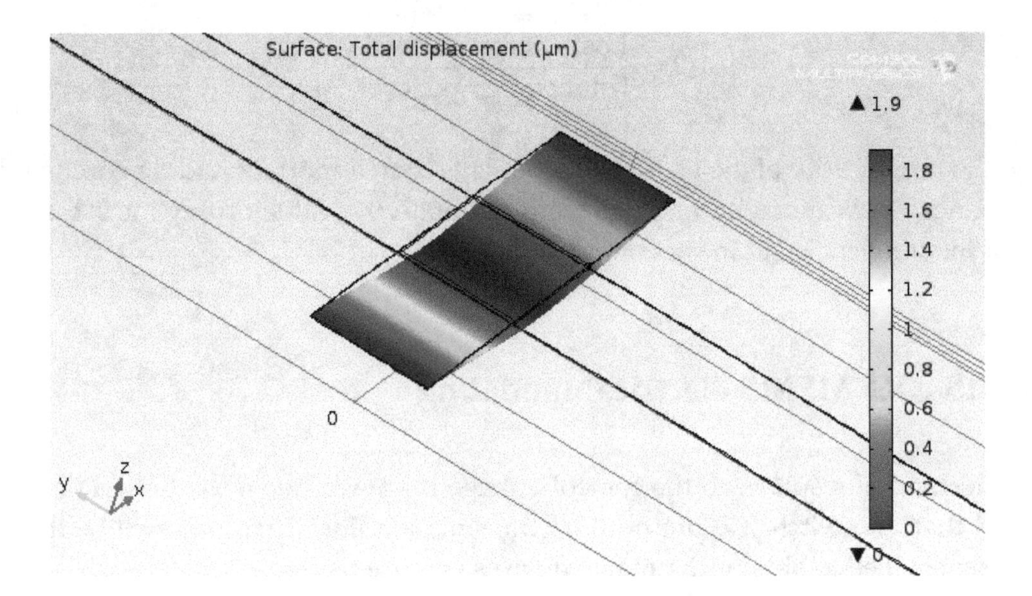

Figure 4. Simulation results of the beam.

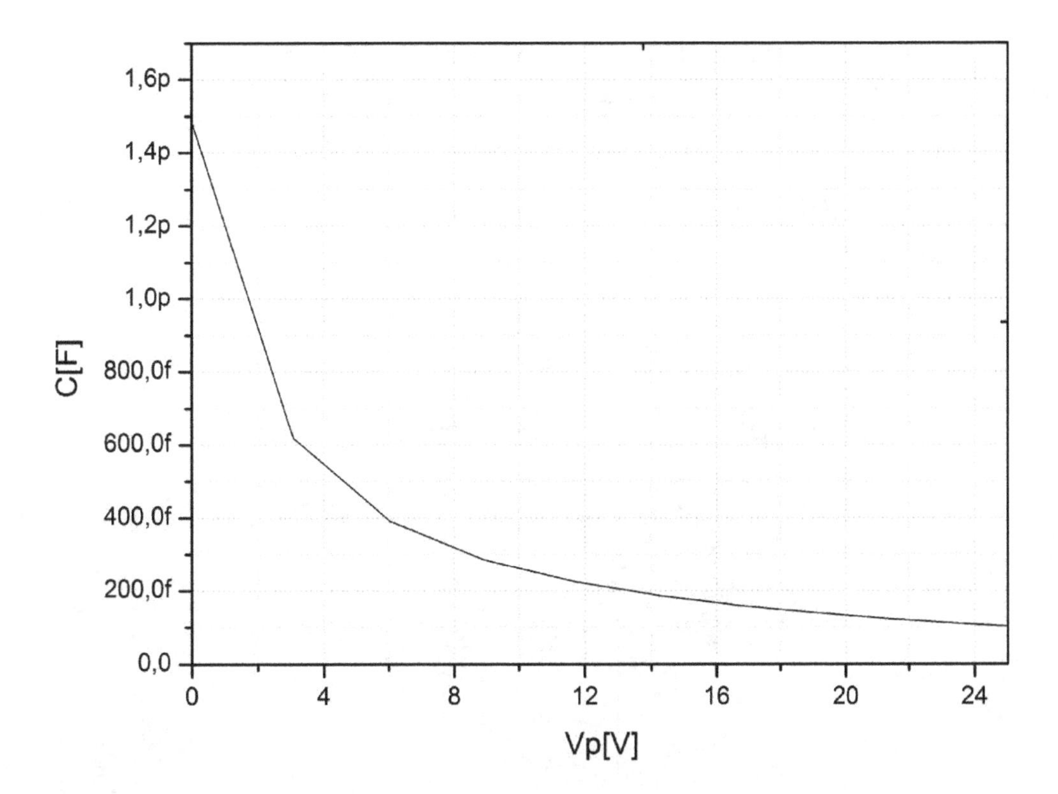

Figure 5. Relationship between the capacity and the voltage applied.

$$TR = \frac{C_{max} - C_{min}}{C_{min}} \tag{9}$$

Quality factor: the quality factor Q of a component is an important parameter. Indeed, it determines the losses of a variable filter or the noise of a VCO using a variable capacity. It is defined by the ratio between the energy stored and the energy lost by the component:

$$Q = \frac{\text{Lost energy per cycle}}{\text{Total energy per cycle}} \tag{10}$$

Linearity: the nonlinearity of the passive components is an important and demanding data for radio-frequency applications. In fact, we want to obtain the value of the linear capacity as a function of the frequency and the actuating voltage.

4. Capacitive RF MEMS electrical modeling

Once the mechanical response to the control voltage is known, the next step will be to model the RF load that the switch will present to a transmission line. That load will be the variable that can be controlled to obtain RF tunable devices or systems.

4.1. RF modeling approach

Figure 6 shows the proposed model of the proposed RF MEMS [19]. The proposed circuit model consists of two CPW lines, separated by a shunt RLC circuit. The RLC is the equivalent bridge circuit.

The MEMS can be modeled by the association of three subsystems in cascade. The ABCD matrix is given by Eq. (11):

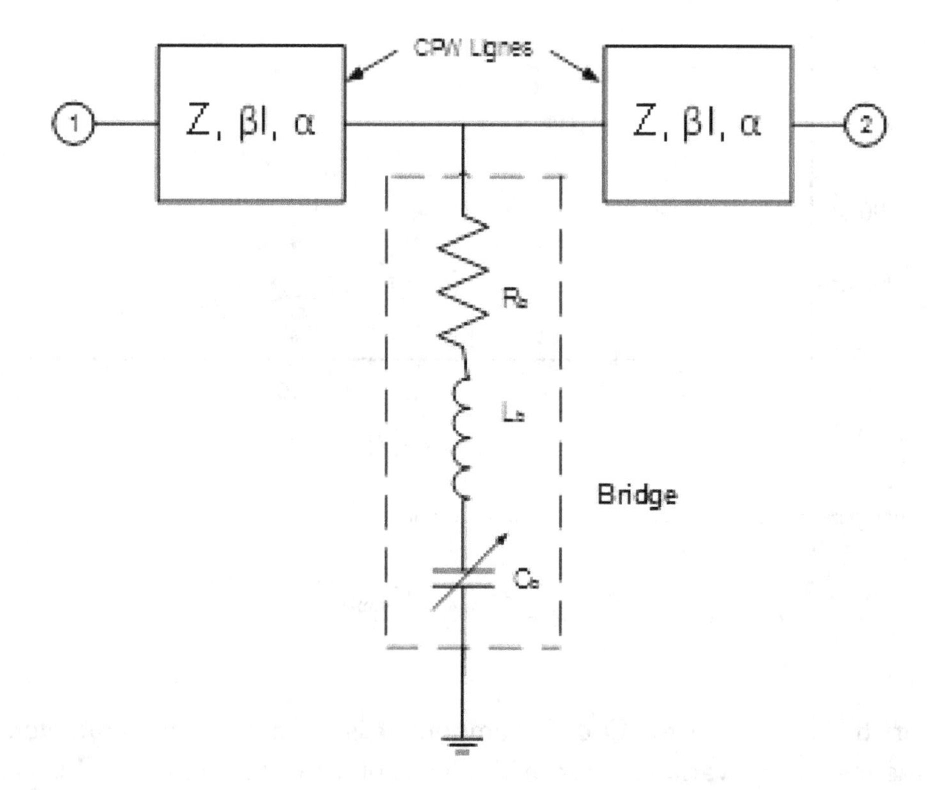

Figure 6. Electromagnetics model.

$$\begin{bmatrix} A & B \\ C & D \end{bmatrix} = \underbrace{\begin{bmatrix} A_1 & B_1 \\ C_1 & D_1 \end{bmatrix}}_{TL1} * \underbrace{\begin{bmatrix} A_b & B_b \\ C_b & D_b \end{bmatrix}}_{Bridge} * \underbrace{\begin{bmatrix} A_2 & B_2 \\ C_2 & D_2 \end{bmatrix}}_{TL2} \quad (11)$$

Considering a system represented by an association of three devices where the ABCD matrix is known, the next step is to determine each subsystem matrix.

4.1.1. Coplanar waveguide modeling

The CPW's most important electrical parameters are the characteristic impedance Z_c, and the effective permittivity ε_{eff}, both given by Eq. (12), where K (k) and K′ (k) present the elliptic integral which essentially depends on the CPW's geometric and physical characteristics. Here, ε_r is the relative permittivity, w is the width of the RF line, s is the gap between the RF line and ground, h is the thickness of substrate, and h_1 is the thickness of the buffer layer:

$$\begin{cases} Z_C = \dfrac{30\pi}{\sqrt{\varepsilon_{eff}}} \dfrac{K'(\mathrm{k})}{K(\mathrm{k})} \\ \varepsilon_{eff} = 1 + \dfrac{\varepsilon_r - 1}{2} \dfrac{K'(\mathrm{k}) K(\mathrm{k}_1)}{K(\mathrm{k}) K'(\mathrm{k}_1)} \end{cases} \quad (12)$$

$$\begin{cases} k = \dfrac{w}{w + 2s} \\ k_1 = \dfrac{sh\left(\dfrac{\pi w}{4h}\right)}{sh\left(\dfrac{\pi(w + 2s)}{2h}\right)} \end{cases} \quad (13)$$

$$\begin{cases} K'(\mathrm{k}) = K(\mathrm{k}') \\ \mathrm{k}' = \sqrt{1 - k} \end{cases} \quad (14)$$

The ratio $K(k)/K'(\mathrm{k})$ approximation is done by Eq. (15):

$$\frac{K(k)}{K'(\mathrm{k})} = \begin{cases} 0 \le k \le \dfrac{1}{\sqrt{2}} \Rightarrow \dfrac{K(k)}{K'(\mathrm{k})} = \dfrac{\pi}{\ln\left(2 \dfrac{1 + \sqrt{k}}{1 - \sqrt{k}}\right)} \\ \dfrac{1}{\sqrt{2}} \le k \le 1 \Rightarrow \dfrac{K(k)}{K'(\mathrm{k})} = \dfrac{1}{\pi} \ln\left(2 \dfrac{1 + \sqrt{k}}{1 - \sqrt{k}}\right) \end{cases} \quad (15)$$

The second parameter for modeling the CPW line is the propagation constant. This parameter is given by Eq. (16), as a function of the attenuation constant and the phase constant. The attenuation constant is due to the conductor as well as to the attenuation in dielectric, both presented in Eq. (17):

$$\gamma = \alpha + j\beta \begin{cases} \alpha = \alpha_c + \alpha_d \\ \beta = \dfrac{2\pi}{\lambda}; \lambda = \dfrac{c}{f\sqrt{\varepsilon_{eff}\mu_{eff}}}; c = 3 \times 10^8 \ \text{m s}^{-1} \end{cases} \tag{16}$$

$$\begin{cases} \alpha_d = 27.3 \dfrac{\varepsilon_r}{\varepsilon_r - 1} \dfrac{\varepsilon_{eff}(f) - 1}{\sqrt{\varepsilon_{eff}(f)}} \dfrac{\tan\delta}{\lambda_0} (\text{dB/lenght}) \\ \alpha_c = \dfrac{8.68R_s}{480\pi K(k_1)K(k_1')\left(1 - k_1^2\right)} \left[\begin{array}{l} \dfrac{1}{a}\left(\pi + Ln\left(\dfrac{8a\pi(1-k_1)}{t(1+k_1)}\right)\right) \\ + \dfrac{1}{b}\left(\pi + Ln\left(\dfrac{8b\pi(1-k_1)}{t(1+k_1)}\right)\right) \end{array} \right] (\text{dB/lenght}) \end{cases} \tag{17}$$

The model of CPW line is presented by ABCD matrix in Eq. (18):

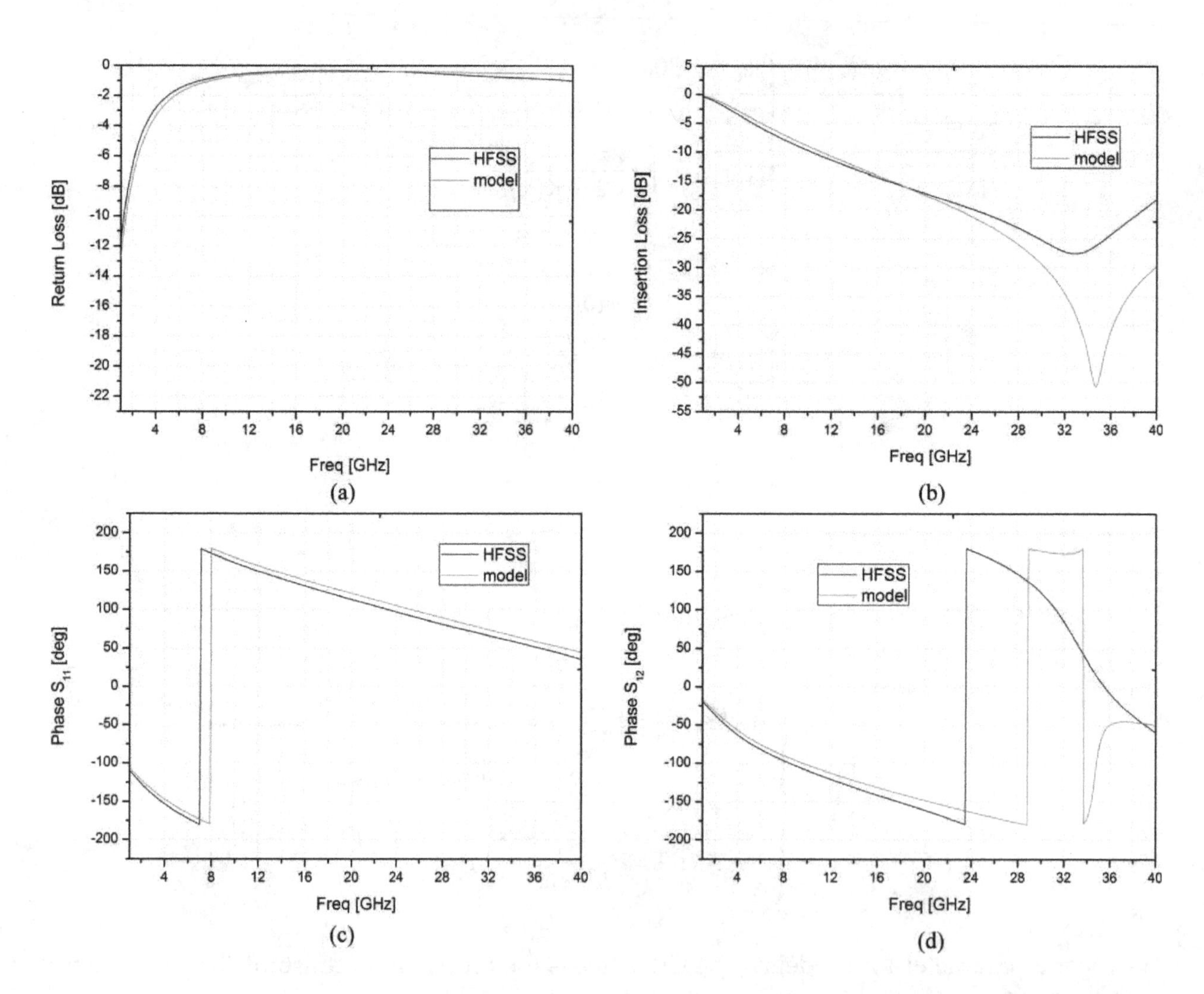

Figure 7. Simulation results of the scattering parameters versus frequency at the OFF state: (a) return loss, (b) insertion loss, (c) phase of S_{11}, (d) phase of S_{12}.

$$\begin{bmatrix} A & B \\ C & D \end{bmatrix} = \begin{bmatrix} \cosh(\gamma L) & Z_c \sinh(\gamma L) \\ \dfrac{\sinh(\gamma L)}{Z_c} & \cosh(\gamma L) \end{bmatrix} \quad (18)$$

4.1.2. Bridge modeling

The bridge is modeled by an inductor (L_b) and a resistance (R_b) in series. These parameters are independent of the substrate. The proposed RF MEMS has a capacitive variable (C_b). The parameters (L_b), (R_b), and (C_b) are given by Eqs. (19) and (20):

$$\begin{cases} R_b = \dfrac{L}{\sigma w t} \\ L_b = 0.2^*[(\ln(L/(w+t))) + 1.193 + 0.2235^*((w+t)/L)]*L \end{cases} \quad (19)$$

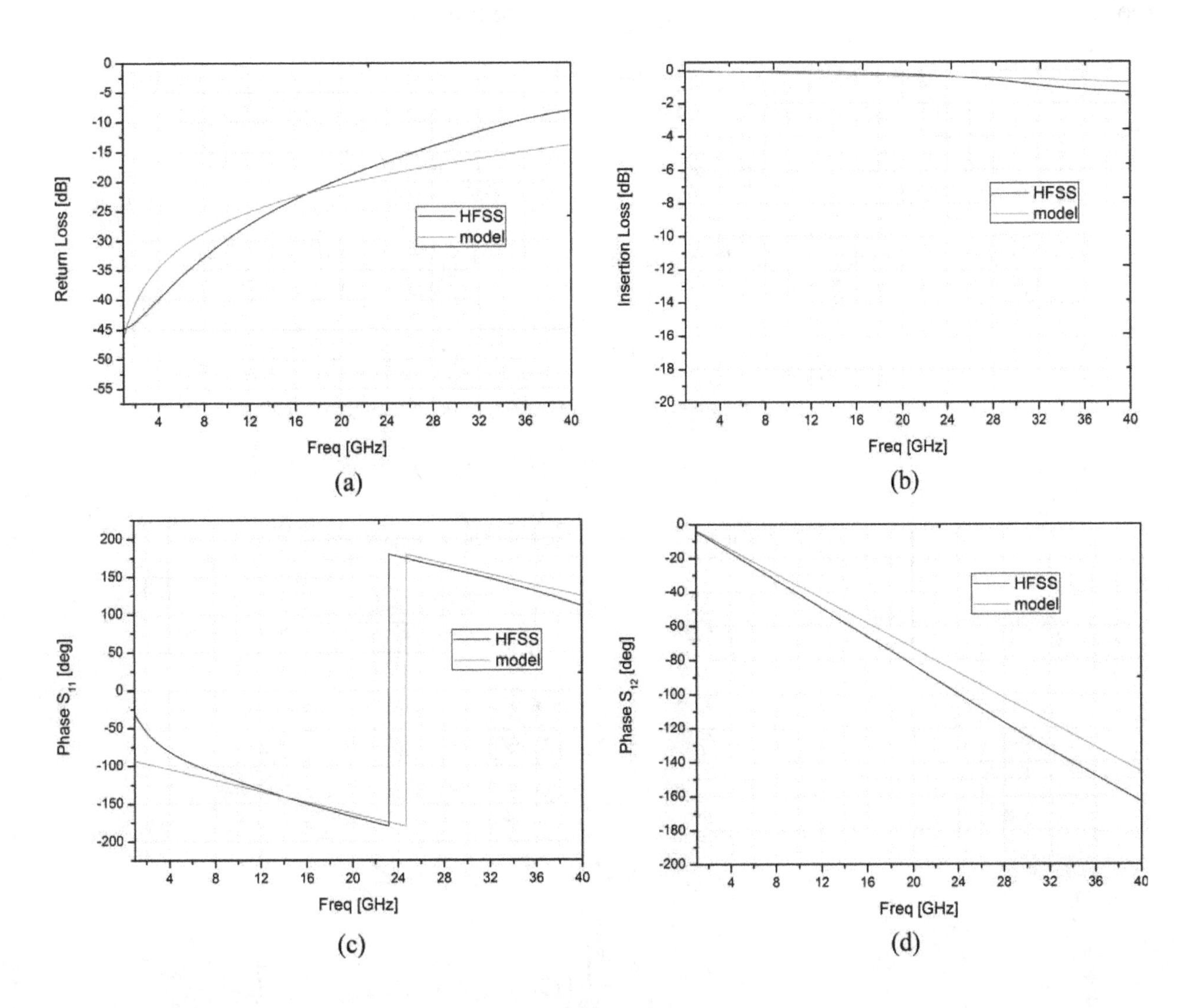

Figure 8. Simulation results of the scattering parameters versus frequency at the ON state: (a) return loss, (b) insertion loss, (c) phase of S_{11}, (d) phase of S_{12}.

$$C_b = \begin{bmatrix} C_{down} = \dfrac{\varepsilon_0 \varepsilon_r A}{th} \\ C_{up} = \dfrac{\varepsilon_0 A}{g_0 + \dfrac{th}{\varepsilon_r}} \end{bmatrix} \tag{20}$$

The bridge model can be presented by the next ABCD matrix:

$$\begin{bmatrix} A_b & B_b \\ C_b & D_b \end{bmatrix} = \begin{bmatrix} 1 & 0 \\ \dfrac{1}{Z_b} & 1 \end{bmatrix} = \begin{bmatrix} 1 & 0 \\ \dfrac{1}{R_b + jX_b} & 1 \end{bmatrix} = \begin{bmatrix} 1 & 0 \\ \dfrac{1}{R_b + j(L_b\omega + 1/C_b\omega)} & 1 \end{bmatrix} \tag{21}$$

4.2. The scattering parameters model

The standard output of simulation tools is the S-parameters. In this way, we need to transform the previous developed model to present it in a convenient way to allow comparison with such tools. The scattering parameters can be expressed by the following form:

$$\begin{bmatrix} S_{11} & S_{12} \\ S_{21} & S_{22} \end{bmatrix} = \begin{bmatrix} \dfrac{A + B/Z_0 - CZ_0 - D}{A + B/Z_0 + CZ_0 + D} & \dfrac{2(AD - BC)}{A + B/Z_0 + CZ_0 + D} \\ \dfrac{2}{A + B/Z_0 + CZ_0 + D} & \dfrac{-A + B/Z_0 - CZ_0 + D}{A + B/Z_0 + CZ_0 + D} \end{bmatrix} \tag{22}$$

where the reflection coefficient and its phase are given by Eq. (23) and the insertion loss and its phase are given by Eq. (24). The scattering parameters are written in the following form:

$$S_{11} = S_{22} = |S_{11}|e^{j\Phi_{11}} = \frac{(Z_c{}^2 - Z_0^2) + 2R_bZ_c) + j2Z_cX_b}{\left((Z_0 + Z_c)^2 + 2R_b(Z_0 + Z_c)\right) + j2X_b(Z_0 + Z_c)}$$

$$=> \begin{cases} |S_{11}| = \left\{ \dfrac{\sqrt{((Z_c{}^2 - Z_0^2) + 2R_bZ_c)^2 + (2Z_cX_b)^2}}{\sqrt{\left((Z_0 + Z_c)^2 + 2R_b(Z_0 + Z_c)\right)^2 + (2X_b(Z_0 + Z_c))^2}} \right\} \\ \Phi_{12} = tg^{-1}\left(\dfrac{2Z_cX_b}{2R_bZ_c}\right) - tg^{-1}\left(\dfrac{2X_b(Z_0 + Z_c)}{\left((Z_0 + Z_c)^2 + 2R_b(Z_0 + Z_c)\right)}\right) \end{cases} \tag{23}$$

$$S_{12} = S_{21} = |S_{12}|e^{j\Phi_{12}} = \frac{2R_bZ_c + j2Z_cX_b}{\left((Z_0 + Z_c)^2 + 2R_b(Z_0 + Z_c)\right) + j2X_b(Z_0 + Z_c)}$$

$$=> \begin{cases} |S_{12}| = \dfrac{\sqrt{(2R_bZ_c)^2 + (2Z_cX_b)^2}}{\sqrt{\left((Z_0 + Z_c)^2 + 2R_b(Z_0 + Z_c)\right)^2 + (2X_b(Z_0 + Z_c))^2}} \\ \Phi_{12} = tg^{-1}\left(\dfrac{2Z_cX_b}{2R_bZ_c}\right) - tg^{-1}\left(\dfrac{2X_b(Z_0 + Z_c)}{\left((Z_0 + Z_c)^2 + 2R_b(Z_0 + Z_c)\right)}\right) \end{cases} \tag{24}$$

The simulation results of the capacitive RF MEMS switch at the two states OFF (the bridge at downstate) and ON (bridge position g = 3 μm) are shown, respectively, in **Figures 7** and **8**.

To validate our model, we simulated the capacitive RF MEMS switch with HFSS. The results are compared in terms of return loss, insertion loss, and phase at the two states of the capacitive RF MEMS switch (ON and OFF). Here, we intend to show the similarity between the results of our model and the software HFSS simulation.

5. RF MEMS-based reconfigurable component design

This section will present the use of RF MEMS switches to obtain different tunable RF devices, namely, a phase shifter, a resonator, and a tunable antenna.

5.1. Reconfigurable phase shifter at 18 GHz based on RF MEMS

The reflection-type phase shifter as shown in **Figure 9** is constituted with hybrid coupler and RF MEMS capacitive switches (i.e., metal-dielectric-metal) [20]. The first RF MEMS is connected between a through-port and ground. The second is linked between the coupled port and ground.

In this design, the tunability is achieved by the use of a capacitive RF MEMS switch acting as a reflection load. The capacitor value, which is controlled by a DC voltage, operates from downstate to upstate. This variable capacitance is used to tune the variable phase shifter.

The reflection-type phase shifter using switch RF MEMS capacitive was implemented in ADS simulation software. The reflection coefficient Γ is given by Eq. (25), where ($X_b = L_b\omega - (1/c_b\omega)$).

$$\Gamma = |\Gamma|e^{j\Phi_{21}} = \frac{Z_s - Z_0}{Z_s + Z_0} = \frac{[Z_c^2 + 2Z_cR_b - Z_cZ_0 - Z_0R_b] + jX_b[2Z_c - Z_0]}{[Z_c^2 + 2Z_cR_b + Z_cZ_0 + Z_0R_b] + jX_b[2Z_c + Z_0]} \quad (25)$$

Figure 10 shows the RF phase shifter performance around 18 GHz, in terms of the return and insertion loss, and the phase shift dependency on the applied voltage. Despite not fully linear, it is possible to observe an almost linear characteristic of the phase shifter in different frequency ranges.

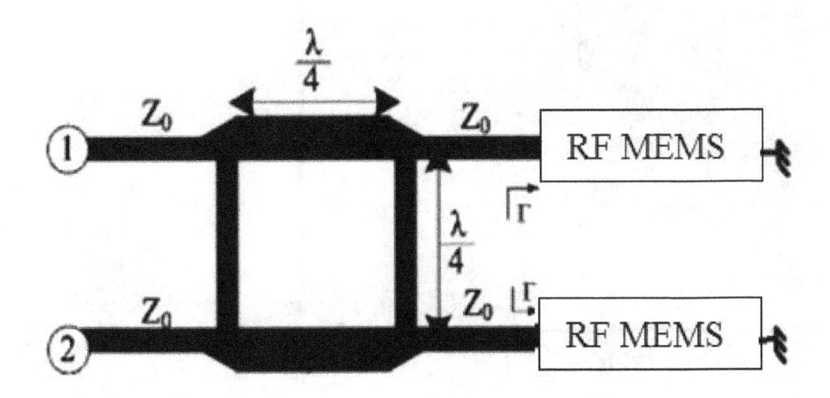

Figure 9. Simulation results of the scattering parameters versus frequency at ON state.

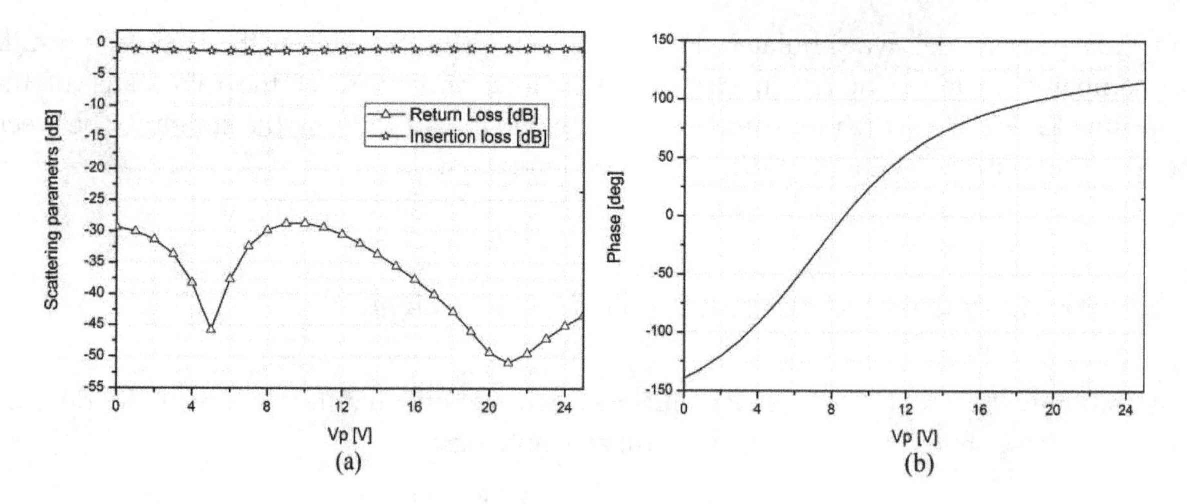

Figure 10. Simulation results of the phase shifter versus the applied voltage at 18 GHz: (a) scattering parameters and (b) phase.

5.2. MEMS-based reconfigurable resonator

There is an important claim that reconfigurable radio-frequency components on a single chip with high performances and multiband characteristics may be a solution for wireless communication [21, 22]. In this study, an improvement of the capacitive RF MEMS structure is proposed in order to obtain a reconfigurable resonator. **Figure 11** shows the suggested RF MEMS resonator structure [23].

The tunable RF MEMS characteristic was designed based on capacitive and inductive effects. The capacitive effect is due to the space between the bridge and the RF line, while the inductive

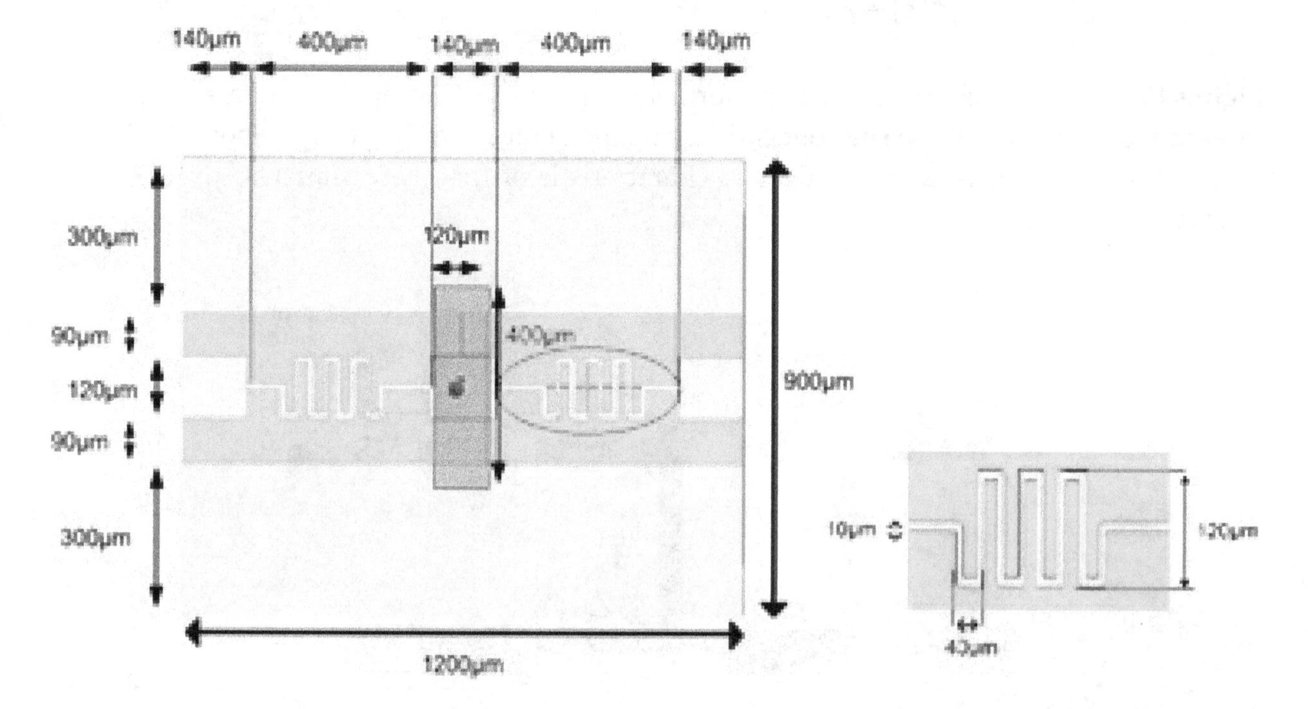

Figure 11. Proposed resonator RF MEMS.

effect is due to the presence of two meander inductors which are integrated in line with the RF waveguide. The combination of these two effects leads to a resonant phenomenon, introducing different resonant frequencies. If the applied voltage Vp is equal to 0 V, the bridge is in the UP state; therefore, the device is at a normally ON state. Moreover, the spacing *g* between the membrane bridge and the RF line affects the resonance frequency. The spacing *g* among bridge and CPW line varies between g = 2 μm at OFF state and g = 3 μm at ON state.

The proposed tunable resonator was simulated with HFSS and CST-MWS tools. **Figure 12** presents the scattering parameters for different bridge positions, in order to achieve a tunability on the frequency band between 10 GHz and 40 GHz.

Figure 12a and **b** shows, respectively, the return loss (S_{11}) and the insertion loss (S_{12}) for g = 2, 2.5, and 3 μm. It is possible to observe that controlling the bridge position level allows to obtain three resonant frequencies: 21.9, 24, and 25.1 GHz.

The S_{12} parameter presents almost constant value equal to −1 dB for all simulated spacing *g* factor when the S_{11} parameter is down to −10 dB. There is a good correspondence between the simulation results on HFSS and CST-MWS simulators.

Table 4 summarizes the spacing *g* factor versus the applied voltage. Moreover, the resonance frequency and the frequency range in different states of the bridge are shown. The RF MEMS is

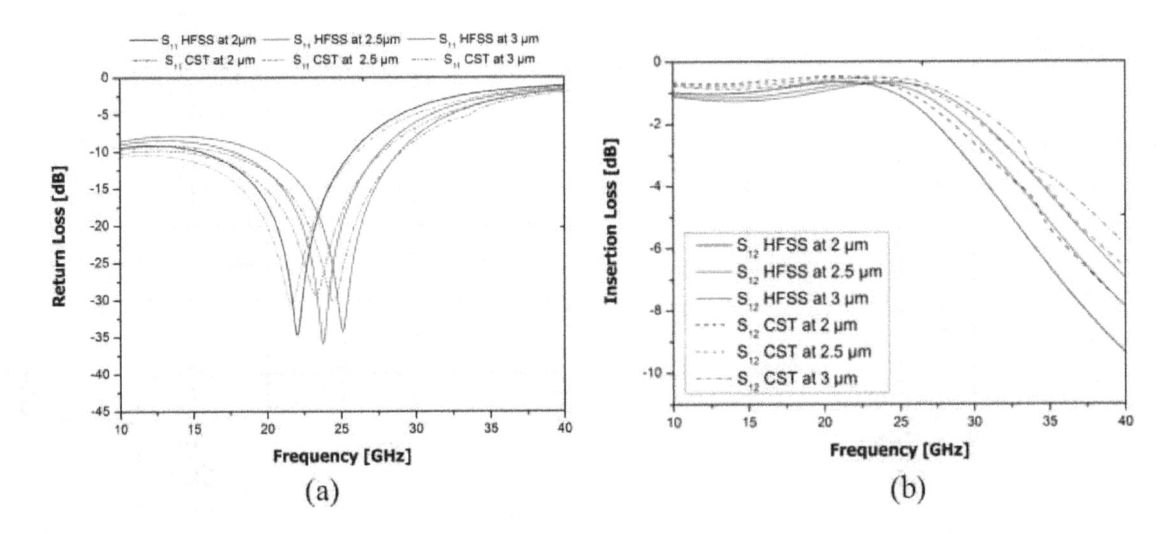

Figure 12. Scattering parameters at g = 2, 2.5, and 3 μm: (a) return loss (b) insertion loss.

Space *g* (μm)	Applied voltage (V)	Cover band			
		Resonance frequency (GHz)		Frequency range (GHz)	
		HFSS	CST	HFSS	CST
2	25 V	21.9	21	15.6–25.7	10–26.1
2.5	19 V	24	23.1	17.8–27.6	14.4–27.8
3	0 V	25.1	24.6	19.5–29	16.8–29

Table 4. Simulation results of the proposed resonator.

normally ON component, i.e., at g = 3 μm, the applied voltage equal to 0 V. This table presents a comparison study of the simulation results of the resonator between HFSS and CST-MWS. The proposed resonator covers three bands.

5.3. Reconfigurable antenna based on a RF MEMS resonator

Figure 13 presents the structure of the proposed reconfigurable CPW antenna [24], which is based on the integration of a resonating RF MEMS device with the CPW antenna on the same substrate. The reconfigurability of this antenna depends on the load provided by the state of

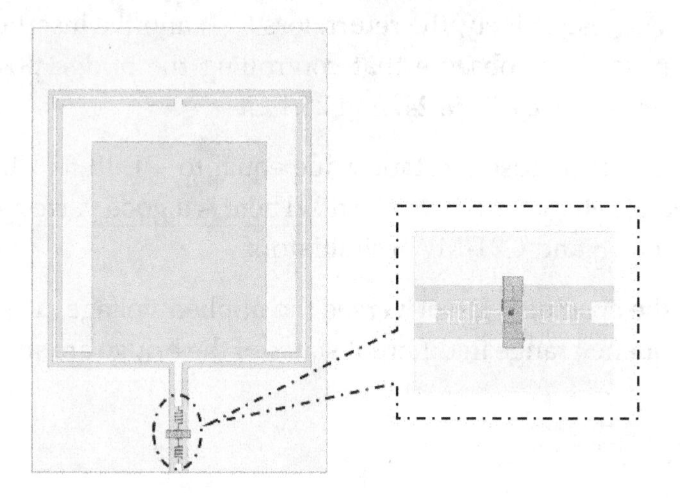

Figure 13. Reconfigurable antenna based on RF MEMS.

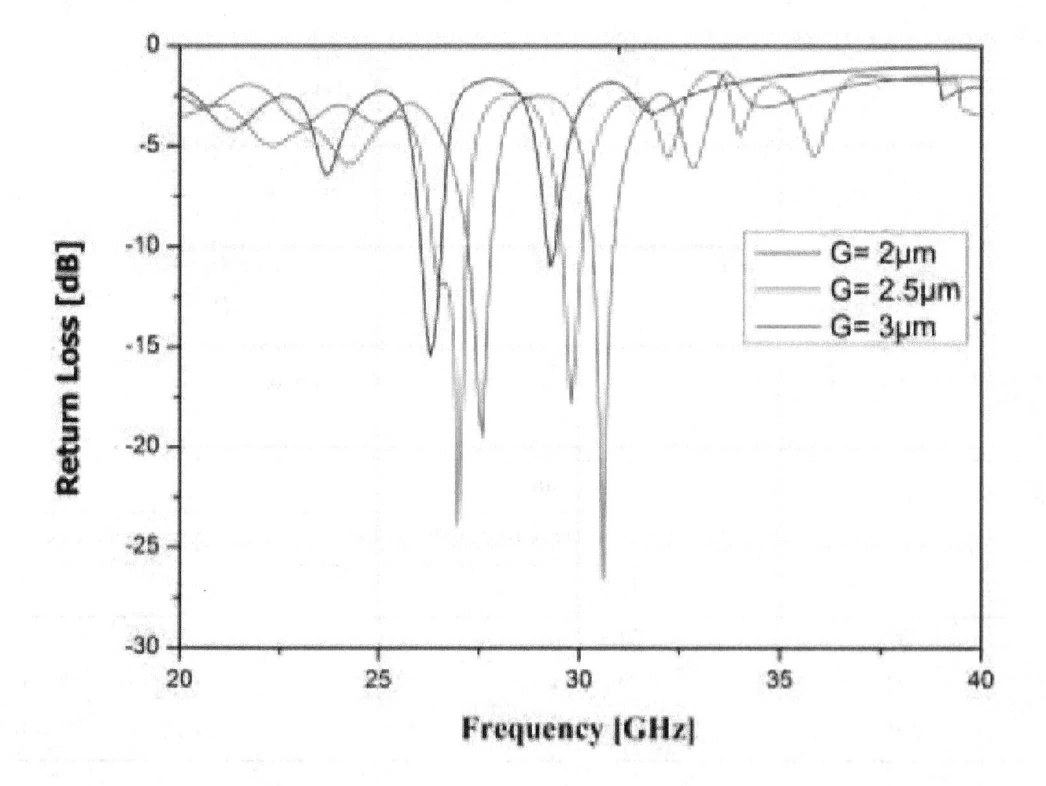

Figure 14. Return loss results versus frequencies.

the RF MEMS resonator. The antenna considered was a modified patch antenna with a printed inverted U-shaped ring resonator.

Figure 14 shows the reflection coefficient simulation results. The resonant frequencies can be observed at three states of the bridge. For g = 2 μm, the device has a single resonant frequency of 26.3 GHz, and the return loss will be 15.1 dB; for g = 2.5 μm, it shows two resonant frequencies: first at 27 GHz with a return loss of 23 dB and, second at 29.8 GHz with a return loss of 18 dB; and for g = 3 μm with also two resonant frequencies: 27.5 and 30.6 GHz, with return loss of 19.84 dB and 26.62 dB, respectively.

Figure 15 shows the radiation pattern for different resonance frequencies at three different antenna configuration states, considering phi = 90°. Firstly, the three states bridge given three resonance frequencies and the main lobe at teta = 310°. Secondly, only for g = 2.5 and g = 3 μm given the resonance frequency and the main lobe at teta = 0°.

Table 5 summarizes the simulation results of the reconfigurable antenna in terms of the spacing *g* factor versus the applied voltage, the resonant frequencies, the return loss, bandwidths, and gains.

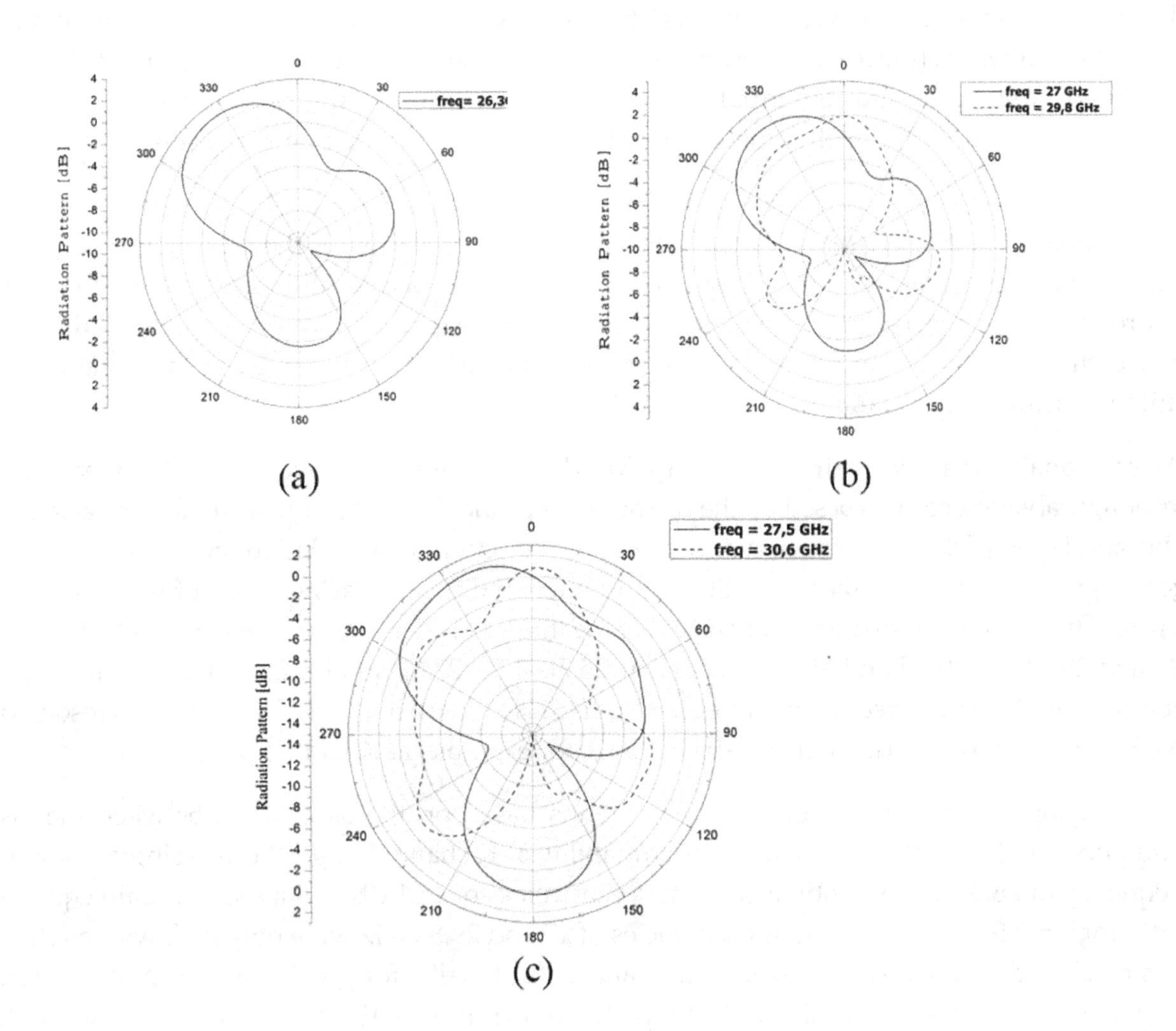

Figure 15. Realized gain of the reconfigurable antenna at different states: (a) at 2 μm, (b) at 2.5 μm, and (c) at 3 μm.

Parameters	Values				
Space g (μm)	2	2.5		3	
Applied voltage (V)	25	19		0	
Resonance frequency (GHz)	26.3	27	29.8	27.5	30.6
RL (dB)	15.1	23	18	18.84	26.62
Frequency range (GHz)	26–26.6	26.4–27.3	29–30.1	27.3–28	30.13–30.7
BW (%)	2.281	3.333	3.691	2.545	1.863
Gain (dB)	3	3	2	2	1

Table 5. The reconfigurable antenna results.

6. Conclusions

In this chapter, we have developed a parametric model for a capacitive RF MEMS switch in the frequency range between 1 and 40 GHz. A comparative study was carried out to analyze the theoretical model's mechanical characteristic with COMSOL and to compare the theoretical electrical properties (return loss, insertion loss, and phase) to those obtained using HFSS. The proposed model computes the input data swiftly and efficiently and produces results similar to those obtained with the aforementioned simulation software, while being much faster and less resource-demanding.

It was also presented the design, the analysis, and the simulation of the reflection-type phase shifter, using integrated RF MEMS switches. The proposed phase shifter at 18 GHz, which has approximately a constant insertion loss (≈−1 dB) with a significant return loss (<–30 dB) and phase shift (−138.822° and 128.15°), can be used in different RF applications, namely, it is suitable for radar applications.

We have analyzed a new contribution for RF MEMS to obtain a tunable resonator. The idea of this reconfigurable resonator is based on the use of two meander inductors and a variable capacitance. The simulation of this component was made by two electromagnetic design tools, and there was good agreement between them for different tuning conditions (spacing states of g = 2, 2.5 and 3 μm). The obtained resonance frequencies for the three considered states were, respectively, 21.9, 24, and 25.1 GHz. The bandwidths were [15.6, 25.7], [17.8, 27.6], and [19.5, 29] GHz, respectively, demonstrated for the three resonant frequencies (|S11| = 35 dB and |S12| = 1 dB). This resonator switches can be used in different RF applications, for example, at K and Ka bands.

We have also proposed a reconfigurable antenna based on the association between the last resonator and the CPW antenna to obtain tunability at Ka band. For g = 2 μm, a single resonant frequency of 26.3 GHz was obtained, with a return loss of 15.1 dB and a realized gain equal to 3 dB; for g = 2.5 μm, two resonant frequencies of 27 and 29.8 GHz were obtained, with a return loss of 23 and 18 dB and a realized gain 3 and 2 dB. Finally, for g = 3 μm, it also allows two resonant frequencies, 27.5 and 30.6 GHz, with a return loss of 19.84 and 26.62 dB, showing a realized gain of 2 and 1 dB, respectively.

Acknowledgements

This work was supported by the Laboratory of Circuit and Electronic System in High Frequency of University of Tunis El Manar and Research Center for Microelectromechanical Systems (CMEMS) of the University of Minho Braga-Portugal. Foundation for Science and Technology (FCT) project PTDC/EEI-TEL/5250/2014, by FEDER funds through POCI-01-145-FEDER-16695 and Projecto 3599-Promover a Produção Científica e Desenvolvimento Tecnológico e a Constituição de Redes Temáticas.

Author details

Bassem Jmai[1]*, Adnen Rajhi[2,3], Paulo Mendes[4] and Ali Gharsallah[1]

*Address all correspondence to: bassem.jmaiesti@gmail.com

1 Department of Physics, FST, Unit of Research in High Frequency Electronic Circuit and System, University Tunis El Manar, Tunis, Tunisia

2 Department Electrical Engineering, National School of Engineering Carthage, Tunis, Tunisia

3 Laboratory of Physics Soft Materials and EM Modelisation, FST, University Tunis El Manar, Tunis, Tunisia

4 Department of Industrial Electronics, Microelectromechanical Systems Research Center, University of Minho, Guimarães, Portugal

References

[1] Gevorgian S. Ferroelectrics in Microwave Devices, Circuits and Systems: Physics, Modeling, Fabrication and Measurements. Springer Science & Business Media; 2009. ISBN: 9781848825062

[2] Sharma M, Kuanr BK, Sharma M, Basu A. Tunable coplanar waveguide microwave devices on ferromagnetic nanowires. International Journal of Materials, Mechanics and Manufacturing. February 2014;**2**(1):9-13. DOI: 10.7763/IJMMM.2014.V2.88

[3] El Cafsi MA, Nedil M, Osman L, Gharsallah A. The design of a 360°-switched-beam-base station antenna. In: Antenna Arrays and Beam-formation. Rijeka: InTech; May 2017. pp. 13-30. ISBN: 978-953-51-3146-5

[4] Mabrouki M, Jmai B, Ghyoula R, Gharsallah A. Miniaturisation of a 2-bits reflection phase shifter for phased array antenna based on experimental realisation. International Journal of Advanced Computer Science and Applications. May 2017;**8**(5):438-454. DOI: 10.14569/IJACSA.2017.080553

[5] Rijks TG, Steeneken PG, Beek JTM, Ulenaers MJE, Jourdain A, Tilmans HAC, Coster JD, Puers R. Microelectromechanical tunable capacitors for reconfigurable RF architectures.

Journal of Micromechanics and Microengineering. February 2006;**16**(3):601-611. DOI: 10.1088/0960-1317/16/3/016

[6] De Los Santos Héctor J. RF MEMS Circuit Design for Wireless Communications. Artech House; 2002. ISBN: 1-58053-329-9

[7] Brown ER. RF-MEMS switches for reconfigurable integrated circuits. IEEE Transactions on Microwave Theory and Techniques. November 1998;**46**(11):1868-1880. DOI: 10.1109/22.734501

[8] Rebeiz GM. RF MEMS, Theory, Design and Technology. John Wiley & Sons; 2003. ISBN: 978-0-471-20169-4

[9] Dussopt L, Rebeiz GM. High-Q millimeter-wave MEMS varactors: Extended tuning range and discrete-position design. IEEE MTT-S International Microwave Symposium Digest. June 2002;**2**:1205-1208. DOI: 10.1109/MWSYM.2002.1011869

[10] Blondy P, Mercier D, Cros D, Guillon P, Rey P, Charvet P, Diem B, Zanchi C, Lappierre L, Sombrin J. Packaged millimeter wave thermal MEMS switches. In: IEEE 31st European Microwave Conference; 24–26 September 2001

[11] Park JY, Yee YJ, Nam HJ, Bu JU. Micromachined RF MEMS tunable capacitors using pizeoelectric actuators. In: IEEE International MTT-S Microwave Symposium Digest; 20–24 May 2001

[12] Ruan M, Shen J, Wheeler CB. Latching micromagnetic relays. IEEE Journal of Microelectromechanical System. December 2001;**10**:511-517. DOI: 10.1109/84.967373

[13] Mafinejad Y, Kouzani A, Mafinezhad K. Review of low actuation voltage RF MEMS electrostatic switches based on metallic and carbon alloys. Journal of Microelectronics, Electronic Components and Materials. May 2013;**43**(2):85-96

[14] Muller Ph. Conception, Fabrication et Caractérisation d'un Microcommutateur Radio Fréquences pour des Applications de Puissance [PhD thesis]. L'université des sciences et technologies de Lille UFR d'électronique, Lille 1; 2005

[15] Lonac JA, Merletti GA. Parametric analysis on the design of RF MEMS series switches. European Scientific Journal. December 2015;**1**:248-257. ISSN: 1857-7431

[16] Brown ER. RF-MEMS switches for reconfigurable integrated circuits. IEEE Transactions on Microwave Theory and Techniques. November 1998;**46**(11):1868-1880. DOI: 10.1109/22.734501

[17] Caekenberghe KV. Modeling RF MEMS devices. IEEE Microwave Magazine. January 2012;**13**(1):83-110. DOI: 10.1109/MMM.2011.2173984

[18] Jmai B, Rajhi A, Gharsallah A. Controllable bridge of the RF-MEMS: Static analysis. In: International Conference on Green Energy Conversion Systems (GECS); October 2017

[19] Jmai B, Anacleto P, Mendes P, Gharsallah A. Modeling, design, and simulation of a radio frequency microelectromechanical system capacitive shunt switch. International Journal

Numerical Modeling, Electronic Networks, Devices and Fields. 2017:e2266. DOI: 10.1002/jnm.2266

[20] Jmai B, Rajhi A, Gharsallah A. Conception of a tunable analog reflection-type phase shifter based on capacitive RF MEMS for radar application. International Journal of Microwave and Optical Technology. September 2016;**11**(5):339-346. DOI: IJMOT-2016-6-1031

[21] Chun YH, Hong JS. Electronically reconfigurable dual-mode microstrip open-loop resonator filter. IEEE Microwave and Wireless Components Letters. July 2008;**18**(7):449-451. DOI: 10.1109/LMWC.2008.924922

[22] Rinaldi M, Zuo C, Vander Spiegel J, Piazza G. Reconfigurable CMOS oscillator based on multifrequency AlN contour-mode MEMS resonators. IEEE Transactions on Electron Devices. May 2011;**58**(5):1281-1286. DOI: 10.1109/TED.2011.2104961

[23] Jmai B, Rajhi A, Gharsallah A. Novel conception of a tunable RF MEMS resonator. International Journal of Advanced Computer Science and Applications. January 2017;**8**(1):73-77. DOI: 10.14569/IJACSA.2017.080111

[24] Jmai B, Gahgouh S, Gharsallah A. A novel reconfigurable MMIC antenna with RF-MEMS resonator for radar application at K and Ka bands. International Journal of Advanced Computer Science and Applications. May 2017;**8**(5):468-473. DOI: 10.14569/IJACSA.2017.080556

4

MEMS Technologies Enabling the Future Wafer Test Systems

Bahadir Tunaboylu and Ali M. Soydan

Abstract

As the form factor of microelectronic systems and chips are continuing to shrink, the demand for increased connectivity and functionality shows an unabated rising trend. This is driving the evolution of technologies that requires 3D approaches for the integration of devices and system design. The 3D technology allows higher packing densities as well as shorter chip-to-chip interconnects. Micro-bump technology with through-silicon vias (TSVs) and advances in flip chip technology enable the development and manufacturing of devices at bump pitch of 14 μm or less. Silicon carrier or interposer enabling 3D chip stacking between the chip and the carrier used in packaging may also offer probing solutions by providing a bonding platform or intermediate board for a substrate or a component probe card assembly. Standard vertical probing technologies use microfabrication technologies for probes, templates and substrate-ceramic packages. Fine pitches, below 50 μm bump pitch, pose enormous challenges and microelectromechanical system (MEMS) processes are finding applications in producing springs, probes, carrier or substrate structures. In this chapter, we explore the application of MEMS-based technologies on manufacturing of advanced probe cards for probing dies with various new pad or bump structures.

Keywords: wafer and package test systems, MEMS technology, interconnects, interposer, wafer probes

1. Introduction

Increased connectivity and functionality is driving the evolution of 2D technology toward 3D technology for integration of silicon devices and system design. This technology is becoming a scaling engine for silicon technology [1] allowing higher packing densities and shorter chip-to-chip interconnects. Shrinking die dimensions and pitch pose challenges on the probing and test side of the equation forces development of newer probes, interposers, interconnects and robust assembly systems [2, 3]. As 3D IC packaging is becoming mature, there is a strong push toward 3D IC Si integration. In a 3D IC integration, some of the chips, a microdisplay, microelectromechanical systems (MEMSs), memory, microprocessor, application-specific IC (ASIC), micro-controller unit, digital signal processor, microbattery and analog-to digital mixed signal are combined and stacked in three dimensions [4, 5]. These system and component level challenges are being addressed by silicon carriers or 3D-stacking, interposers, substrates and newer probe materials by MEMS processes. Developing a common intermediate board for a substrate or space transformer and probe card assembly will help solve technical challenges and reduce cost of test in both wafer and package level testing. An optimal design, which includes the IC design, the automated test equipment (ATE) test cell and the probe card solution, of the test flow between wafer sort and final test can yield benefits. Standard vertical probing technologies use microfabrication technologies for probes, templates and substrate-ceramic packages [6]. Pitches below 50 μm pose enormous challenges on fabrication of probe card components and nanotechnology and MEMS processes are required for producing probes, carrier or substrate structures for precision requirements. Probe structures must be designed with precision and their power delivery properties must be optimized. Advanced probe cards must be able to support high-speed testing and cold and hot temperature cycle testing with precision contact capability. They also need to address contact challenges for multi-row pads/bumps, full array Cu-pillar micro-bumps with various solder-bump metallurgies at temperature. Application of various technology approaches in test systems against the test requirements of silicon logic or memory or mixed signal devices is discussed.

2. Trends in silicon and systems for test

The cost of scaling is rapidly increasing and the expected development cost for system-on-chip (SoC) for 10 nm is 400M USD and for 7 nm it is projected to be approaching 600M USD [2]. This means that it requires multibillion dollar lifetime revenue to be economically feasible per design. System solutions need to balance performance, power and cost. The industry of moving to 3D architectures adds challenges in variability in manufacturing next generation devices, requires more stringent variability control by data analytics and Industry 4.0 applications [7]. Advanced packaging also adds multiple levels and variability can happen across multiple die, as memory chip stack with through-silicon vias (TSVs) placed on a logic device which is integrated to a substrate with copper pillar bumps, SnAg bumps or micro-bumps. At 10 nm process node, 3D TSVs are projected to be at 6 μm diameter with depth of 55 μm [2]. Logic-memory integration improves the bandwidth and provides higher performance per watt while SoC partitioning increases yield and helps cost optimization. In a total package stack-up, thin silicon layers become an issue due to low-k modulus reliability while the substrate can become subjected to a thermal mismatch stress and induced warpage problems, as well as routability issues.

The cost is increasing with decreasing pitch, increasing probe count and increasing parallelism.

The area-array type of logic test is challenging below 100 μm bump pitch and push for MEMS type of probe solutions are required to scale with the technology. Design for tests (DFTs) with wrappers are targeted to reduce number of I/O's that need to be contacted during test. Also, the ability to reuse testers is also studied to lower the total cost of test. A test system architecture with vertical style probe card is shown is **Figure 1**. In the system, ST stands for space transformer, multilayer ceramic substrate (MLC) and device under test (DUT).

When the roadmaps for probe card requirements are reviewed, there are many critical test system parameters that must be considered especially for large-sized highly parallel cards. They are mainly:

- Controlled overdrive
- Reduced temperature drifts
- Planarity self-adjustable function
- Low voltage test operation
- Reduced pad damage and increased uniformity
- Less particle generation
- Smart repair concepts
- Expanded temperature range
- Diagnostic functions on probe cards
- Smart alignment features
- Cost efficiency
- Lower lead times

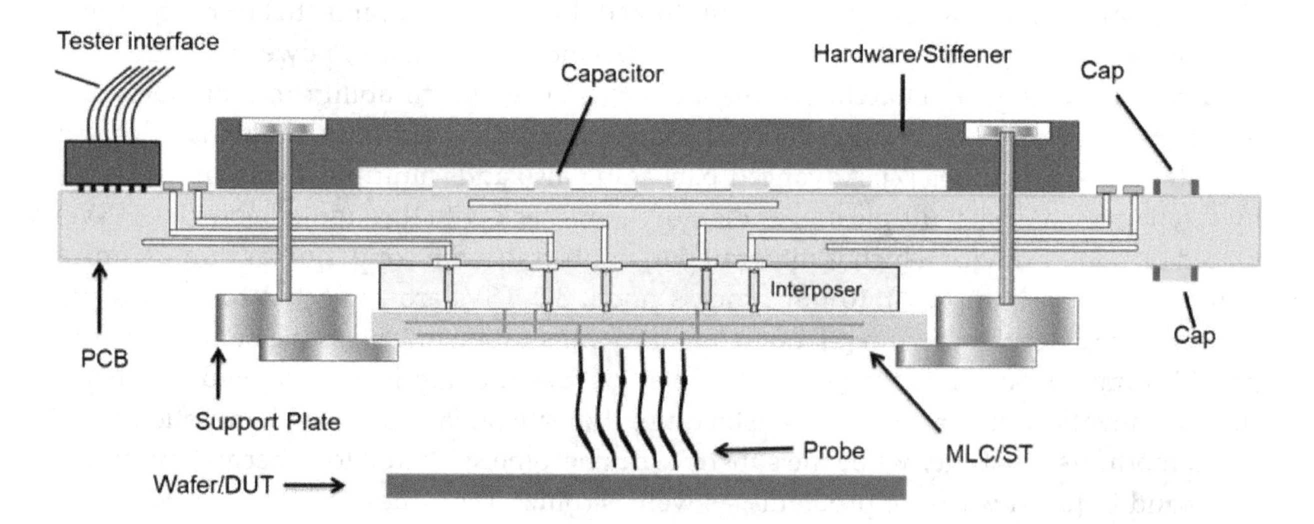

Figure 1. Probe card system architecture is shown.

3. Probe technology and designs for fine-pitch probing

Cantilever probing technologies, both traditional and MEMS-style cantilever, have limitations for multi-DUT probing at 50 μm or below. Wafer test becomes challenging because of design complexity of devices. For instance, one limit is the number of rows of bond pads that can be tested at one time, dependent heavily on pad density. Another parameter of a design test limitation with cantilever-style technologies is the corner keep-out in device layouts. Yet another requirement of this mode of technology is the need for skip DUT configurations, compromising test stepping efficiency.

Vertical style technology approaches allow more rows of peripheral pads and array patterns for contacts. Images of probe cards of a traditional cantilever, vertical and MEMS-memory types are illustrated in **Figure 2**. The market for devices with multiple peripheral pads is moving to finer pitches and the demand for higher levels of parallel testing is increasing for such logic configurations. These requirements are driven by higher I/O requirements, smaller device dies, longer test times and more challenging cost of test economics. It is required to probe devices at higher levels of parallelism and finer pad pitches. Pads can be arranged inline, dual or multi inline rows or staggered pads. This design space is typically not addressable by standard vertical, advanced memory cards or standard cantilever cards but a new segment for advanced fine-pitch MEMS type probe technologies. Major product families in this space at increasingly higher parallelism requirements are high-end ASICS, SoCs/high level digital signal processor (DSPs) and low-end DSPs/low-end microcontrollers.

Cantilever probe cards are used in addressing 1-row peripheral multi-DUT or 1–2 row peripheral layouts of pads on devices, as shown in **Figure 3**. Probe card with 2-row cantilever probes

Figure 2. Cantilever, standard vertical and MEMS type probe cards.

is illustrated on the right. There are significant limitations to standard cantilever probes technologies as the number of rows or DUTs rises. For three rows of pads, a vertical style technology is needed for efficient probing, as shown in **Figure 4**. Traditional vertical buckling beam style probes of three different diameters (4-mil, 3-mil and 2-mil) which address different device pitch requirements in wafer test are also shown. MEMS-vertical technologies enable probing of full arrays, as shown in **Figure 5**, that are typically not feasible with conventional vertical probe technologies.

Probe action, scrub mark size and depth must be precisely controlled to prevent damage to bond pads, typically Al or Cu, and low-k dielectrics during wafer probe. Fine-pitch probing requires precise control of alignment at pad sizes smaller than 40 μm.

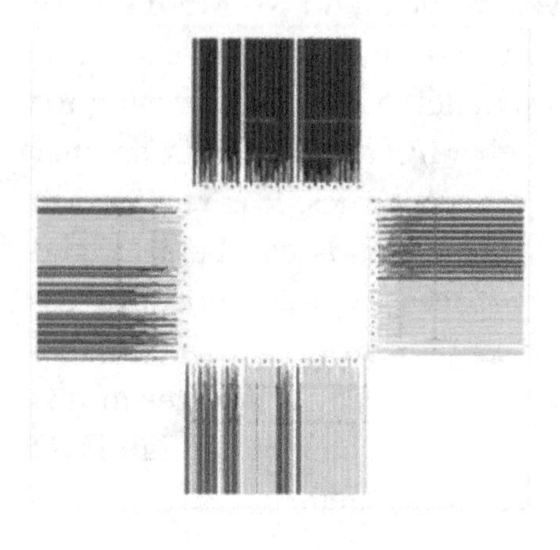

Cantilever probe and pads 1-row peripheral 2-row peripheral 2-row probes

Figure 3. Cantilever design and contact pad layouts.

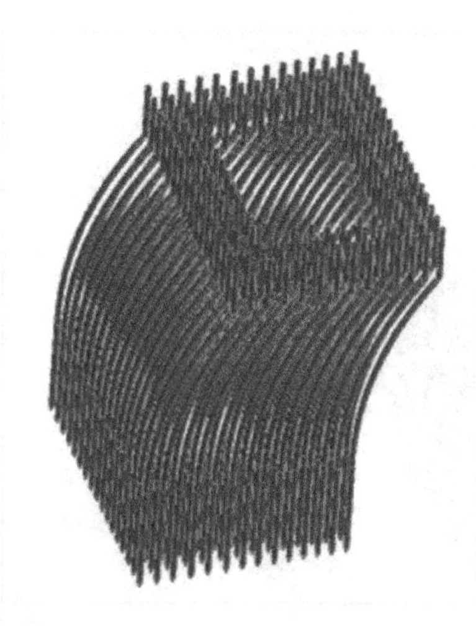

4 mils 3 mils 2 mils

Vertical probes 3-row peripheral layout 3 different vertical probe styles

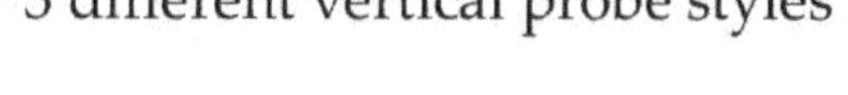

Figure 4. Vertical probes, 3-row peripheral layouts on a device and illustrations of three different vertical probe designs on a wafer.

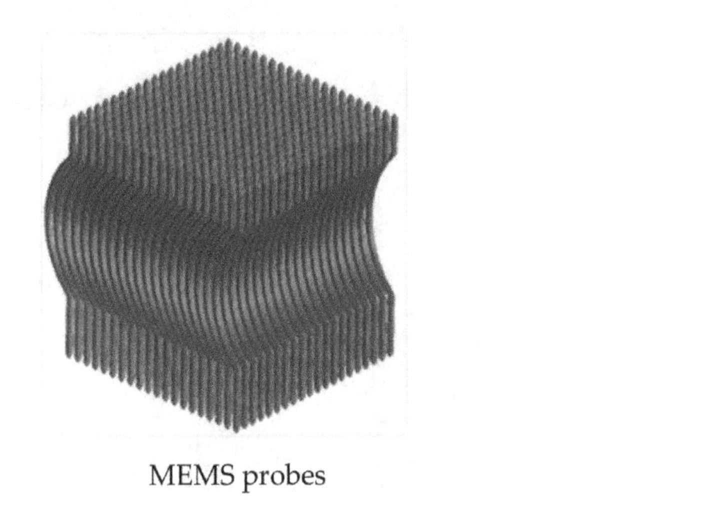

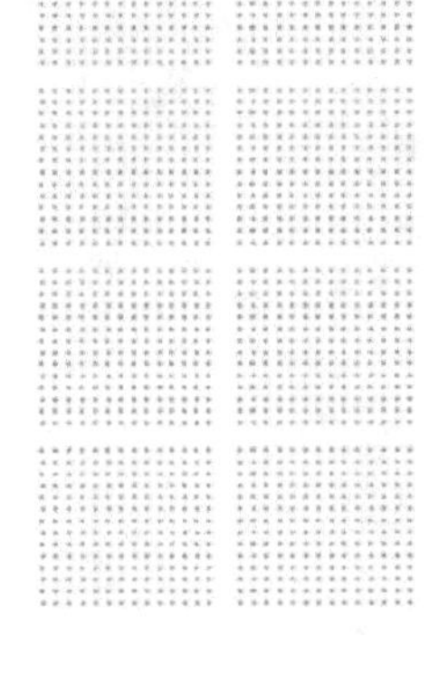

MEMS probes Array layout

Figure 5. MEMS-vertical probes for contacting an array of bumps.

The contact model for vertical probe contacts is different than cantilever-style beams. Scrub marks generated by cantilever beams by design are typically longer than marks by vertical probes. Accurate guiding of probes permits finer controls and precise scrub marks for vertical. The tolerances on guiding holes as well as probes are critical for positions.

Figure 6 illustrates results of deflection and stress simulations for the models of cantilever probe designs and MEMS-cantilever probe designs exhibiting deflection upon pad contact and generating scrub motion on probe tips. MEMS-cantilever type designs are well suited for memory device testing up to 1–4 touchdowns for 300 mm wafers with probe counts up to 60,000 probes.

Vertical buckling beam model and MEMS-fine pitch vertical probe design contacts and simulations of deflection under load are shown in **Figure 7**. Vertical probes are typically manufactured

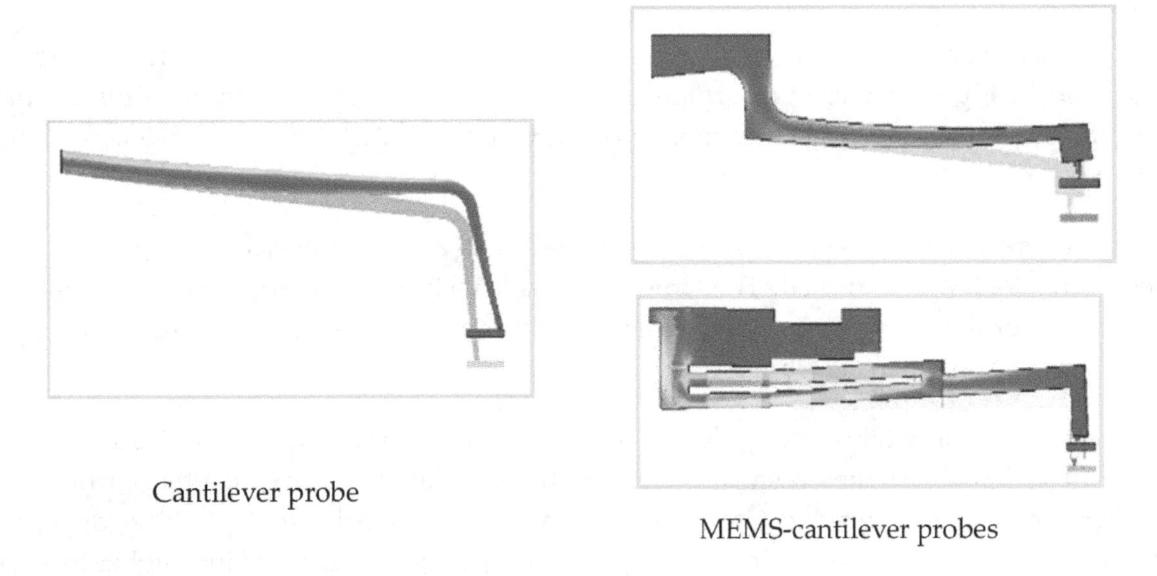

Cantilever probe

MEMS-cantilever probes

Figure 6. Cantilever probe design (conventional) and MEMS-cantilever probe designs showing deflection and scrub on pad during contact.

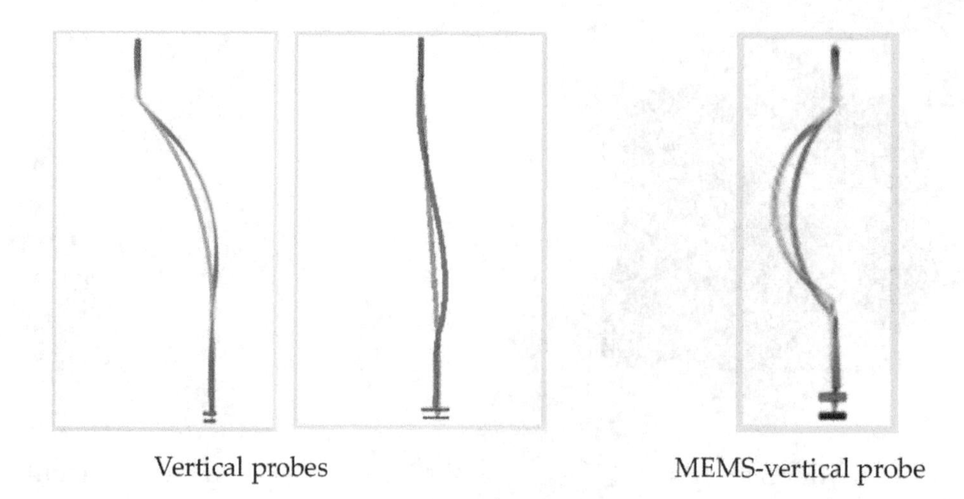

Figure 7. Vertical buckling beam probe design and MEMS-vertical probe designs showing deflection and contacting a pad/bump.

from Paliney 7™ or BeCu materials by stamping a wire version followed by a final finishing process. MEMS-vertical or cantilever probes are lithographically produced and involves many process steps typical in MEMS technologies. Different types of nickel alloys (Ni-Co and Ni-Mn) are commonly used for MEMS spring or probe manufacturing. Probe tips may be coated with harder alloys for better lifecycle, which may involve Pd and Pt alloys such as PdCo, PtIr, PtNi, Rh or hard gold and other alloys. It should be noted that MEMS-based vertical technology has an edge over buckling beam technologies for design flexibility for highly parallel peripheral devices as well as the accuracy of scrub signatures required for smaller pad sizes.Flip chip type area-array applications such as microprocessors, graphics chips and microcontrollers, are addressed by traditional vertical or MEMS-style vertical probe technologies. **Figure 8** shows MEMS probe products, advanced vertical probe technologies for testing full area-array (A) or testing multi-row peripheral or partial arrays (B) and advanced cantilever types for testing memory devices (DRAM or flash).

The electrical contact resistance measurements for MEMS-vertical probe technology as illustrated in **Figure 5** were performed on various emerging bump types. **Figure 9** provides the eutectic bump resistance measurements done by MEMS-vertical probes on a test system.

The contact resistance (Cres) was indicated to be stable at 25 μm overdrive (3 gf) for all tip sizes (9, 12, 16, 36 μm) studied. The Cres is the path resistance including connections from tester to the probe tip. The effective contact resistance of just the bump and the probe tip is estimated to be less than 0.2 Ohms.

The contact resistance on copper pillar bumps is illustrated for MEMS-vertical technology in **Figure 10**. It that shows the Cres with 12 μm probe tips is much higher than those from 9 μm tips. For probes with 9 μm-tips, Cres was stable at 50 μm overdrive (5–6 gf). The copper pillars are much harder than eutectic or Sn-Ag type bumps, therefore it requires higher forces to establish good contact. However, the probe tips remain much cleaner in life testing on copper pillars compared to solder-based bumps.

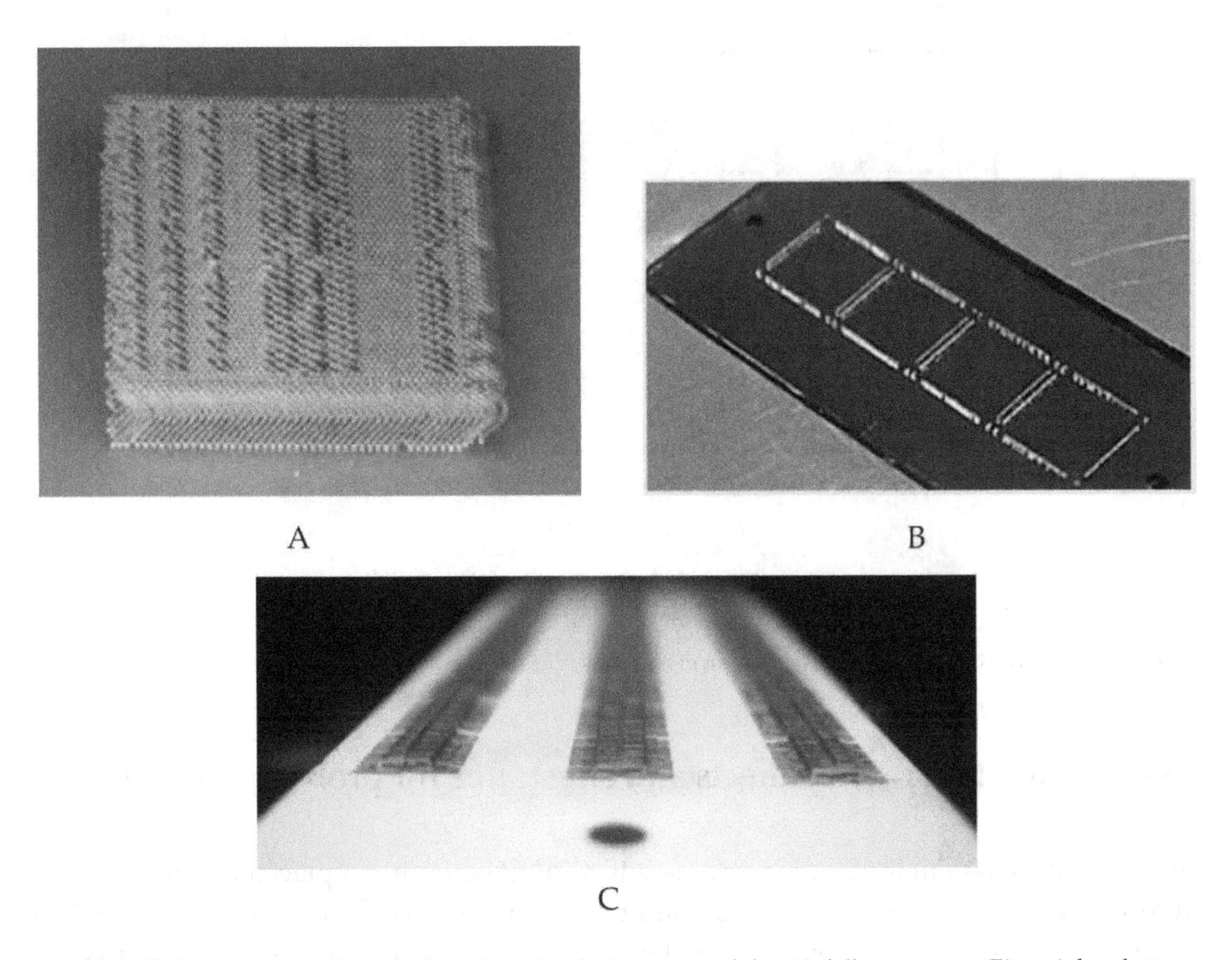

Figure 8. MEMS type advanced vertical probe technologies are used for (a) full area-array, (B) peripheral-rows or partial-array and (C) advanced cantilever probes for inline memory testing.

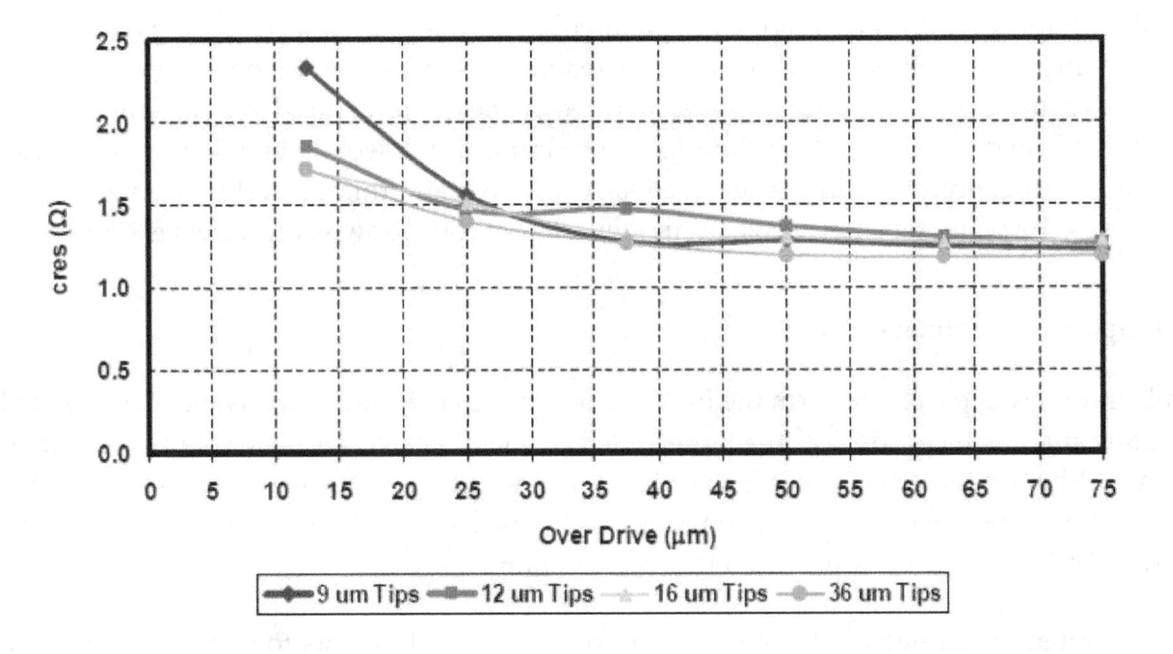

Figure 9. The results from the eutectic bump resistance measurements.

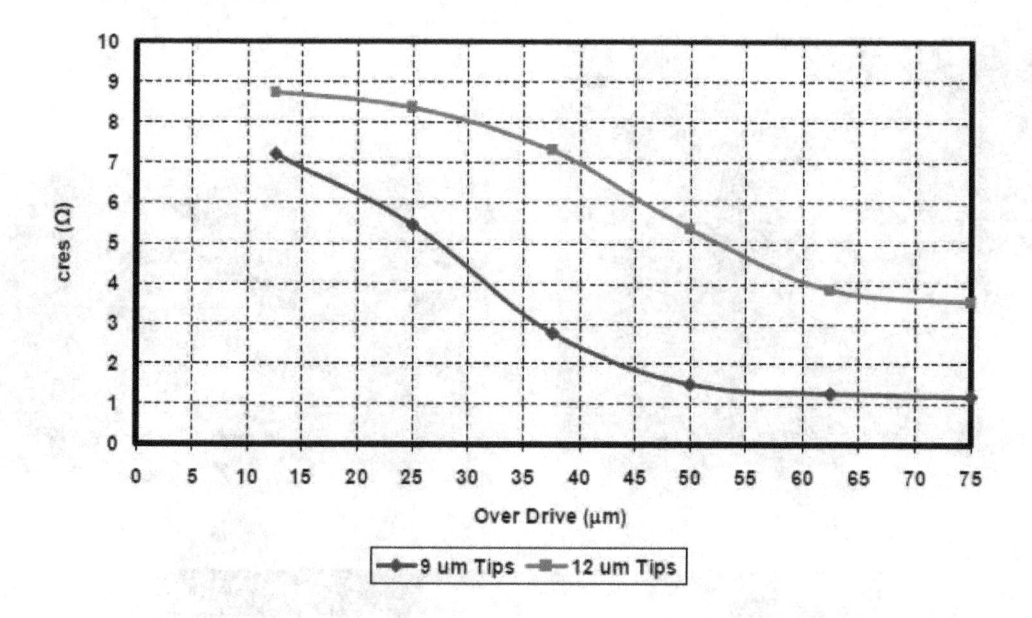

Figure 10. The eutectic bump resistance measurements done by MEMS-vertical probes on a test system are shown.

4. Next generation interconnects and substrates for probing systems

Higher levels of system integration and new IC technologies allow placement of significant test resources on the probe card, such as caps or resistors to very close proximity to the DUT, device supplies, digital channels and analog test circuitry, to improve signal integrity and performance. This capability helps overcome some test limitations, and make it possible to add RF test structures, and circuits enabling high-speed loop-back solutions. These solutions help in the cost-effectiveness of the test strategy. Advanced probe cards have to be designed to support high-speed testing and cold and hot temperature testing (from −55 to 150°C). Providing robust precision contact capability enables reliable contacts on smaller die sizes with better signal fidelity. Probe structures can be manufactured in a cost-effective way by MEMS methods to enable scaling to a finer bump pitch well below 50 μm area-arrays. Probe repair concepts are available on a restricted basis and this capability usually strongly requested by wafer test houses when high number of touchdowns on wafers is required.

4.1. Space transformers

Substrates are typically perform the function of space transformers in advanced probe cards, routing fine pitch of a device to a larger pitch of a PCB and tester boards in wafer test systems. Although the probe count is very large, memory type probe cards can handle 200mm or 300mm wafers due to device geometries with 1 or 2 row peripheral layouts. Space transformer in this case is typically a single-layer thin film on MLC.

Space transformers need to be able feature following requirements to support next generation advanced probe cards: (1) very low pitch fanout (30 μm), (2) high frequency operation with a high bandwidth of 3 GHz, signal length matching, low crosstalk for analog and digital signal, shielding, (3) high pin counts for dense device designs (> 5000), (4) large arrays, (5) path resistance <2 Ohms, (6) no skip DUT and (7) peripheral device test with ore than 3 rows of pad per side and array configuration. Some of these requirements may not possible with MLC ceramic manufacturing with extra polyimide (PI) layers. Multilayer organic substrate (MLOs) is lower cost versions, but also have similar geometric and process limitations along with some thermal test restrictions.

MLS (multilayer substrate) is proposed as a type of silicon interposer manufacturable using MEMS technology to reach these target requirements. Space transformer technology comparison is provided for fine-pitch probing applications in **Figure 11**. WST stands for wired space transformer used in standard vertical probe cards. These are quickly changing with various capability enhancing feature every year. WST, MLC and MLO are well established while MLS and other high density ST scenarios are emerging. BGA and LGA stands for ball-grid array or land grid array versions of MLC. CTE is the coefficient of thermal expansion of a material.

Substrate interconnect process flow is shown in **Figure 12(A, B)** for creation of a silicon interposer with fine pitch and its bonding onto a MLC carrier. This type of a silicon interposer allows for fine pitch top surface routing capability for fanout on a MLC. Silicon interposer will have TSVs for connecting the top to bottom side. It features a probing contact pad on the surface and a bump connection to 200 μm-pitch MLC on the other side.

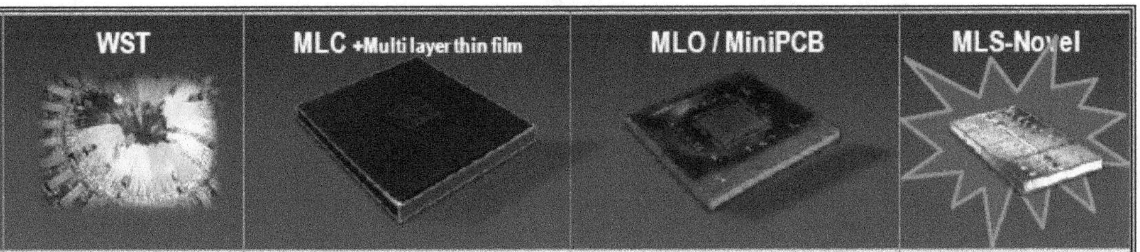

	WST	MLC +Multi layer thin film	MLO / MiniPCB	MLS-Novel
Material	•Copper alloy wire	•Ceramic (Alumina or HiTCE) •Tungsten or Cu conductor •Cu/PI Multilayer thin film	•Epoxy/PI/FR4 •Copper	•Si core •Cu/PI/SiOx Multi layer
Pitch mini & Configuration (density)	•In line : 60μm •Peripheral (1row) : 60μm •Peripheral (2row) : 60μm •Peripheral (3row) : 60μm*	•In line : 60μm •Peripheral (1row) : 60μm •Peripheral (2row) : 80μm (60μm w Skip) •Peripheral (3row) : 120μm (80μm w Skip) •Array : 120μm (PAD size is 35x35μm)	•In line : 100μm •Peripheral (1row) : 100μm •Peripheral (2row) : 200μm (100μm w Skip) •Peripheral (3row) : 200μm (200μm w Skip) •Array : 200μm (PAD size is about 50x50μm)	•target is 50μm or below at any configuration !
Pin count	1000	>5000 (for pitch >120μm)	>5000	>5000
Pros	•Cycle time •Cost •Probe depth adjustable •Generic PCB	•High pin count •BGA/LGA •Good mechanical resistance •Low CTE •Good electrical performance	•High pin count •Hard Gold possible •NRE lower than MLC •Shorter lead time than MLC •Good electrical performance	•Very low pitch potential •Hard gold •Low CTE •Good electrical performance
Cons	•Risky above 1000 wires •Electrical performance limitation	•Cost at low qty •Cycle time •Probe Depth limitation •Hard Gold none std	•Density & pitch limitation vs MLC/MLO •Probe Depth limitation •No PGA •Low mechanical stiffness (deflection) •Planarity after reflow	•New technology •Cost at low qty •Probe Depth limitation

Figure 11. The space transformer technology capabilities for advanced probe cards.

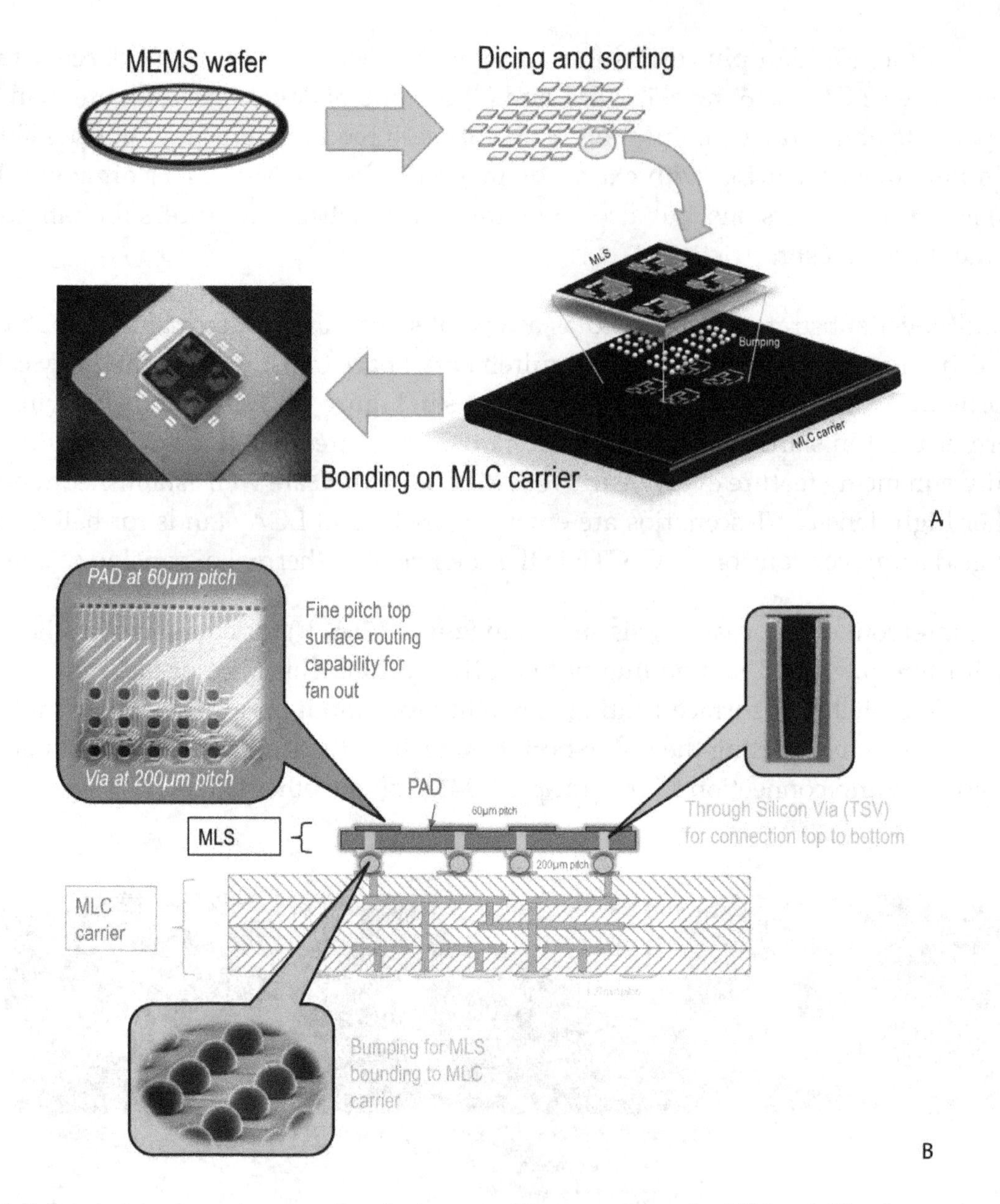

Figure 12. Substrate interconnect process flow is shown in (A) process steps for MLS and (B) substrate interconnect features.

MLC process may involve ceramic manufacturing process and additional polyimide (PI) process layers. Fine pitch on the surface is reached by stacking up three layers of PI. Vias on ceramic layer are routed to pad locations. As the pin count is raised, thin film layer count increases to match the required probing pads. In some cases as row count increases, then a test scheme requires skipping a DUT due to such substrate density restrictions. 3-row pad structure is shown in **Figure 13** and the pin count versus the pad density is also shown.

4.2. Copper pillar bumps

The increasing requirement for more functionality in smaller packages forces trends toward 3D packaging approaches for portability [8]. Higher routing density enabled by finer pitch flip chip technology using copper pillars is highly desirable for lower costs and

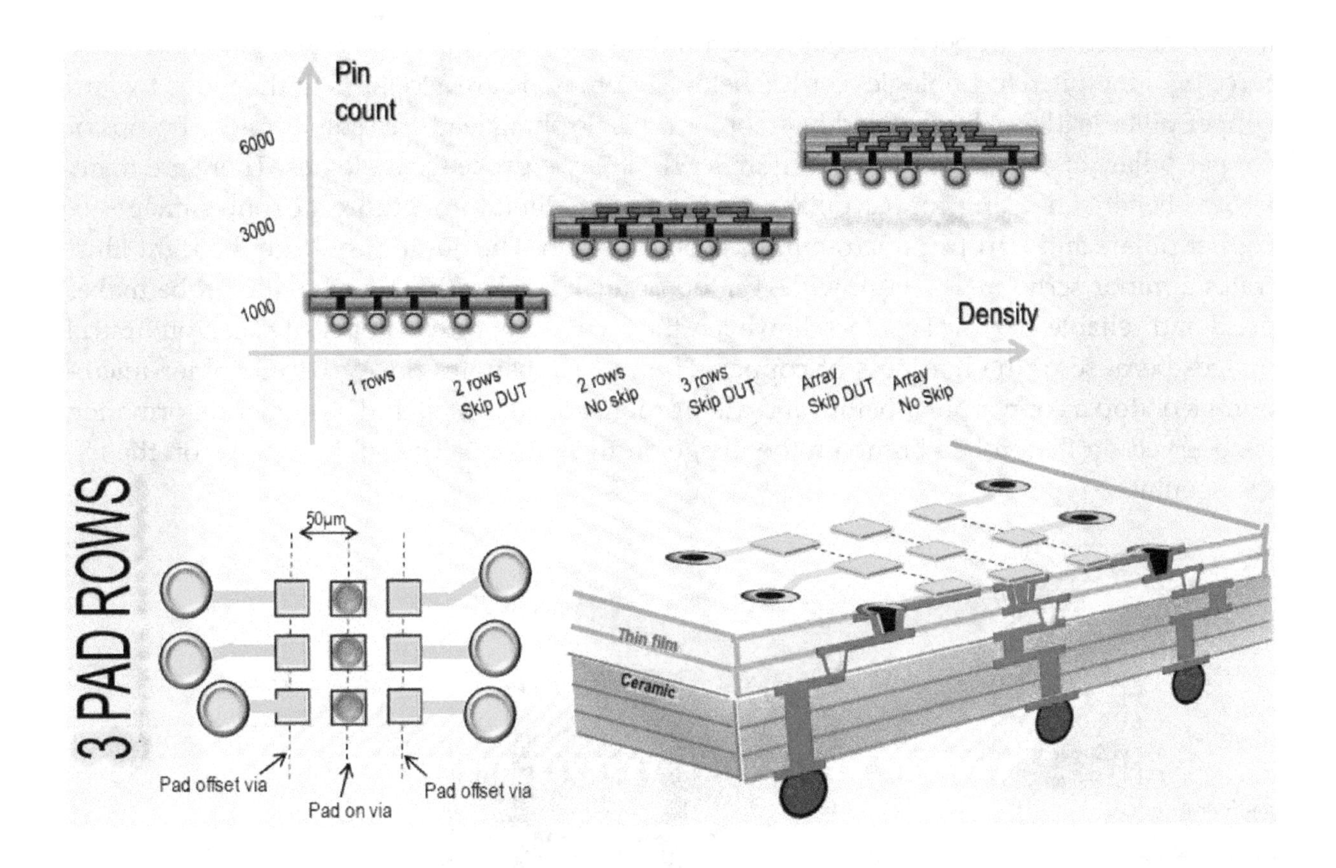

Figure 13. Density on substrates is illustrated for various configurations.

scalability. Copper pillar bumps typically consist of a copper base and a solder capped top [6]. These copper pillars, sometimes called high pillars or micro-bumps, act as an interconnect structure which lowers stress on low-k layers in finer silicon nodes and increase reliability. Use of such micro-bumps simplifies substrates for packages, thereby decreasing cost, and allows natural migration toward TSV technologies of the future. On the other hand, as the metallurgy of the bump structure changes from eutectic to lead-free solders and more importantly to solder-cap on copper pillars with varying contact geometries, probing very fine-pitch bumps presents new test, process and precision challenges [8]. There is also an increasing trend toward performing the final test in wafer level to reduce both cycle time and cost of test while moving to environment friendly manufacturing processes. It is important for IC design and packaging development and test engineers to understand the material impacts of new wafer bumping system and technology. They need to address both reliability and manufacturability of the entire process, which includes test process development early in the cycle so that the overall system level cost is optimized [8].

Probing of traditional solder bumps at 120 μm pitch or above, whether eutectic, high-lead or lead-free solder balls are performed typically by buckling beam/vertical technologies. The contact area formed on the top of a round bump after a probe contact is related to the metallurgy and the mechanical properties of bump materials as well as the probe tip geometry and probe force [6]. Fine-pitch technologies for ICs below 40-nm node are accelerating the move toward copper pillar lead-free bumps and interconnections. Probing lead-free solder micro-bumps or copper pillars at 40 μm-array pitch requires MEMS-style probe technologies. There are many known benefits of using copper pillars reported in the literature. **Figure 14** shows images of copper pillars and lead-free micro-bumps at 50 μm pitch. The bump profile on the right illustrates a minor scrub mark, 9 μm wide, on top of the Cu-pillar. In this case, the probe makes good and reliable electrical contact, however, the scrub signature is not easily seen on optical images because of the hardness of copper. **Figure 15** illustrates Sn-Ag based solder micro-bumps on top a copper pillar before and after probing at 50 μm pitch. The solder deformation is observed on the probed bump on the image on the left side. The solder bumps on the left are of eutectic type.

Figure 14. Images of copper pillars and lead-free micro-bumps at 50 μm pitch. The profile shows the pillar bump after probing with vertical MEMS probe technology.

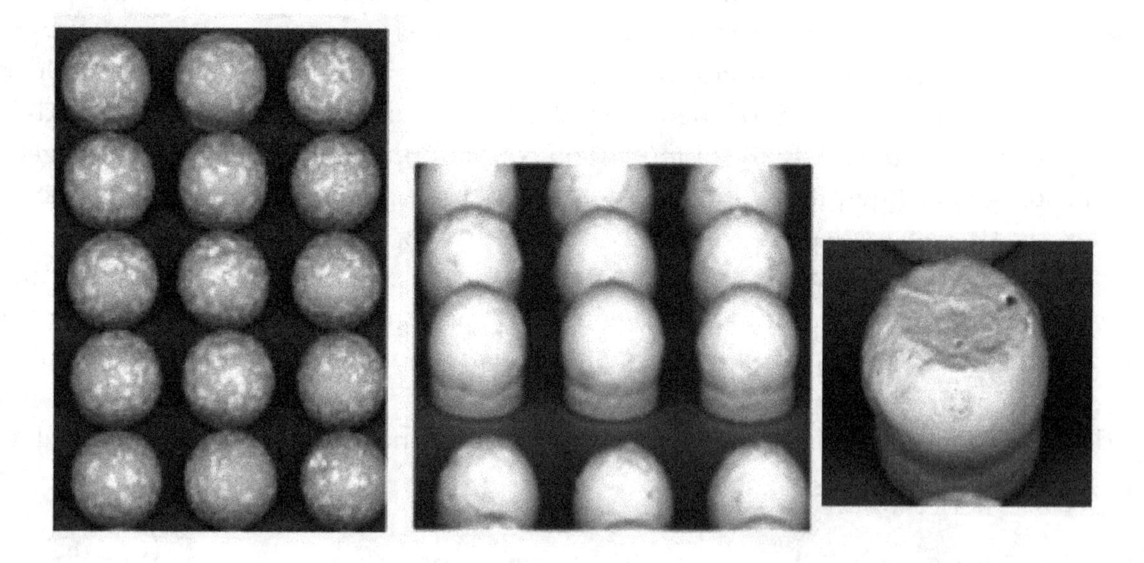

Figure 15. Solder micro-bumps illustrated on the left have no copper pillar-base. Solder micro-bumps on top a copper pillar before and after probing (on the right) at 50 μm pitch.

5. Final test and spring pins

The decision on how to partition test between wafer and package tests and where to focus efforts to increase parallel testing is always on the agenda of test practitioners in the industry. A wafer test probing is a short-cut in addressing both wafer test and package test, if it can be a bundled solution for productivity of the test floors. This potentially reduces total cost of test substantially. The trend nowadays is to focus on wafer sort in high parallelism mode.

The wafer level chip scale package (WLCSP) format has been rising and in the final test, there is strong push for cost-effective RF testing solutions [9, 10]. The spring-pin technology for the final test still inexpensive workhorse of the package test industry. The system board level functionality is moving into package-level (SiP) or chip-level (SoC) implementations. The spring pins, are not scalable at fine pitches and will not support test speeds necessary.

A socket-contactor design is proposed for reliable electrical contact and allows testing for best wafer yields. This type of approach must replace known vertical probe technology or membrane probe technology for testing wafer level packages. A novel contactor and socket were designed for high performance and low-cost for use in wafer probe or final test [11].

Figure 16 shows a socket test system overview showing a load-board and device under test (DUT) with a ball-grid-array (BGA) in contact with traditional spring pins, that is, pogo pins™. The DUT can be packaged as BGA with bumps or land grid array (LGA) and the contactor pin geometry will change depending on the pad/bump materials and contact surfaces [11].

The proposed design of new contactor is illustrated in **Figure 17** including a plunger, spring wire and the socket with a retaining plate. The contactor consists of a plunger pin made of beryllium copper and a braided stainless steel spring wire. The spring wire is typically copper over-plated. The socket materials with retaining plates were made of FR4. The overall

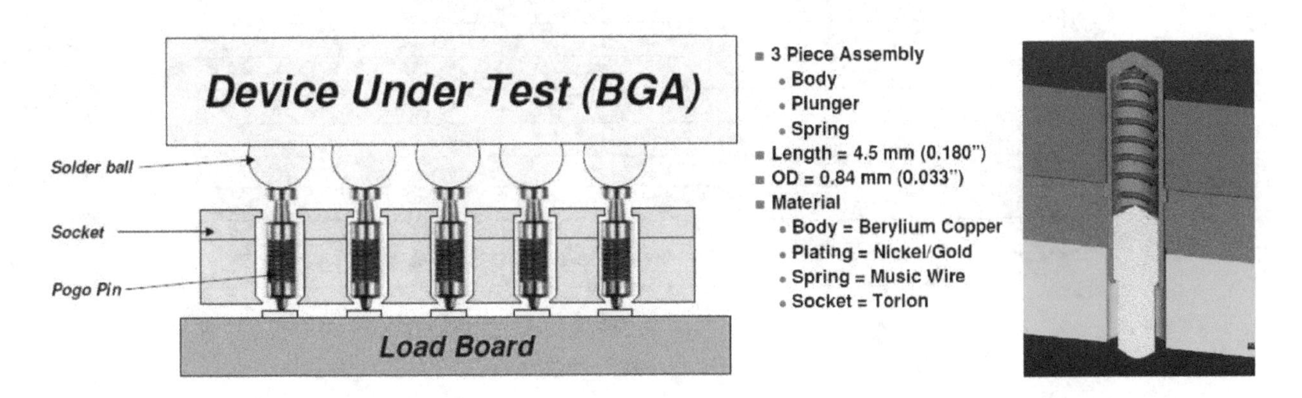

Figure 16. A pogo pin socket system overview.

diameter of the spring wire section was 0.51 mm. The contactor has a 5 mm in uncompressed total length including the plunger and the spring. The length of the spring wire section was 0.27 mm.

The measurement results in **Figure 18** show the contact resistance behavior of the spring assembly for SS304V/Cu plated with Ni/Au contacts in a 36-Pin test socket. SS304 stands for stainless steel spring wire and Cu, Ni and Au are overplating applied to the spring to improve the electrical performance characteristics. S parameter characterization has shown better results than those of traditional spring pins. **Figure 19** shows electrical simulation results for pitches of 0.8, 1.6 and 2.5 mm are shown. Insertion loss was estimated to be at −1 dB bandwidth as 5.025 GHz (A). Return loss of −16 dB at 4 GHz is illustrated at (B).

The plunger pin and stainless steel spring wire can be manufactured with MEMS processes to make them scalable to much finer pitches than these versions can support.

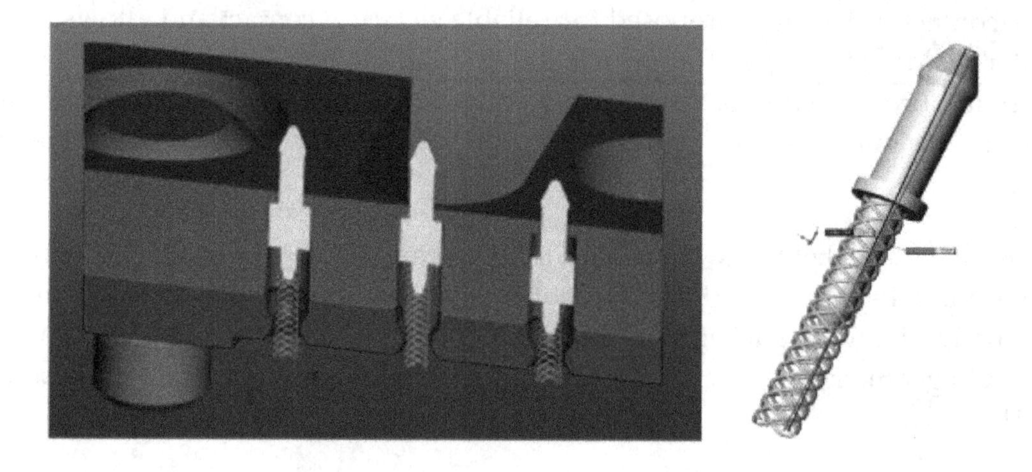

Figure 17. Prototype of new contactor design showing the plunger and the spring sections held in a retaining plate.

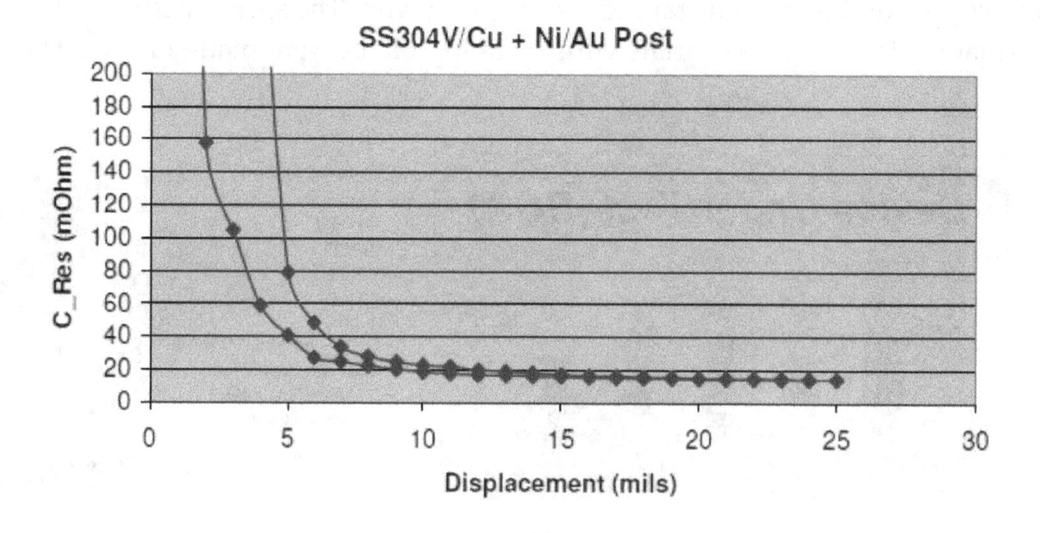

Figure 18. The contact resistance behavior if the plated spring contact.

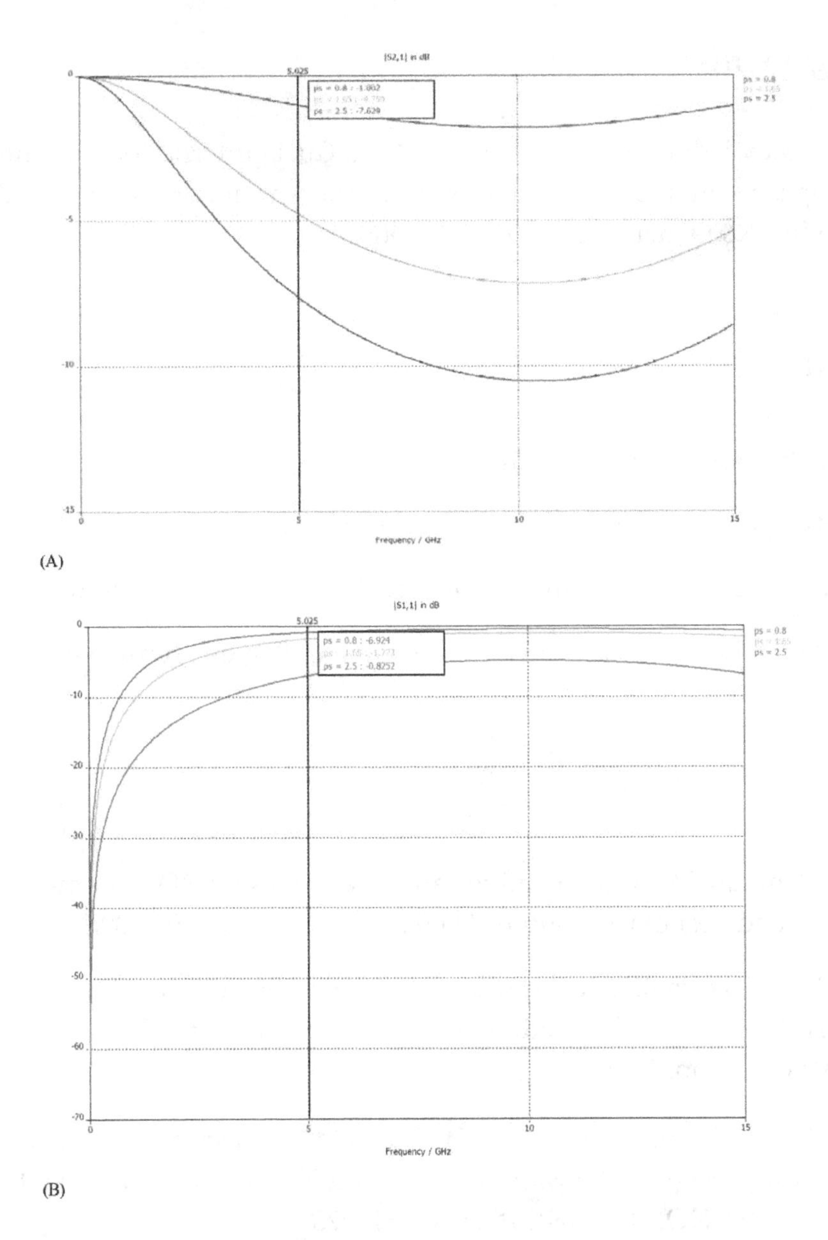

Figure 19. Simulation results for pitches of 0.8, 1.6 and 2.5 mm are shown. (A) Insertion loss, −1 dB bandwidth is 5.025 GHz (top). (B) Return loss of −16 dB at 4 GHz.

6. Conclusions

Wafer test systems and enabling requirements for effective testing of mixed signal, logic and memory ICs were reviewed. TSVs and 3D packaging are evolving and making silicon interposers available and high performance stacked die packages without wire-bonding. Silicon interposers using TSV technology based on MEMS processes can be utilized in probe card assemblies to enable next generation fine-pitch vertical probing. MEMS technologies are being developed for manufacturing of novel high density substrates and fine-pitch probes for cantilever as well as vertical probing. MEMS technologies already dominate the memory test market. It is clear though the overall market is trending toward MEMS technologies and purely vertical, cantilever, blade technology or others will shrink in probe card market and advanced MEMS technologies will win.

Acknowledgements

We gratefully acknowledge the support by a Marie Curie International Reintegration Grant within the European Union Seventh Framework Program under Grant No. 271545. We also thank SV Probe Inc. R&D members for discussions.

Author details

Bahadir Tunaboylu[1]* and Ali M. Soydan[2]

*Address all correspondence to: btunaboylu@sehir.edu.tr

1 Istanbul Sehir University, Department of Industrial Engineering, Istanbul, Turkey

2 Gebze Technical University, Institute of Energy Technologies, Gebze-Kocaeli, Turkey

References

[1] Emma PG, Kursun EM. Opportunities and challenges for 3D systems and their design. IEEE Design and Test of Computers. 2009;**26**(5):6-14. DOI: 10.1109/MDT.2009.119

[2] McCann D. Trends, challenges, and directions for silicon beyond 28 nm that drive interconnect development. In: Proceedings of IEEE Semiconductor Test Workshop, June 2015; San Diego. S01-00. pp. 1-49

[3] Li J, Liao J, Ge D, Zhou C, Xiao C, Tian Q, Zhu WM. An electromechanical model and simulation for test process of the wafer probe. IEEE Transactions on Industrial Electronics. 2017;**64**:1284-1290. DOI: 10.1109/TIE.2016.2615273

[4] Lau JH. Overview and outlook of through-silicon via (TSV) and 3D integrations. Micro-Electronics International. 2011;**28**:8-22. DOI: 10.1108/13565361111127304

[5] Zhang Y, Wang H, Sun Y, Wu K, Wang H, Cheng P, Ding GM. Copper electroplating technique for efficient manufacturing of low-cost silicon interposers. Microelectronic Engineering. 2016;**150**:39-42. DOI: 10.1016/j.mee.2015.11.005

[6] Tunaboylu B. Testing of copper pillar bumps for wafer sort. IEEE Transactions on Components, Packaging and Manufacturing Technology. 2012;**2**:985-993. DOI: 10.1109/TCPMT.2011.2173493

[7] Sperling E. Variation spreads at 10/7 nm. Semiconductor Engineering, Nov 17, 2017. Available form: https://semiengineering.com/author/esperling [Accessed: Nov 28, 2017]

[8] Tunaboylu B, Theppakuttai S. Wafer-level testing of next generation pillar solder bumps, test, assembly and packaging times. 2010;**1**(6):1-7

[9] Armendariz N. A Cost-effective approach for wafer level chip scale package testing. In: Proceedings of IEEE Semiconductor Test Workshop, June 3-6, 2007. San Diego, CA, USA. S08-04. pp. 1-31. http://www.swtest.org/swtw_library/2007proc/PDF/S08_04_Armendariz_SWTW2007.pdf

[10] Zhou J, Diller J. Are spring contact probes valid at fine pitch. In: Burn-in & Test Strategies (BiTS) Workshop, March 4-7, 2012. Mesa, AZ, USA. S5P1. pp. 1-16. https://www.bitsworkshop.org/archive/archive2012/2012s5.pdf

[11] Tunaboylu B. Electrical characterization of test sockets with novel contactors. IEEE Transactions on Device and Materials Reliability. 2014;**14**:580-582. DOI: 10.1109/TDMR.2012.2209888

5

Micropatterning in BioMEMS for Separation of Cells/Bioparticles

Rajagopal Kumar and Fenil Chetankumar Panwala

Abstract

Biofluids remain a difficult issue in some drug delivery processes for separation of bioparticles through microchannels. This chapter reviews the techniques which have been substantiated and proven helpful for the separation of particles depending on mass and size with some constraints of high throughput. In this study, a key focus will be on separation criterion by patterning of a microchannel and utilize sieve type channels based on spherical bioparticles. The first part focuses on the designing of the pattern for issues of the network like clogging and theoretical experiments by both hydrodynamic and other passive methods for sorting/separation. The second part focuses on the simulations for separation for small and larger bio particles depending on mass and size, samples of blood and other Klebsiella infected fluidic samples for the experiment. The theme provided for mass and size-based separation is simple and can accomplish operations in microfluidics for several biological experiments, diagnosis approaches and zoological researches.

Keywords: microfluidics particle sorting, patterning, Klebsiella and other bioparticle, COMSOL Multiphysics 5.2a

1. Introduction

A portion of MEMS, that is, micro electro mechanical system technology has contributed in various applications of sensors and actuators, BioMEMS applications [1] in which it has played a crucial role for Micro/Nano fluidic devices and a key role for validating a factor in integration of multiple functions for different microdevice and miniaturization. These technologies of Microsystem are used widely in biomedical, disposability, low power consumption, low cost as well as it incorporates multiple phenomena physically and due to its design complications it is difficult to deploy these devices than other sensors [1].

BioMEMS has different types of devices which determines that device for this term have some manipulations in chemicals for about smallest part in form of microlitre for bacteria or proteins separation purpose of different cells (spherical and non-spherical) drug delivery and detection of contaminant with other manipulations necessary. However, some of the micro electro mechanical system (device instrument), which is attached to normal surgical instruments, is also called BioMEMS type but it is not included in such normal devices due to restriction honored and considered as technical type instrument. There are some other devices which involve itself under BioMEMS to accentuate the idea and perform all tasked from input of samples to the detection of the cells or proteins named *micro-total-analysis system*. Operations at chip scale can replace some familiar works of laboratory process known as *lab-on-a-chip* as well as for the approach for conducting measurements parallel an *array processor* are been included. There are some other term which does not belongs to the BioMEMS but it emerges with it are defined with self as *microfluidics* [1].

A splendid addition has been exhibited which allows for the sorting depending on selection and different interests in analytes [2–4]. Thus this chapter deals with the different techniques used for separation and for micropatterning of array channels used in different applications of biomedicals and other BioMEMS terms. First part of this chapter deals with the active and passive approaches for bioparticles and cellular separation depending on the fluid velocities and its concentration depending on applications and designing. The second half of the chapter focuses on the simulations of the sorting and the detection techniques [5].

2. Patterning, separation and detection in microfluidics

In Microfluidics system technology there is an separation methods which can manipulate individual cells having the potential and empowers the experiment sets larger with lower reagent costs and allow for faster reaction work compared to conventional methods [6] for separation and modeling area using micropatterning for the samples like blood and other species like mammalian cells (*K. pneumoniae*) focused on the size and mass of the bioparticle (Spherical and Non-spherical). Nonetheless in recent years Klebsiella species has now become one of the important antibody in infections like nosocomial even there are some important members of Klebsiella breed of Entero-bacteriaceae and some have been exhibited in human laboratory specimens like *K. rhinoscleromatis* and *K. oxytoca* [7]. However separation and detection for the manipulations from the fluidic sample of bioparticles using microfluidics are sensible and challenging issues in laboratories depending on the flow rates, throughput and clogging on microscale [5].

Separation of size-based particle and aborting is occupied in many filtration systems of commonly used tap water filters to a system with complex size separation chromatography systems. As separation of size-based particle has various intent which consists of distillation of fluids or air and analytes concentration (macromolecules or proteins, DNA and others) in separation of components and biological cells. Depending on the size and mass based separation the modeling like patterning or sieving methods are commonly used where it will allow those bioparticles (spherical and non-spherical) to flow through gaps according to hydrodynamic flow rate and based on fluid velocity and viscosity measurement the other particles can be sorted individually for precise outcome [8].

2.1. Bioparticles of spherical and non-spherical shape

Ideally for geometry, it has a definite shape associated with the index of shape for adequate characterization. Depending on the characteristics of the bioparticles and its recognition visually is straightforward as defining in words or numerical value of a shape can be in a limited portion. Despite the efforts made in previous works in evolving a sphericity or index have created many hurdles for distributing values on the 3-dimensional complex irregular shape [9–11]. For the shape factor there is no perturbation with it for changes in the drag force of particle for spherical and non-spherical as there has been a development in the dynamic shape factor relatively. This ratio of drag force is calculated for the forces experienced by the spherical or non-spherical particle over the ratio of other particles with the volume equivalently flowing with the medium and same flow velocity. Other than spherical particle there is a an phenomenon change in the non-spherical particle as understanding the indexing of non-spherical particle is important with the chemical, physical, biological functions and other phenomenon by which the changes in the effect of bioparticles cannot be denied. However the anatomy of some proteins molecule to the utmost dimensions of galaxies and stars neglecting the effect of shape can be made. Interaction with the analysis of particles deeply requires some considerations of shape for behavior in the system of the particle in applications practically to conclude [8] (**Figure 1**).

2.2. Techniques for sorting

In recent years for studies of application on cell separation there were many techniques extended which emerged in the field of microfluidics for biomolecular analysis which are initially driven as per requirement. While manipulation and Cell separation is a crucial sample step for processing in many medical assays and low Reynolds numbers as well as predictable flows, small fluid volumes, small dimensions, materials and the established microfabrication

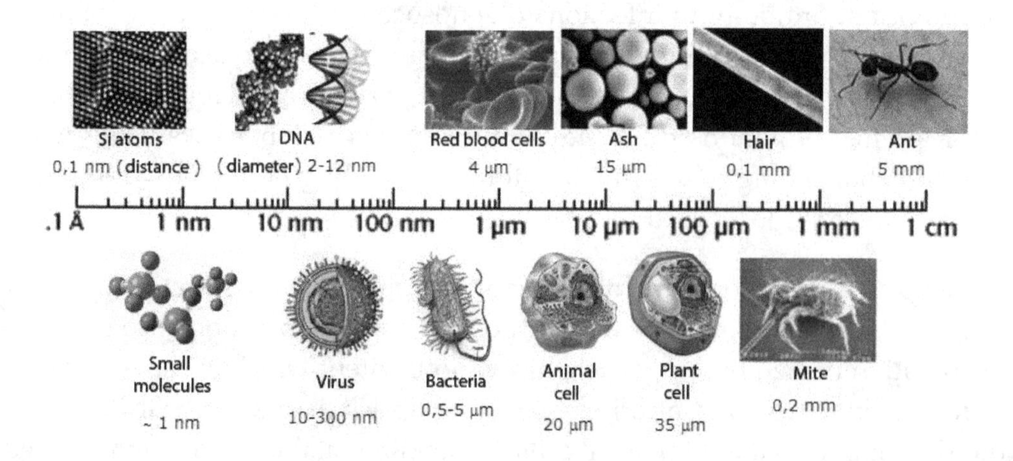

Figure 1. Types of bacteria present with its size of shape [12].

techniques which are typical microfluidic device factors allows user to work with cells on microscale. Existing microfluidic separation methods are categorized into two methods as active and passive methods where active methods incorporate an external force and passive methods rely on carefully designed channel geometries and internal forces to sort different particles. Classification is as shown in **Figure 2**.

In this chapter, we introduce various principles and related methods including some common separation methods for active type include immunomagnetic separation (IMS), acoustophoresis, electrophoresis, dielectrophoresis, optical force and flow cytometry or FACS [14]. There are different technologies or methods in microfluidics for precise, passive and continuous sorting systems invented including Hydrodynamic inertial force, Deterministic Lateral Displacement (DLD), Pinched-Flow Fractionation (PFF) and Hydrodynamic Filtration (HDF) [15, 16].

Microfluidic technologies that can manipulate individual cells with the potential enable larger experiment sets, lower reagent costs and allow for faster reaction work compared to conventional methods [6]. PFF is a separation technique where a field is applied to a fluid suspension pumped through channel which is narrow and long, flow in perpendicular direction depending on their differing "Flexibility" under the exerted force by the field and to cause separation of the particles present in the fluid. In recent years a number of microfluidic devices have been advanced for the continuous separation of bioparticles by employing unique techniques. One of the prominent separation method is DLD in which particles drives through the post to post arrays positions. The interaction of the different particles with different size and post to post array directions leads to different bacteria to drift in different directions with respect to the arrays thus causing the continuous fractionation and two dimensional of the sample mixture.

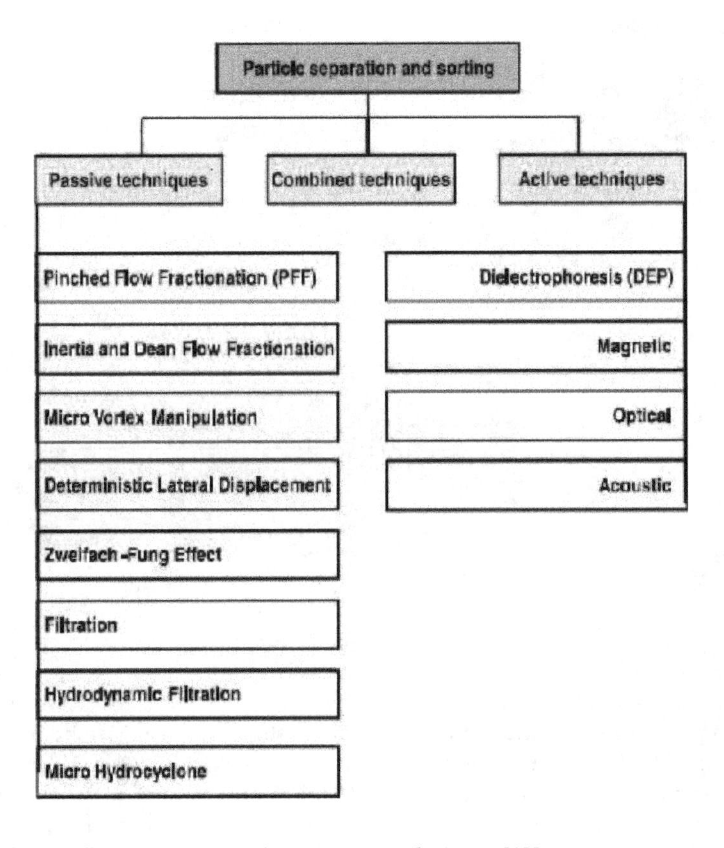

Figure 2. Classification of microfluidic sorting and separation techniques [13].

Hydrodynamic force is one of the inherent physical principle and basic in system of microfluidics where inertial force and dean rotation force has been utilized, though hydrodynamic force based separation system can be created by fluid dynamic theory based microchannel network. An Acoustic method utilizes the ultrasonic standing waves and allowing a manipulation of cells then other separation methods like dielectrophoretic, magnetic and others. The basic principle of acoustic sorting is to use pressure gradients generated by ultrasonic standing waves since most microfluidic system uses liquid medium as the working fluid. DEP is the electro kinetic motion which occurs when polarisable particle is placed in non-uniform electric fields and particle motion induced by DEP force is influenced by the ambient electric field and the properties of electric particles or solutions. Magnetic sorting system can be of two different categories for separation (I) Attaching magnetic particles to the cells to react with the magnetic field and (II) Utilizing the native magnetic properties of the cells in laminar flow for some basic motion of particles expressed as Newton's second law [17] schematics shown in **Figure 3**.

Microfiltration is a method of basic concept for separating and sorting microparticles which utilizes the size of micropores and the gap between microposts as a lattice or sieve. Accordingly, microfiltration is highly dependent on the sizes of the microparticles. The advantages of this methods discussed are easy for understanding the separation principles for engineer or scientists to implement for an application to existing methods [17]. Thus by this study we have introduced a developed technique HDF which focuses on the shear flow of input associated with buffer streams which constricts the stream flow within the middle of the microchannel

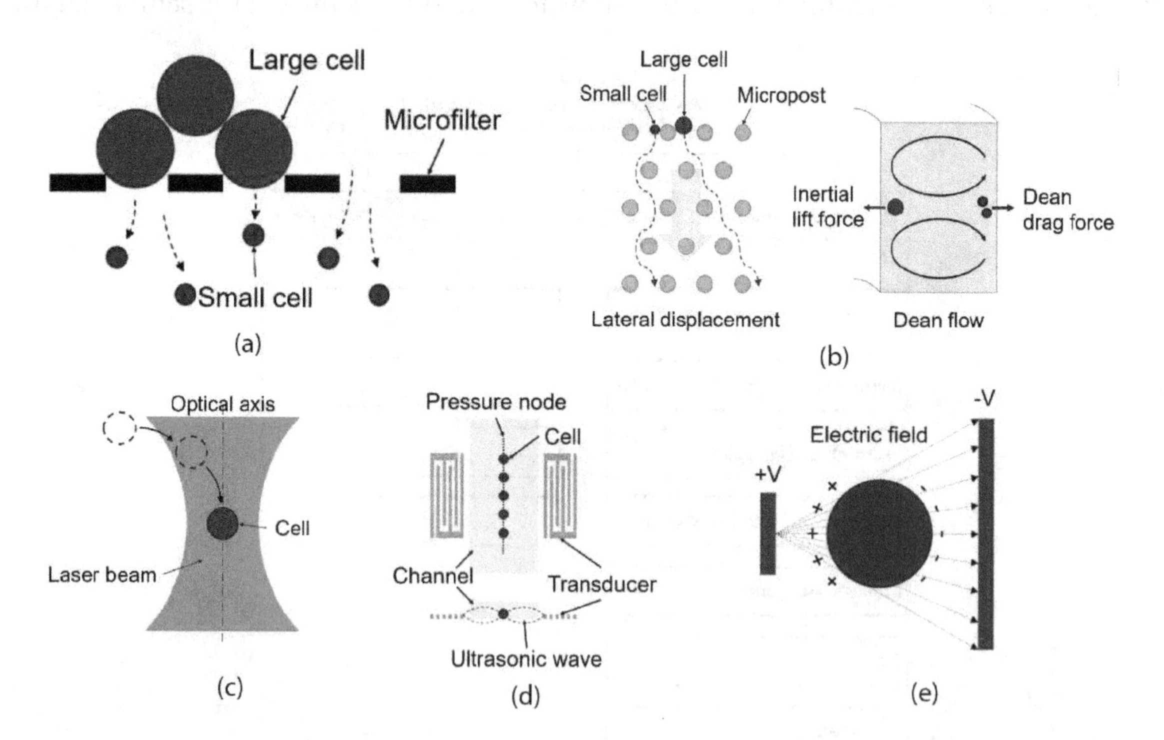

Figure 3. Cell Separation principle schematics (a) Geometrical constraints for cell size in microfiltration technique (b) hydrodynamic forces applied to the cells in the network of microchannel (c) Between the magnetic force field and the magnetic particles attached to the cells they are separated by the attraction force (d) The ultrasonic waves generated by the transducers from which acoustic cell separation utilizes acoustic primary radiation force (e) The electro kinetic motion of polarisable cells in non-uniform electric fields is utilized by Dielectrophoretic force (DEP) [17].

accommodated with particles in the plane of sidewalls with narrow width [18] where both concentration and classification of particles can be examined at the same interval of time by introducing a solution consisting of particle beads of various mass and size.

We preceded here with a method that has been adopted to invent a separation mechanism, Continuous processing, high precision for particle separation and high throughput [17]. When a particle flows in a microchannel the center position of the particle cannot remain present on a certain distance from sidewalls which is equal to the particle radius. The method of filtration utilizes this fact and is performed using a Sieve type-shaped microchannel network having multiple side branch channels/sieve-shaped networks. According to flow rates in microchannel by fluid flow (μl) it withdraws a small amount of liquid continuously on intervals from the main stream through the side microchannel network and particles are concentrated and aligned in the network according to size and mass. However the concentrated and sorted particles can be collected according to size and mass through all other output channels in the stream of the microchannel network. Therefore continuous introduction of a particle suspension into the microchannel enable particle sorting, concentration and classification at the same time with precision [19]. Whenever there is a difference in particle size and mass then separation becomes difficult for the result and mesh clogging is inevitable [20].

Hydrodynamic Filtration is one of the most frequently used technique to classify particles suspended in fluid flow due to sedimentation. Existing filtration methods performed either in batch or continuous manner and large-scale treatment can be easily achieved. Biological entities such as rod-shaped bacteria and disc-shaped red blood cells (RBCs), disproportional length and width which complicate the separation process designed for spherical particles and the narrowest width has to be considered for the separation criteria within the design parameters of the microfluidic devices. Some of the examples for cell separation include (**Figures 4** and **5**):

- Blood cells (WBC) isolation from tissue
- Circulating Tumor Cells (CTC) from blood
- Separation of some bacteria from food which are pathogenic to health and other systems
- Resistant cells isolation from peripheral blood

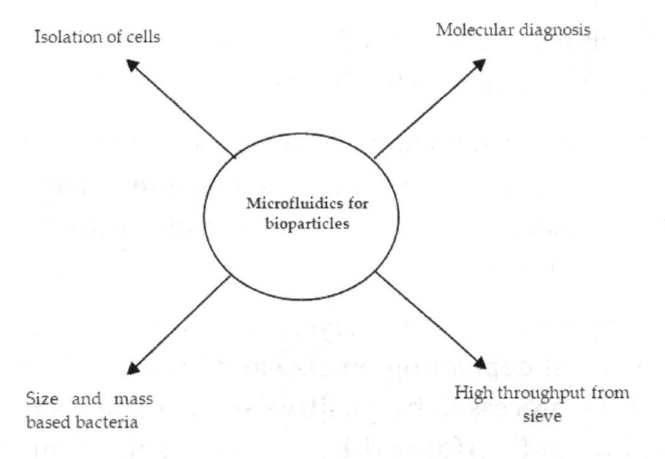

Figure 4. Role of microfluidics for the separation.

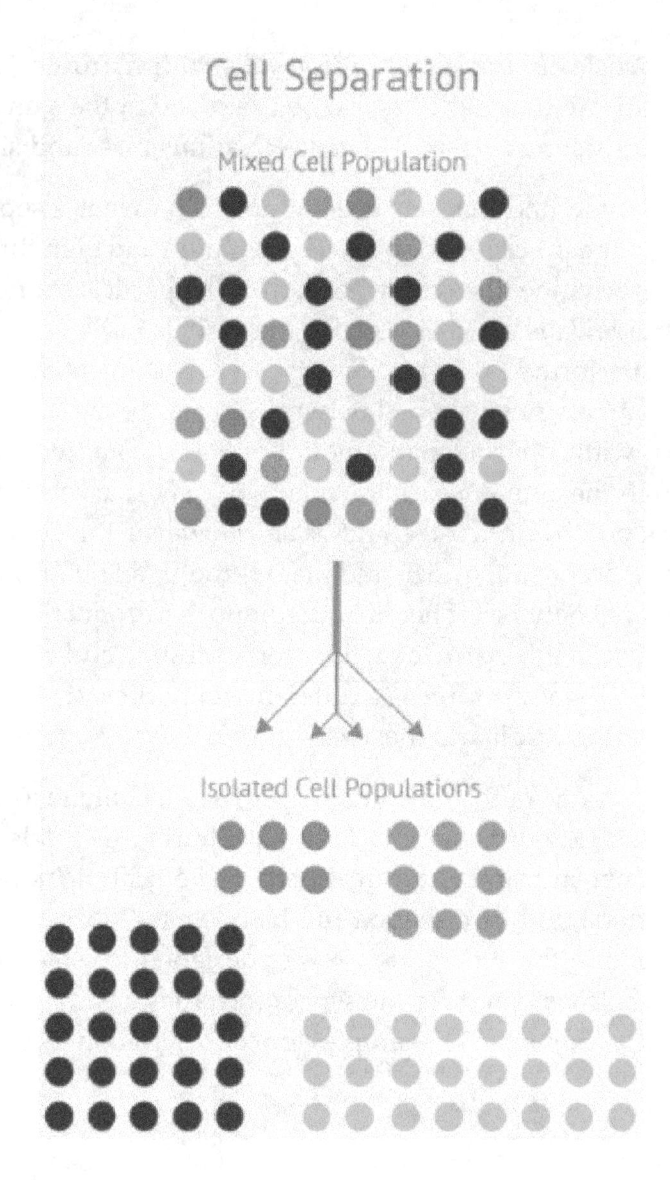

Figure 5. Sorting and separation for different cells from mixed population to isolated populations [20].

However there are some essential approaches through which separation takes in microfluids,

- From a biological sample if a single type cell is removed it can be called as **Depletion region.** Example mononuclear cells from which RBCs can be removed.
- Similarly when there is need of removing cells other then the single cell like RBC a **Negative selection** can be done where it will leave single cell type and other packet of cell needed to be removed can be separated. Example like bone marrow or whole blood sample, removal of cells varying with size and others.
- Whenever there is any downstream analysis then a mechanism of removal cell type is targeted whichever cell depending on size and mass can be separated this type of typical selection can be processed by **positive selection**. This type can be possible in monoclonal antibodies and performed by aiming a surface marker of cells (bioparticles) [20].

3. Analysis and modeling for microchannel

3.1. Sieve or lattice type microchannel network design

The sieve-shaped micro channel network is assured of two types of microchannels (I) The main channel and (II) Separation channel, In microchannel main channels are deeper than the separation channels as the main channels are at different angle positions of channel to lower right/left and perpendicularly-crossing are separation channels. Particle/cell is introduced continuously from the inlet at intervals, whereas a buffer solution without particles/cell is introduced from two side inlets shown in **Figures 6–8**. Smaller particles can reach the ceiling (upper region) of the main channels but the larger particles cannot because of the effect of hydrodynamic filtration. Consequently the repetition of the larger particles entering the separation channels will be higher than that of the smaller particles and the positions of the larger particles will shift in the direction according to flow rate more greatly than those of the smaller particles depending on the mass and size of the particle achieving continuous separation [21].

As discussed about the techniques employing the effect of hydrodynamics in microfluidic applications has been prominent in the decades to be fruitful in terms of efficiency, throughput and continues to develop in the future through some more improvements in separation processing rate and resolution. However there are many other new areas of hydrodynamic microfluidic phenomena for an application which demand further investigations and promisingly both explain and informs researchers theoretically about basics on physics and

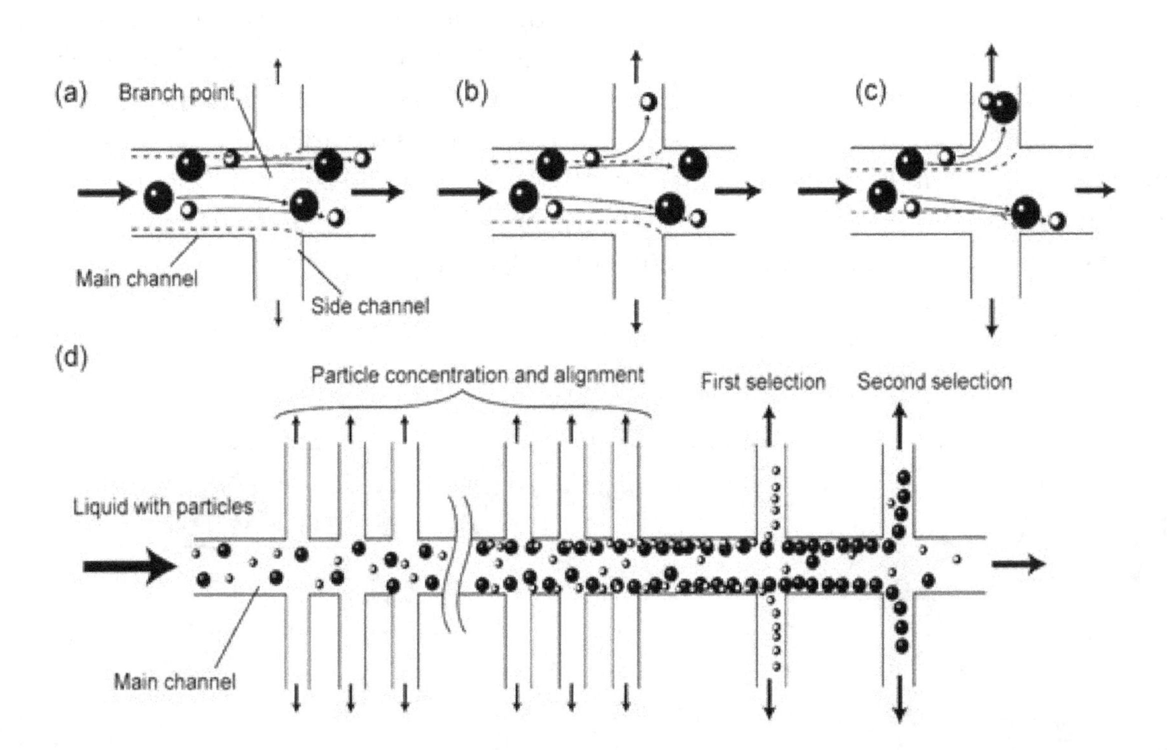

Figure 6. Hydrodynamic filtration principle showing behavior of particle at branch point according to different flow rates which are high, medium and low at multiple channels [19].

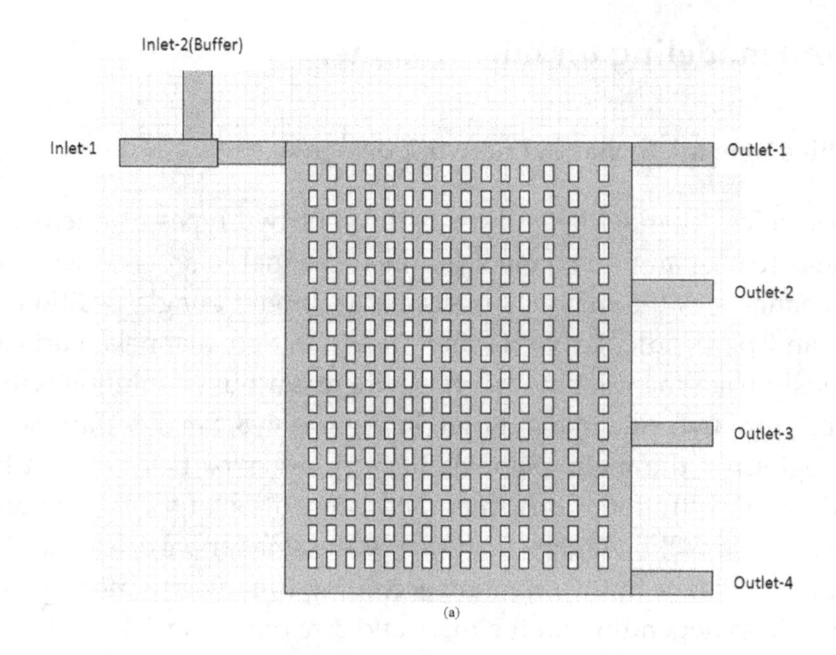

Figure 7. Schematic of Sieve type – shaped Microchannel network.

to exploit them experimentally in applications of biological terms. In spite of a few active separation techniques have been developed to conform the demand growing in these new area [22]. Presently a continuous particle/cell separation system utilizes a Sieve type-shaped micro channel network has been shown ahead. The difference in the densities, velocity, pressure and viscosity in sample it generates the asymmetric and symmetric flow distribution at each intersection with intervals resulting in the separation of large size particles through the streamline [23]. A modified mechanism of particle sorting using sieve type microchannel patterning is presented where it potentially enables the throughput separation highly and can prevent clogging problem of micro channel at some extent. The presented system would become a simple but valuable unit operation in the microfluidic apparatus for medical and biological experiments [21]. The presented network system would be highly useful because of sorting microparticles and cells with a high precision and would become an important useful tool for general chemical/ biological experiments in laboratories.

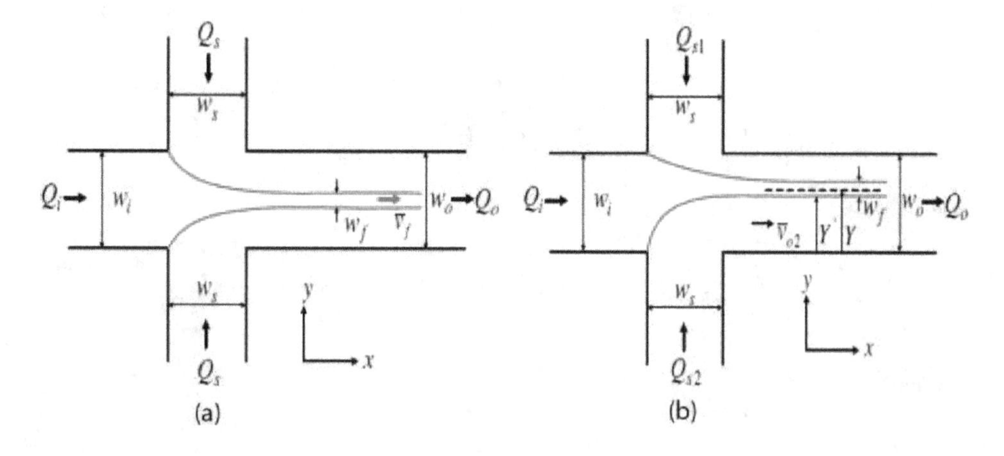

Figure 8. Schematic of: (a) Symmetric hydrodynamic flow focusing and (b) Asymmetric hydrodynamic flow focusing [27].

Here two inlets are employed from which one is used to introduce fluid without containing particles so that the particles flow along the sidewall. Multiple side channels are used in area so that particles larger than a certain size cannot pass through. As a result, such particles are concentrated and focused onto the sidewall. Microchannel network can be as shown in figure using COMSOL Multiphysics 5.2a software. In the downstream area the side channels are made gradually wider or shorter so that the particles are removed from the main stream in ascending order of size. Thus, the particles are sorted by size and concentrated with cytometry of flow with a basic laminar flow that focuses particles in same dimension, while at High flow rate inertial forces on particles cause are used to manipulate particles and inertial forces dominates when the particles Reynolds number is >1.

It is well known that a microchannel acts as a resistive circuit when an incompressible Newtonian fluid is continuously introduced into the channel. The micro device was therefore designed according to the concept that the volumetric flow rate *Q*, applied pressure *P*, and hydrodynamic resistance *R* are analogs of *I*, *V* and *R* in Ohm's law, respectively. In this study the following equation was used to estimate the hydrodynamic resistance *R* of each segment of the microchannel [24].

$$R \propto \frac{L}{l_1^3 l_2}\left[1-\frac{192\,l_1}{\pi^5 l_2}\sum_{n=1,3,5}^{\infty}\frac{\tanh(n\pi l_2/2l_1)}{n^5}\right]^{-1} \tag{1}$$

where l_1 and l_2 are either the width or depth of the microchannel however, l_1 is the larger of these two values.

3.2. Theoretical and numerical analysis

For theoretical discussion prediction made with the width of two-dimensional hydrodynamically streams in rectangular shape micro channels is designed. Here critical diameter of bacteria particles will sustain a stable detonation for minimum diameter while the spacing that is center to center between the post is known as λ and d is known as the relative shift between adjacent posts, thus to measure the parameter λ ¸t can be measure with relative shift and tangent of the angle with respect to the vertical objects through the array as shown in (**Figure 8**) [25].

$$\varepsilon = \frac{d}{\lambda} \tag{2}$$

"Unconfined" and "confined" critical diameter was determined directly by inspecting the experimental distance/time data and examining the lattice-shaped microchannel at different angle [26].

Generally, ε (smaller) gives an result for smaller critical size in an array. However the clogging problem and large particle densities in an array do not occur easily. Therefore for a simple object [25],

$$D = \frac{K_B T}{6\prod \eta a} \tag{3}$$

where K_BT is thermal energy at temperature, D is diffusion coefficient of particle of radius 'a' inflow that we wish to separate from flow streamlines.

In this sieve type-shaped design a single region of posts will have a single threshold and particle flow will be in two directions. According to the principle of mass conservation, the supply of fluid flow passing through the dimension of the stream should be equal to the fluid passing through the inlet channel, i.e. [27]

$$w_f = \frac{Q_i}{\overline{v}_f h} \tag{4}$$

Moreover, the total fluid flowing through the outlet channel must equal the total amount of fluid supplied from the inlet and side channels, i.e. [27]

$$\overline{v}_o = \frac{Q_i + Q_{s_1} + Q_{s_2}}{w_o \times h} \tag{5}$$

Therefore, the relation between the width of the hydrodynamic focused stream (w_f) and the volumetric flow rates of the inlet channel (Q_i) and the side channels (Q_{s1} and Q_{s2}) can be as [27].

$$\frac{w_f}{w_o} = \frac{Q_i}{\gamma\left(Q_i + Q_{s_1} + Q_{s_2}\right)} \tag{6}$$

where,

$$\gamma = \frac{\overline{v}_f}{\overline{v}_o}$$

where the velocity ratio γ to be found, w_o is width of the outlet channel and v_f and v_o are the average flow velocities in the focusing stream and the outlet, respectively shown in **Figure 6** [27].

Thus the performance of the device using sieve type-shaped microchannel can be improved with the faster flow rate and clogging problem can be reduced. A critical hydrodynamic diameter can be designed easily with G between posts of the microchannel shown in **Figure 9**. Some assumptions on this study were made are:

1. Flow within the micro channels is steady and laminar. But, because of the smaller characteristic dimensions are involved the flow in microchannel is laminar.

2. The fluids are Newtonian.

3. The fluid has constant density in the inlet channel, side channels and outlet channel.

4. Inlet, outlet and all the channels are of the same measurement.

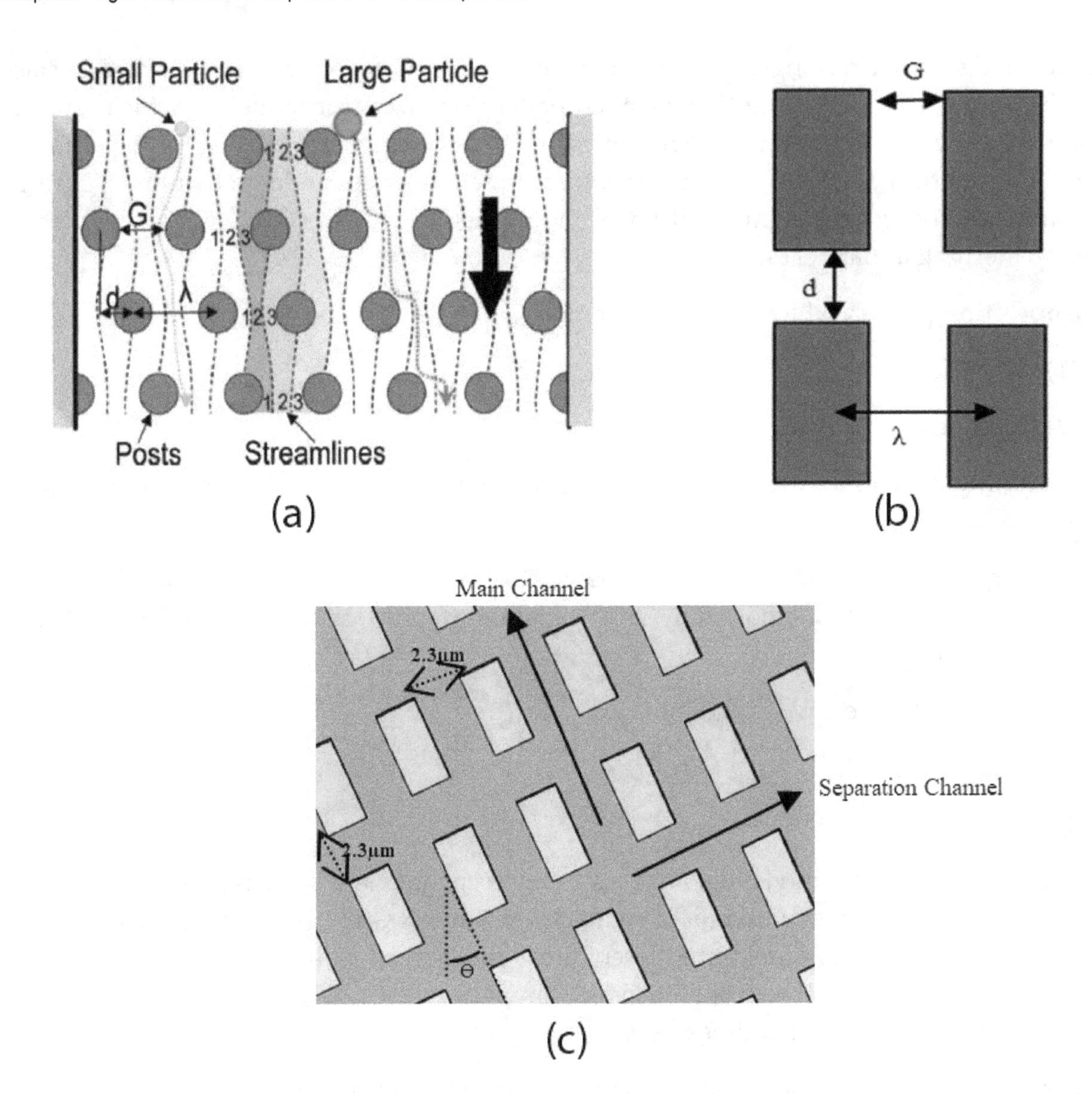

Figure 9. (a) Separation by a DLD in an array of micro posts with streamlines shown with G is spacing between the gaps in structure [28] (b) and (c) Structure and rectangular shape have and x and y distance of 4.3 and 2 μm showing main channel and separation channel.

For the particle movement depending on the volumetric flow rate was measured in COMSOL multiphysics 5.2a software by the Navier-stokes equation in compressible fluid flow [29].

$$\left\{\left(\rho\left(\frac{\partial u}{\partial t}\right) + u \cdot \nabla u\right)\right\} = \{-\nabla p\} + \left\{\nabla \cdot \left(\mu(\nabla u + (\nabla u)^T)\right) - \frac{2}{3}\mu(\nabla \cdot u)I\right\} + F \tag{7}$$

There is a unique term that corresponds to the inertial forces, viscous forces, pressure forces and external forces which are applied to the fluid flow as Eq. (7) plays a vital role for the flow and to predict the movement for the particle according to the volumetric flow through the channel. For the different velocity magnitude at various streamlines for the particle in microchannel-1 is calculated on the basis of Eq. (7) precisely in COMSOL multiphysics 5.2a. By solving equation for specific conditions include inlets, outlets and walls predictions for the velocity and pressure in geometry can be observed and using sieves or grooves in a microchannel can be simulated.

However instead of rectangle type shape in microchannel there are many related surfaces which can be used for sieving type or micropatterning of microchannel by which a precise result can be generated and can be modified using microchannel network for experimental work after fabrication of the microchannel for different surfaces are as shown in (**Figure 10**). For fabrication in micropatterning it has some processes used in BioMEMS and carried for micropatterning a channel as,

- **Deposition process** which is subdivided into,
  (1) Physical
  (2) Chemical
- **Patterning process** in MEMS is the transfer of pattern into a material lithography is widly used process which are framed as,

(1) Lithography where some types of categories are available with its process as Photolithography, Electron beam lithography, Ion beam lithography, Ion track technology and X-ray lithography

- **Etching process** are subdivided into,
  (1) Wet etching
  (2) Dry etching

Depending on different modeling, analysis of microchannel and sieve type patterns the fabrication can be proceeded with reliable material which can withstand some parameters for a patterned chip and can be sued in different areas for sample analysis like separation of particle and detection of bacteria from the given sample like CTCs, DNA, Klebsiella bacteria like *E. coli*, *K. pneumoiea* and other types.

The main task which can be carried for separation are the size and mass of the bacteria ranging between 0.3 and 10 μm and mass weight can vary from 1 to 10^{-12} Kg depending on the sample in biological terms precisely. Similarly for Klebsiella species one of the bacteria cells are *K. pneumoniae* whose particle size varies around 0.5–2.5 μm. However the mass of the

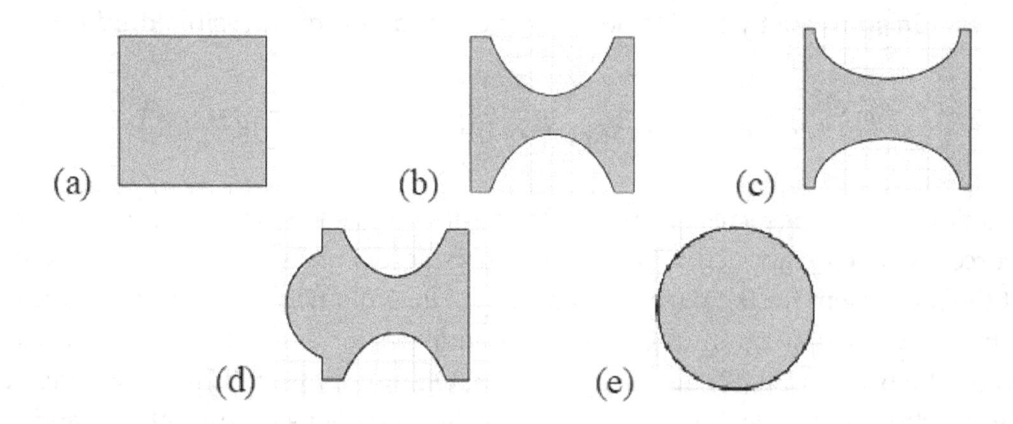

Figure 10. Different shape for microchannel (grooves).

bacteria particle of *K. pneumoniae* changes with the size of the particle and remains in a range of 10^{-12} to 10^{8}Kg. Depending on the volumetric flow rates in microfluidic system for a hydrodynamic filtration method the flow rates for the separation of *K. pneumoniae* bacteria from the sample can be precisely separated. Similarly the flow rates depending on the fluid viscosity and density can test at different flow rates in microliters (µl) numerous times for the inlets. According to model designed in (**Figure 7**) can be tested with two or multiple types of particles differing in size and shapes for the separation however it is possible for experimentation and simulation in software like COMSOL Multiphysics 5.2a, Intellisuite and other related tools. Whereas the clogging problem in the channel can be reduced using this structure which is a major issue for separation in micropatterned channels. Thus can achieve higher throughput sorting when a separation occurs at every intersections and robust against the problem of clogging in microchannel.

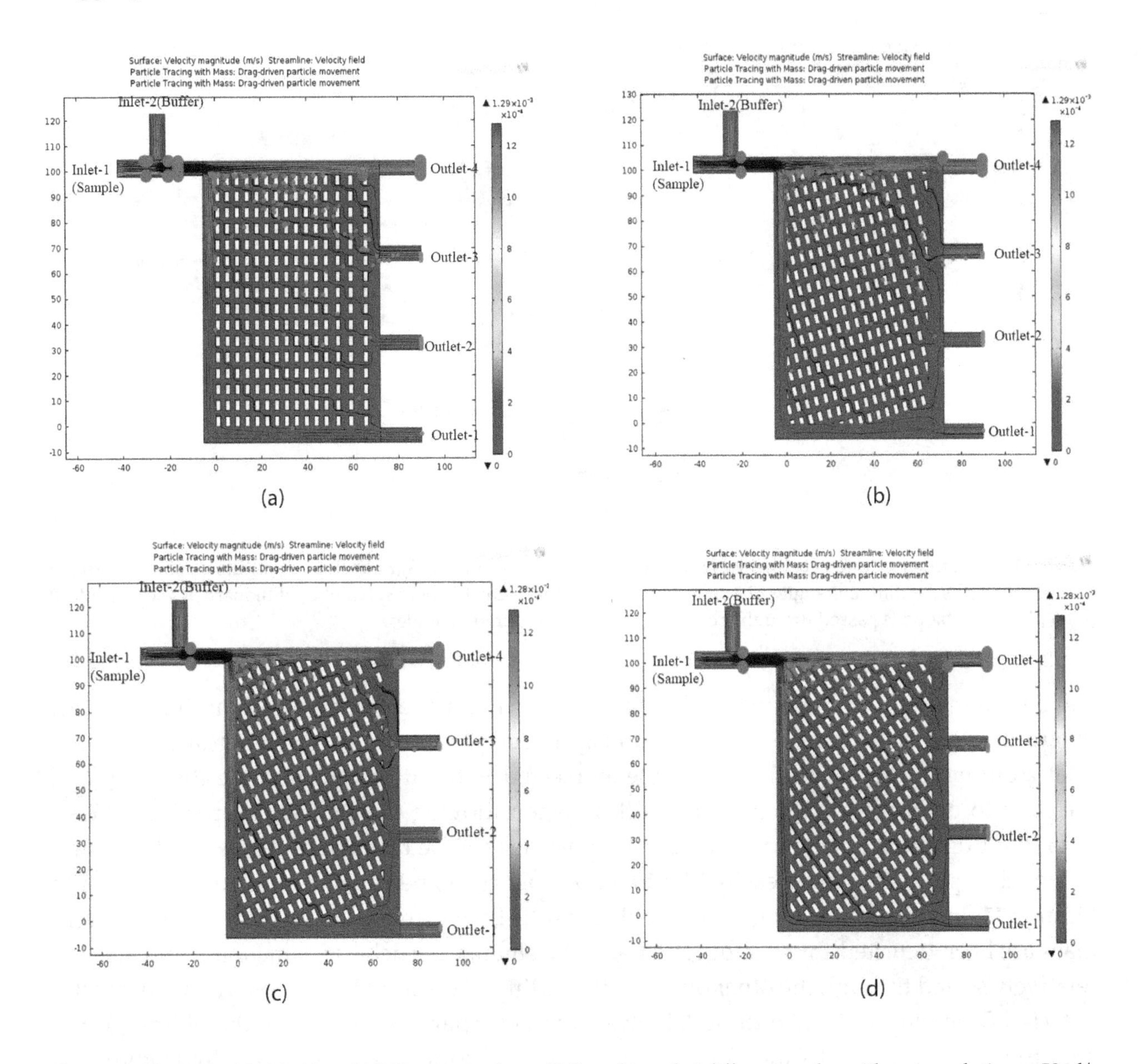

Figure 11. Shown in (a), (b), (c) and (d) Sieve type-shaped Microchannel at different angles with water solution at 50 µl/s of fluid flow rate in water. Similarly tested at different flow rate for different angles with measureable flowrates and simulated with blood sample properties also with the software COMSOL Multiphysics 5.2a.

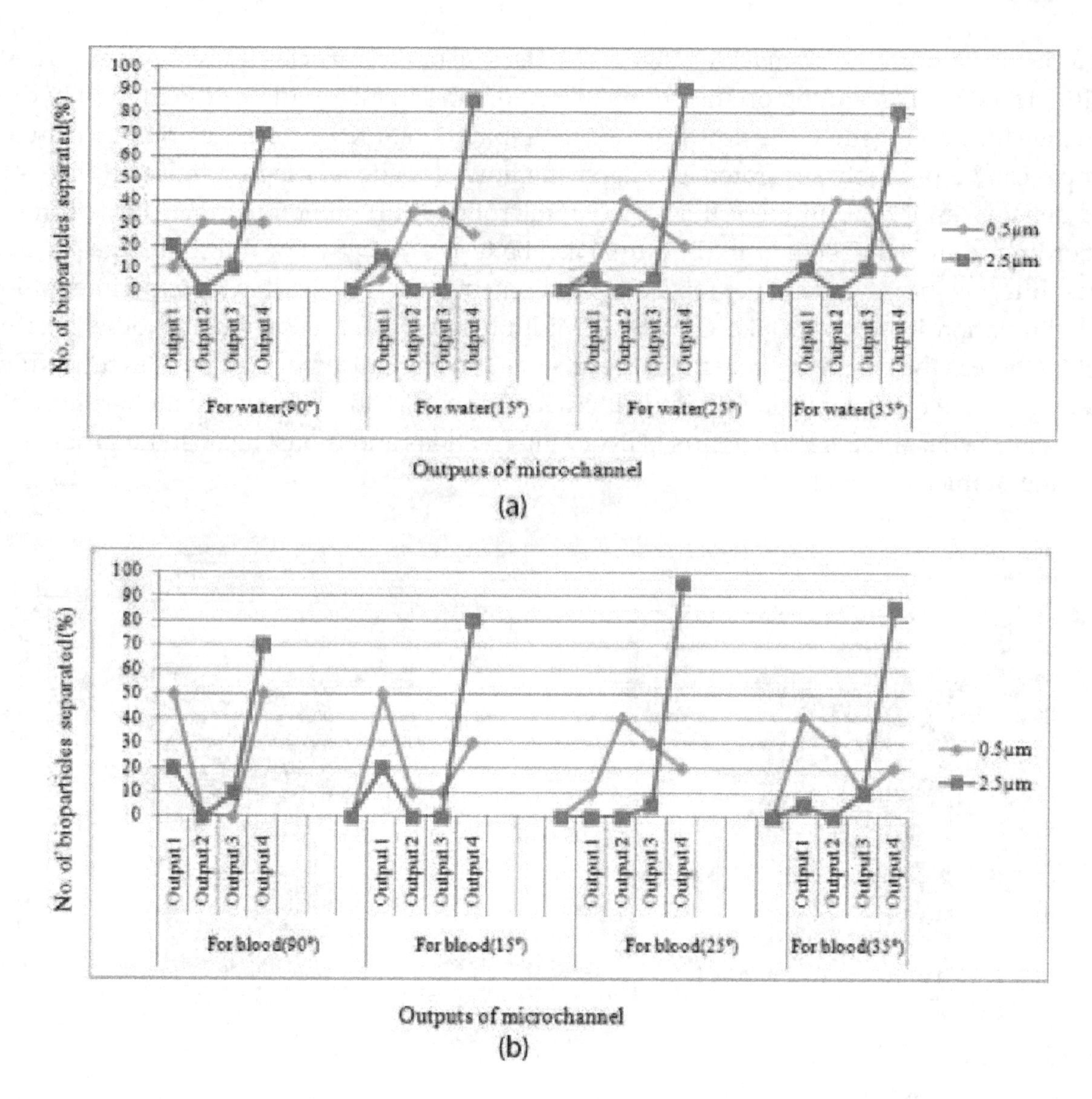

Figure 12. (a) Outputs for microchannel vs. Number of bioparticles seprated(%)(m/s) plot when water is passed through grooves at 50 μl/s for different angles, (b) Similarly, Outputs for microchannel vs. Number of bioparticles separated(%)(m/s) plot when blood is passed through grooves at 3000 μl/s for different angles.

Sieve type microchannel network is designed with main channel and a separation channel where the cells with sample are continuously introduced through an inlet-1 and the other inlet-2(buffer) through which the sample is injected without cells to precede the sample in the flow to decrease the fluctuations in flow when moving through the chip. Overall outputs (Four outputs) through which bacteria particles can be counted at it rates and pressure by which separation can be easily identified. As per the experimental simulations shown in **Figures 11** and **12** (Graphs) it give a probable results by which the larger particles of different mass are been detected through output-4 and the smaller particles of same mass and size are relatively sorted through the other outputs depending of the fluid flow rates which are mostly detected from output-3 and output-2. However Smaller particles will reach the ceiling (upper region) of the primary channel But the larger particles cannot because of the hydrodynamic effect. Owing to the change in frequency of particles they gets shifted toward right direction according to the fluid rate and buffer effect by which paticles are individually identified through outputs. Similarly different shapes of particles like non-spherical or rod-type shape can be utilized and detected through microchannel network for sorting. By this system of sieve network it achieves a high throughput sorting while separation occurred at every intersection and is robust against the issue of clogging in microchannel network [6].

Micro device having sieve-shaped micropatterned channel has been designed and simulated for a distinct slanted angles at 90, 15, 25 and 35° through which separation of mass and size-based particles from a sample was carried for *K. pneumoiae* beads (by using its properties). The model for the Sieve type microchannel has been done based on the equation for critical diameter, Navier-stokes equations analysis and accordingly depending on the mass and size of the particle it was separated through the main and separation channels of micropatterned chip of sieve type model. The fluid flow simulation was tested with blood sample and distilled water properties in simulation at different angles of channels by which a bigger particle as well as smaller particles were separately identified via outputs for the precise separation (**Figures 11** and **12**). The main channel and the separation channel were crossing at right angle and also the particles with the size of 3 and 8 μm (aspect ratio = 0.3) were separated through outputs at various flow rates in μl/s for blood and water mixed with properties of *K. pneumoniae* beads by micro channel network as the size of the *K. pneumoniae* varies from 0.5 to 2.5 μm shown in **Figure 11**.

We examined the model of separation at different input flow rates concerning about 13, 20, 25, 50, 66, 80 μl/s with water and 133, 666, 800, 1000, 3000, 5000 μl/s with blood by which the separation behavior of the particles using COMSOL Multiphysics 5.2a were simulated and calculation is done. Similarly, at various angles of the channels the samples through input were simulated for the quick and precise performance for the sorting of bacteria at different fluid flow rates mentioned shown in **Figure 12**.

Finally a discussion can be made for the *K. pneumoniae* particles that it can be examined with different size of the spherical or non-spherical shape cells for separation which has been a big issue in hospitals and routine lifestyle. Since infections *K. pneumoniae* are air borne type which is controlled by mucosal vaccination through pulmonary route and nasal can be an assured approach in order to resemble natural infection. Other than traditional vaccines new vaccines like subunit vaccines are safer but they have less immune response [30]. Therefore to enlarge their immunogenicity, dynamic and delivery systems and safe adjuvant are necessary to be developed. Thus by using micropatterning system the sorting of particle from samples can be easily invented [30]. Furthermore this microchannel network can be carried for different purposes for biological or industrial work and even for the separation of rod-type particles from the *K. pneumoniae* by improving separation efficiency and change or modification in design for the better cause.

4. Conclusion

There was an good development in past, where various microfluidic techniques were developed which involves application in enrichment, particle sorting and detection. However expected high throughput and clogging problem are the facts which has to be properly experimented for higher purpose of biological testings, their conditioning and patterning related fabrication techniques for developement with some characteristics and precise separation of cells should be one of the top priorities. Thus sieve-typed microchannel network can be a precise separation chip network which is efficient for hydrodynamic cell in microfluidic device of size and mass based separation. Effective for mucosal vaccine delivery systems, while flow can be related for immune protection against *K. pneumoniae* infection and suitable for group vaccination programs resulting in considerable health as well as economic benefits. Therefore to contribute to the different successful collaborations at this stage where the stability and the robustness of the microfluidic viability should be taken into care to steadily contribute to successful collaborations between different biomedical, clinicians, biologists, and other the microfluidic association.

Acknowledgements

The authors would greatly acknowledge the support from National Institute of Technology Nagaland, Chumukedima, Dimapur, India and MEMs Design Centre from SRM university, Chennai, Tamil Nadu, India for their assistance.

Author details

Rajagopal Kumar and Fenil Chetankumar Panwala*

*Address all correspondence to: fenilpanwala68@gmail.com

National Institute of Technology Nagaland Chumukedima, Nagaland, India

References

[1] Korsmeyer T, Zeng J, Greiner K. Design tools for BioMEMS. In: Proceedings of the IEEE International Conference on Design Automation; 7-11 July 2004. San Diego: IEEE; 2005. Print ISSN: 0738-100X. pp. 866-870

[2] Arora A, Simone G, Salieb-Beugelaar GB, Tae Kim J, Manz A. Latest development in micro total analysis systems. Analytical Chemistry. 2010;**82**:4830-4847

[3] Gervais L, de Rooij N, Delamarche E. Microfluidic chips for point-of-care immunodiagnostics. Advanced Materials. 2011;**23**:H151-H176

[4] Tran NT, Aved I, Pallandre A, Taverna M. Recent innovations in protein separation by electrophoretic methods : An update. Electrophoresis. 2010;**31**:147-173

[5] Roy E, Pallandre A, Zribi B, Horny M-C, Delapierre F-D, Cattoni A, Gamby J, Haghiri-Gosnet A-M. Molecular microfluidic bioanalysis: Recent progress in preconcentration, separation, and detection. In: Yu X-Y editor. Advances in Microfluidics-New Applications in Biology, Energy And Material Sciences; 2016. ISBN 978-953-51-2786-4, print ISBN 978-953-51-2785-7. Centre for Nanoscience and Nanotechnology, CNRS, University Paris Sud, University Paris Saclay, Marcoussis, France

[6] Kim M-C, Klapperich C. A New Method for Simulating the Motion of Individual Ellipsoidal Bacteria in Microfluidic Devices. Department of Biomedical Engineering. Boston University, Lab Chip. pp. 2464-2471. DOI: 10.1039/c003627g. Epub 9 Jun 2010

[7] Sunahiro S, Senaha M, Yamada M, Seki M. Pinched flow fractionization device for sizes and density de pendent separation of particles utilizing centrifugal pumping. In: 12th International Conference on Miniaturized Systems for Chemistry and Life Sciences; San Diego; 12-16 October 2008

[8] Zeming KK. Rotational separation of non-spherical particles using novel I-shaped pillar arrays [thesis]. Department of Biomedical Engineering, National University of Singapore; 2014

[9] Kulkarni P, Baron PA, Sorensen CM, Harper M. Aerosol Measurement. Chicester: John Wiley & Sons, Inc; 2011. pp. 507-547

[10] Lau R, Chuah HKL. Dynamic shape factor for particles of various shapes in the intermediate settling regime. Advanced Powder Technology. 2013;**24**:306-310

[11] Hakon W. Volume, shape, and roundness of quartz particles. The Journal of Geology. 1935; **43**:250-280

[12] Haick H. Introductipn to Nanotechnology, Israel Institute of Technology. http://www.davidfunesbiomed.eu/2015/06/nanotechnology-introduction.html [Accessed: 23-January-2018]

[13] Sajeesh P, Sen AK. Particle Separation and Sorting in Microfluidic Device: A Review. Department of Mechanical Engineering. India: Indian Institute of Technology Madras/ Berlin Heidelberg: Springer; 2013

[14] McGrath J, Jimenez M, Bridle H. Deterministic Lateral Displacement for Particle Separation: A Review. Riccarton, Edinburgh : Heriot-Watt University, Microfluidic Biotech Group, Institute of Biological Chemistry, Biophysics and Bioengineering (IB3); 2014

[15] Srivastav A, Podgorski T, Coupier G. Efficiency of size- dependent particle separation by pinched flow fractionation. Microfluidics and Nanofluidics. 2012;**13**(5):697-701

[16] Seko W, Yamada M, Seki M. Slanted lattice-shaped microchannel networks for continuous sorting of microparticles and cells, Chiba University. In: Japan 17th International Conference on Miniaturized Systems for Chemistry and Life Sciences; Freiburg; 27-31 October 2013. 978-0- 9798064-6-9/_TAS 2013/$2013CBMS-0001

[17] Lee G-H, Kim S-H, Ahn K, Lee S-H, Park JY. Separation and sorting of cells in mcirosystems using physical principles. Journal of Micromechanics and Microengineering. 2015; **26**(1). IOP Publishing. DOI: 10.1088/0960-1317/26/1/013003

[18] Zhao Q, Zhang J, Yan S, Du Danyuan H, Alici G, Li W. High throughput sheathless and three-dimensional microparticle focusing using microchannel with arc-shaped groove arrays. Scientific Reports. 2017;**7**:41153

[19] Yamada M, Seki M. Hydrodynamic filtration for on-chip particle concentration and classification utilizing microfluidics. Lab on a Chip. 2005;**5**(2005):1233-1239

[20] McNaughton B. Akadeum Life Sciences. https://akadeum.com/2016/11/cell-separation-cell-sorting/ [Accessed: 26-January-2018]

[21] Yanai T, Yamada M, Seko W, Seki M. A New Method for Continuous Sorting of Cells/Particles Using Lattice-Shaped Dual-Depth Microchannels. Department of Applied Chemistry and Biotechnology. Chiba University 1-33 Yayoi-Cho, Inage-Ku, Chiba 263-8522, Japan; 2015

[22] Karimi A, Yazdi S, Ardekani AM. Hydrodynamic Mechanisms of Cell and Particle Trapping in Microfluidics. Department of Aerospace and Mechanical Engineering, University of Notre Dame, Notre Dame, Indiana 46556, USA, Department of Chemical Engineering, The Pennsylvania State University, University Park, Pennsylvania 16802, USA; 2013

[23] Seko W, Yamada M, Seki M. Slanted lattice-shaped microchannel networks for continuous sorting of microparticles and cells, Chiba University. In: Japan 17th International Conference on Miniaturized Systems for Chemistry and Life Sciences; Freiburg: 27-31 October 2013. 978-0- 9798064-6-9/_TAS 2013/$2013CBMS-0001

[24] Yamada M, Seki M. Microfluidic particle sorter employing flow splitting and recombining. Analytical Chemistry. 2006;**78**(4):1357-1362

[25] Davis JA, Inglis DW, Morton KJ, Lawrence DA, Huang LR, Chou SY, Strum JC, Austin RH. Deterministic Hydrodynamics: Taking Blood Part. Princeton Institute of Science and Technology of Materials, and Department of Electrical Engineering and Physics, Princeton University; Proceedings of the National Academy of Sciences of the United States of America. 2006;**103**(40):14779-14784

[26] Wei J, Song H, Shen Z, He Y, Xu X, Zhang Y, Li BN. Senior Member. In: Numerical Study Of Pillar Shapes in Deterministic Lateral Displacement Arrays for Spherical Particle Separation. IEEE transactions on Nanobioscience. 2015;**14**(6):660-667

[27] Lee G-B, Chang C-C, Huang S-B, Yang R-J. The Hydrodynamic Focusing Effect inside Rectangular Microchannels. Tainan: Department of Engineering Science, National Cheng Kung University; 2006

[28] Kumar V, Sun P, Vamathevan J, Li Y, Ingraham K, Palmer L, Huang J, Brown JR. Comparative genomics of *Klebsiella pneumoniae* strains with different antibiotic resistance. Antimicrobial Agents and Chemotherapy. 2011;**55**(9):4267-4276

[29] COMSOL Multiphysics cyclopedia. https://www.comsol.co.in/multiphysics/navier-stokes-equations [Accessed: 02-February 2018]

[30] Jain RR, Mehta MR, Bannalikar AR, Menon MD. Alginate microparticles loaded with lipopolysaccharide subunit antigen for mucosal vaccination against *Klebsiella pneumonia*. 2015;**43**(3):195-201. DOI: 10.1016/j.biologicals.2015.02.001

6

Dual-Mass MEMS Gyroscope Structure, Design, and Electrostatic Compensation

Huiliang Cao and Jianhua Li

Abstract

Dual-mass MEMS gyroscope is one of the most popular inertial sensors. In this chapter, the structure design and electrostatic compensation technology for dual-mass MEMS gyroscope is introduced. Firstly, a classical dual-mass MEMS gyroscope structure is proposed, how it works as a tuning fork (drive anti-phase mode), and the structure dynamical model together with the monitoring system are presented. Secondly, the imperfect elements during the structure manufacture process are analyzed, and the quadrature error coupling stiffness model for dual-mass structure is proposed. After that, the quadrature error correction system based on coupling stiffness electrostatic compensation method is designed and evaluated. Thirdly, the dual-mass structure sensing mode modal is proposed, and the force rebalancing combs stimulation method is utilized to achieve sensing mode transform function precisely. The bandwidth of sensing open loop is calculated and experimentally proved as 0.54 times with the resonant frequency difference between sensing and drive modes. Then, proportional-integral-phase-leading controller is presented in sensing close loop to expand the bandwidth, and the experiment shows that the bandwidth is improved from 13 to 104 Hz. Finally, the results are concluded and summarized.

Keywords: MEMS gyroscope, dual-mass structure, mode analysis, quadrature error, bandwidth expansion, electrostatic compensation

1. Introduction

The precision of micro-electro-mechanical system (MEMS) gyroscope improves a lot in this decade, and achieves the tactical grade level. On the benefits of the small size, low costs, and light weight the MEMS gyro is applied in more and more areas, such as inertial navigation, roller detection, automotive safety, industrial controlling, railway siding detection, consumer electronics and stability controlling system [1–5]. During use, the acceleration along the sense axis causes great error in MEMS gyroscope output signal, and dual-mass gyroscope structure restrains this phenomenon well by employing differential detection technology; so, a lot of research institutes are interested in this structure [6–8].

1.1. Development of dual-mass MEMS gyroscope structure quadrature error compensation

Most of the literature informs that the dominate signal component in output signal is quadrature error, which is generated in the structure manufacture process, and brings over several 100° S^{-1} equivalent input angular [9–14]. The original source of quadrature is the coupling stiffness, which is modulated by drive mode movement and generates quadrature error force. The force has same frequency but has a 90° phase difference with Coriolis force and stimulates sense mode [11]. Most previous works utilize phase-sensitivity demodulation method to pick Coriolis signal from sense channel [9, 13], which requires accurate phase information and long-term, full-temperature range stability. However, the demodulation phase error and noise usually exist (sometimes more than 1° [9, 12]), which bring undesirable bias. The coupling stiffness drift (the drive and sense modes' equivalent stiffness vary with temperature and generate the drift of coupling stiffness [6, 11]) causes the quadrature error force drift, which is considered to be one of the most important reasons leading to bias long-term drift, and is proved by [11, 14] experimental work.

The previous works provide several effective ways to reduce quadrature error and are concluded into three aspects after the structure is manufactured [9, 11]: the quadrature signal compensation, the quadrature force correction and coupling stiffness correction. In work [9], the quadrature error is reduced by dc voltage based on synchronous demodulation and electrostatic quadrature compensation method, and the sigma-delta technology is employed in ADC and DAC. The research in Ref. [14] also employs coupling stiffness correction method to improve the performance of "butterfly" MEMS gyroscope. The bias stability and scale factor temperature stability enhance from 89°/h and 662 ppm/°C to 17°/h and 231 ppm/°C, respectively, which achieves the correction goal. The quadrature error correction in dual-mass tuning fork MEMS gyro structure is investigated in literature [11], and this work also proves the quadrature stiffness are different in left and right masses. A quadrature error correction closed loop is proposed in the work, and utilizing the coupling stiffness correction method, the masses are corrected separately. The stiffness correction combs utilize unequal gap method with dc voltages [15]. The bias stability improves from 2.06 to 0.64°/h with Allan Deviation analysis method, and the noise characteristic is also optimized [11]. Another coupling stiffness correction work is proposed in literature [7]; in this work, coupling stiffness correction controller uses PI technology, and the quadrature error equivalent input angular rate is measured as 450°/s. The experiment in the work shows that the bias stability and ARW improve from 7.1°/h and 0.36°/√h to 0.91°/h and 0.034°/√h, respectively. In Ref. [16], quadrature signal is compensated based on charge injecting technology in the sense loop, the compensation signal has same frequency, amplitude and anti-phase with quadrature error signal. The quadrature error correction method proposed in literature [8] employs both the quadrature force and stiffness correction methods, the modulation reference signal is generated by PLL technology and the correction loop uses PI regulator; the two masses are controlled together. A novel quadrature compensation method is proposed in literature [17] based on sigma-delta-modulators (ΣΔM), the quadrature error is detected by utilizing a pure digital pattern recognition algorithm and is compensated by using DC bias voltages, and the system works beyond the full-scale limits of the analog ΣΔM hardware. The quadrature error is compensated by open-loop charge injecting circuit in Ref. [18], the circuit is implemented on application specific integrated circuits (ASIC) and the experimental results show that the quadrature error component is effectively rejected.

1.2. Development of dual-mass MEMS gyroscope bandwidth expansion

High precision MEMS gyros are reported in literatures, and the bias drift parameters are even better than the tactical grade requirement. But the bandwidth performance always restrains the MEMS gyro application (100 Hz bandwidth is required in both Tactical and Inertial Grade) [4]. For most linear vibrating MEMS gyro, the mechanical sensitivity is determined by the difference between drive and sensing modes' resonant frequencies Δf. It means smaller difference achieves higher mechanical sensitivity (such as higher scale factor, higher resolution and smaller output noise) [19]. It is also proved that the mechanical bandwidth of the gyro is about $0.54\Delta f$ [20], so smaller Δf causes worse bandwidth characteristic. Some works employ mode-matching technology to make $\Delta f \approx 0$ Hz, and the best mechanical sensitivity can be acquired but bandwidth is sacrificed. It seems like that the bargain should be made between mechanical sensitivity and bandwidth in sensing open loop. So the sensing closed loop is required to improve MEMS gyro dynamic characteristics and bandwidth. The work in [21] employs a ΣΔ closed loop to reduce the frequency difference between the drive and sensing mode to less than 50 Hz. The research in work [22] utilizes PI controlling technology to make Δf tunable, and the bandwidth is optimized to 50 Hz. The sensing closed loop for single mass MEMS gyroscope based on automatic generation control (AGC) technology is proposed in paper [23]. The force rebalance controller extends the bandwidth, but it only contains a pure integral section which makes sensing closed loop with high Q value structure unstable. The bandwidth for another single mass MEMS gyroscope is introduced in work [24]; the bandwidth is improved from 30 to 98 Hz with notch filter and lead–lag compensator. The notch filter is designed for the "peak response" caused by the conjugate complex poles at Δf. But the resonant frequencies of the drive and sensing modes usually drift with temperature [6], so the notch filter method cannot satisfy the temperature-changing environment. The work in [25] proposes a method to avoid problems caused by notch filter and expands the bandwidth from 2.3 to 94.8 Hz. However, its left and right sensing modes are not coupled, and the bandwidth is only determined by the low pass filter (different with left and right sensing modes coupled structure). Its dual-mass structure can be considered as two independent gyroscopes with different Δfs, and two sensing loops should be designed separately to expand the bandwidth. The left and right sensing modes coupled structures are investigated in [26, 27, 28]. The work in Ref. [26] illustrates the vibration characteristics of dual-mass and spring structure. The work in Ref. [27] analyzes the energy of gyroscope under in-phase and anti-phase modes. And literature [28] proposes that the real working sensing mode of gyroscope is formed by sensing in-phase and anti-phase modes. However, these works do not focus on the bandwidth method of this kind of sensing loop. The gyro bandwidth optimized by mechanical method is proposed in Ref. [29], but the bandwidth varies with the temperature. Literature [30] employs eight drive mode units to form a drive band (not a single frequency). It provides a high mechanical sensitivity band for the gyroscope, but the bandwidth is determined by sensing mode frequency. And the phase margin of the closed loop is difficult to design. In Ref. [31], the bandwidth estimation methods of MEMS gyroscope in open loop and closed loop are proposed, but the bandwidth expanding method is not discussed clearly. The precise test method of bandwidth is introduced in Ref. [32]. The method uses vibration velocity of virtual drive mode multiplied by virtual angular rate to generate Coriolis force and substitutes angular turntable. The vibration velocity of drive mode is produced by signal source, not by drive mode, which ignores the frequency drift of drive mode and the mechanical coupling between drive and sensing mode.

This chapter focuses on the investigation of the dual-mass MEMS gyroscope structure, and the electrostatic compensation method for the structure, including quadrature error correction and bandwidth expansion technology. Through these technologies, the static and dynamic performance of the MEMS gyroscope is improved.

2. Dual-mass MEMS gyroscope structure design and analysis

2.1. Dual-mass MEMS gyroscope structure design

The fully decoupled linear vibrating gyroscope structure's ideal movement model can be found in many papers; the model can be described as two "spring-mass-damping" systems: drive mode and sense mode. In x-axis direction, drive frame is connected with Coriolis mass which is linked by sense frame in y direction. The drive frame and sense frame can be stimulated by their own mode's effective stiffness and damping, as shown in **Figure 1**. This chapter employs a dual-mass fully decoupled structure as mentioned in [6, 33], and it works in linear vibrating principle, with slide-film combs in both drive and sense modes. The whole

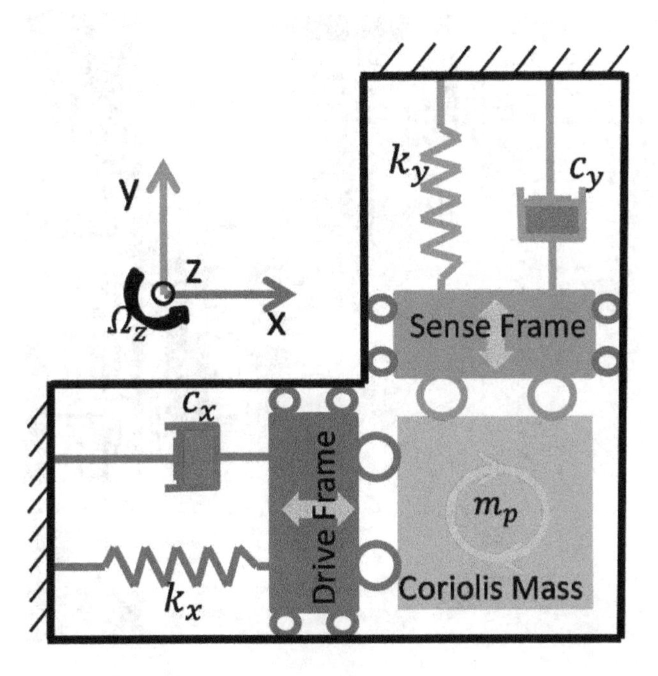

Figure 1. Schematic of ideal gyroscope fully decoupled model.

structure is shown in **Figure 2**. This structure model is constituted by two symmetrical parts, which are connected by two center U-shaped connect springs whose parameters are the same with drive U-shaped spring. The left and right Coriolis mass are sustained by 2 drive U-shaped springs (DS-A,B) and 4 sense U-shaped springs (SS-C,D,E,F), respectively, and these springs are linked by drive and sensing frames. The moving comb fingers are combined with frames while the static ones are fixed with the substrate. The whole structure is suspended and supported by 8 drive (DS-C,D,E,F for left part) and 4 sense (SS-A,B) U-shaped springs [6].

The drive U-shaped springs' stiffness coefficients are large along y-axis and very small in x direction while sense springs have the adverse characteristic. When the structure works, the drive combs support the electrostatic force to stimulate the drive frame and the Coriolis mass (together with SS-C,D,E,F) to move along x direction, and no displacement is generated in sense direction because of SS-A,B. When there is an angular rate Ω_z input around the z-axis, the Coriolis mass together with sense frame and DS-A, B will have a component motion in y orientation under the influence of Coriolis force (the drive frame will not move in y-axis because of the effect of DS-C,D,E,F), and then this displacement involved with Ω_z is detected by the sense combs.

The drive mode of the structure bases on tuning for k theory. The left and right masses are coupled by connect U-shaped spring, when two sensing masses are coupled by the x-axis warp of drive springs. The mode analyses of the first four order modes are shown in **Figure 3** [34]. The first-order mode is drive in-phase mode, the left and right masses together with drive frames vibrate toward same direction along drive axis (x direction in **Figure 2**), as shown in **Figure 3(a)**. The second- and third-order modes are sensing in-phase and anti-phase modes, as shown in **Figure 3(b)** and **(c)**. The left and right masses together with sense frames vibrate in same and inverse directions along sensing axis (y direction in **Figure 2**), respectively. And sensing anti-phase mode is the expected working mode. Drive anti-phase

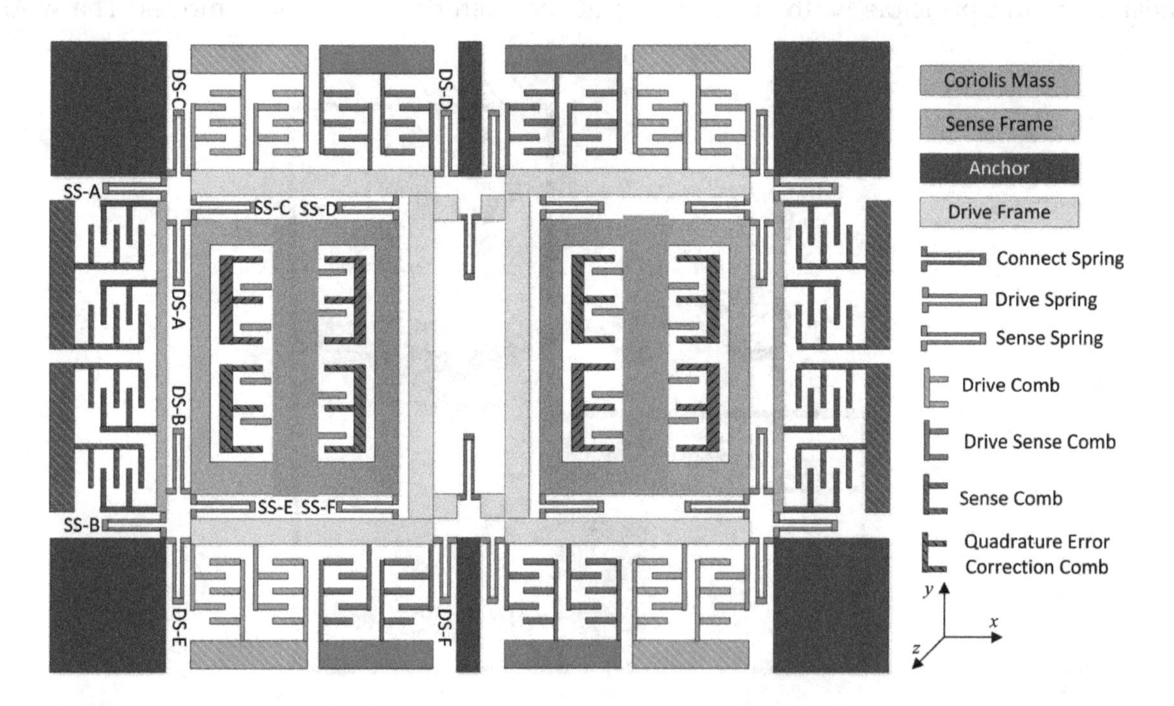

Figure 2. Schematic of the structure.

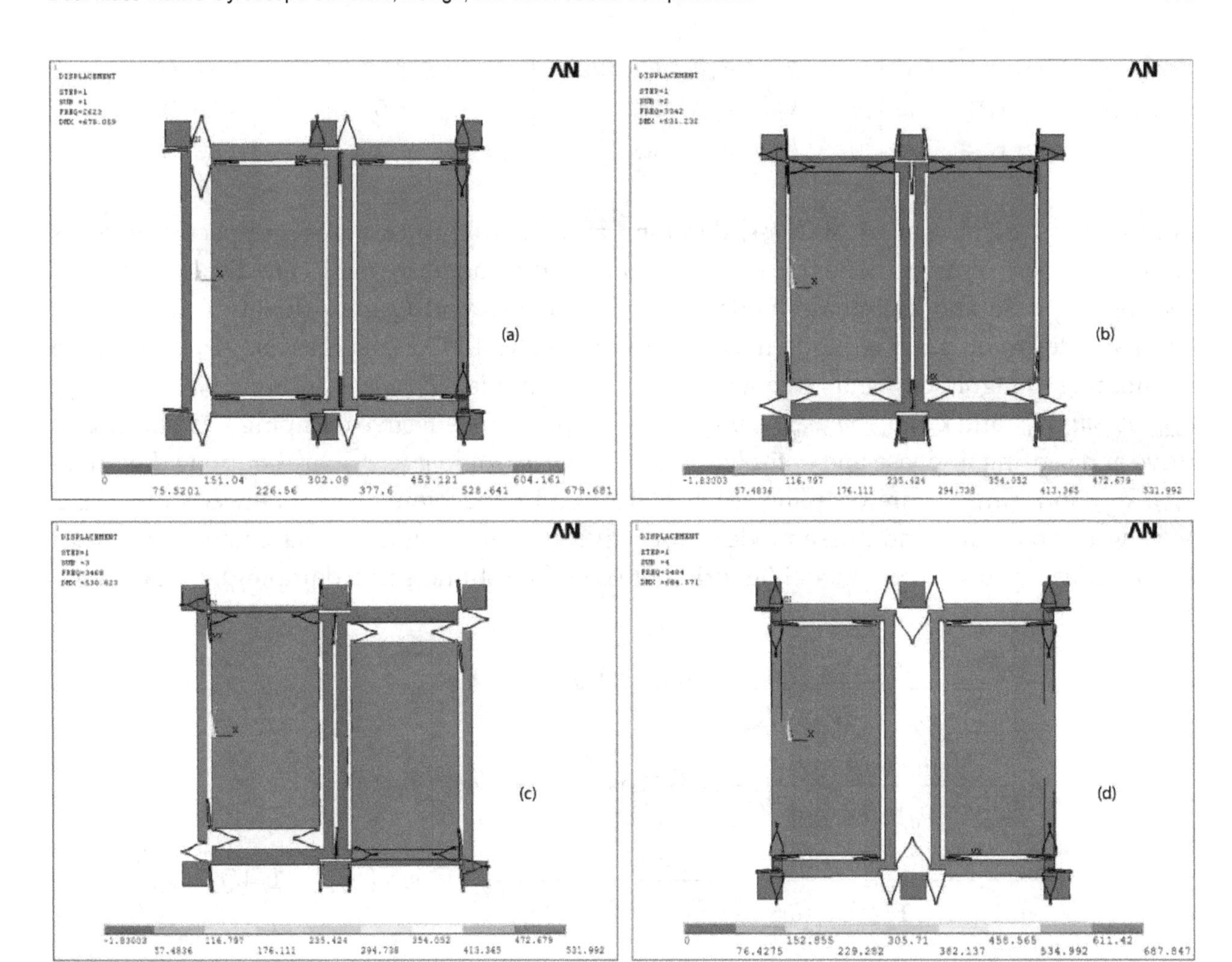

Figure 3. (a) Drive in-phase mode (first mode) with frequency $\omega_{x1} = 2623 \times 2\pi$ rad/s; (b) sensing in-phase mode (second mode) with frequency $\omega_{y1} = 3342 \times 2\pi$ rad/s; (c) sensing anti-phase mode (third mode) with frequency $\omega_{y2} = 3468 \times 2\pi$ rad/s; (d) drive anti-phase mode (fourth mode) with frequency $\omega_{x2} = 3484 \times 2\pi$ rad/s.

mode is the fourth order, in which mode, the left and right masses move in inverse directions along drive axis as **Figure 3(d)** shows. The drive anti-phase is another expected working mode.

2.2. Dual-mass MEMS gyroscope structure working principle analysis

Due to a large difference (>1000 Hz) between the in-phase (the first mode) and anti-phase drive modes frequencies, the quality factor of drive anti-phase mode is $Q_{x2} > 2000$ and the stimulating method of the drive mode. The real working drive mode is considered to be pure anti-phase drive mode (the fourth mode). Considering the real working sensing mode is formed by second and third modes, the motion equation of gyroscope structure can be expressed as (ideal condition) [35]:

$$m\ddot{D} + c\dot{D} + kD = F \tag{1}$$

where $m = \begin{bmatrix} m_x \\ m_y \\ m_y \end{bmatrix}$, $D = \begin{bmatrix} x \\ y_1 \\ y_2 \end{bmatrix}$, $k = \begin{bmatrix} k_{xx} & k_{xy1} & k_{xy2} \\ k_{y1x} & k_{y1y1} & 0 \\ k_{y2x} & 0 & k_{y2y2} \end{bmatrix}$, $c = \begin{bmatrix} c_{xx} & c_{xy1} & c_{xy2} \\ c_{y1x} & c_{y1y1} & 0 \\ c_{y2x} & 0 & c_{y2y2} \end{bmatrix}$, $F = \begin{bmatrix} F_d \sin(\omega_d t) \\ -2m_c\Omega_z\dot{x} \\ -2m_c\Omega_z\dot{x} \end{bmatrix}$.

are the mass, displacement, stiffness, damping and external force matrix, respectively; m_x is equivalent mass of drive mode; x is displacement of drive mode; y_1 and y_2 are displacements of sensing in-phase and anti-phase modes, respectively; Q_{y1} and Q_{y2} are quality factors; Ω_z is angular rate input; sensing mode mass m_y approximates to Coriolis mass m_c; F_d and ω_d are stimulating magnitude and frequency of drive mode; $c_{xx}=\omega_{x2}m_x/Q_{x2}$, $c_{y1y1}=\omega_{y1}m_y/Q_{y1}$, $c_{y2y2}=\omega_{y2}m_y/Q_{y2}$ and $k_{xx}=\omega_{x2}^2 m_x$, $k_{y1y1}=\omega_{y1}^2 m_y$, $k_{y2y2}=\omega_{y2}^2 m_y$ are effective damping and stiffness of drive and sensing in-phase and anti-phase modes; c_{xy1}, c_{xy2} and c_{y1x}, c_{y2x} are coupling damping, k_{xy1}, k_{xy2} and k_{y1x}, k_{y2x} are coupling stiffness (caused by machining error) between drive and sensing in-phase and anti-phase modes, respectively; sensing mode displacement $y = y_1 + y_2$, and for ideal gyro structure model (ignoring the coupling stiffness and damping), we get [35]:

$$\begin{aligned} x(t) = {} & \frac{F_d/m_x}{\sqrt{\left(\omega_{x2}^2-\omega_d^2\right)^2+\omega_{x2}^2\omega_d^2/Q_{x2}^2}}\sin\left(\omega_d t+\varphi_{x2}\right) \\ & +\frac{F_d\omega_{x2}\omega_d/m_xQ_x}{\left(\omega_{x2}^2-\omega_d^2\right)^2+\omega_{x2}^2\omega_d^2/Q_{x2}^2}e^{-\frac{\omega_{x2}}{2Q_{x2}}t}\cos\left(\sqrt{1-1/4Q_{x2}^2}\omega_{x2}t\right) \\ & +\frac{F_d\omega_d\left(\omega_{x2}^2/Q_{x2}^2+\omega_d^2-\omega_{x2}^2\right)/m_x}{\omega_{x2}\sqrt{1-1/4Q_{x2}^2}\left[\left(\omega_{x2}^2-\omega_d^2\right)^2+\omega_{x2}^2\omega_d^2/Q_{x2}^2\right]}e^{-\frac{\omega_{x2}}{2Q_{x2}}t}\sin\left(\sqrt{1-1/4Q_{x2}^2}\omega_{x2}t\right) \end{aligned} \tag{2}$$

$$\begin{aligned} y_{1,2}(t) = {} & \frac{F_c}{\sqrt{\left(\omega_{y1,2}^2-\omega_d^2\right)^2+\omega_{y1,2}^2\omega_d^2/Q_{y1,2}^2}}\sin\left(\omega_d t+\varphi_{x2}+\frac{\pi}{2}+\varphi_{y1,2}\right)- \\ & \frac{F_c\left[\omega_{y1,2}\omega_d\sin\varphi_{x2}/Q_{y1,2}+\left(\omega_{y1,2}^2-\omega_d^2\right)\cos\varphi_{x2}\right]}{\left(\omega_{y1,2}^2-\omega_d^2\right)^2+\omega_{y1,2}^2\omega_d^2/Q_{y1,2}^2}e^{-\frac{\omega_{y1,2}}{2Q_{y1,2}}t}\cos\left(\sqrt{1-1/4Q_{y1,2}^2}\omega_{y1,2}t\right)+ \\ & \frac{F_c\left[\omega_{y1,2}\left(\omega_{y1,2}^2-3\omega_d^2\right)\cos\varphi_{x2}/\left(2Q_{y1,2}\right)+\omega_d\left(\omega_{y1,2}^2/\left(2Q_{y1,2}^2\right)+\omega_{y1,2}^2-\omega_d^2\right)\sin\varphi_{x2}\right]}{\omega_{y1,2}\sqrt{1-1/4Q_{y1,2}^2}\left[\left(\omega_{y1,2}^2-\omega_d^2\right)^2+\omega_{y1,2}^2\omega_d^2/Q_{y1,2}^2\right]} \\ & e^{-\frac{\omega_{y1,2}}{2Q_{y1,2}}t}\sin\left(\sqrt{1-1/4Q_{y1,2}^2}\omega_{y1,2}t\right) \end{aligned} \tag{3}$$

where $\varphi_{x2} = -tg^{-1}\left(\frac{\omega_{x2}\omega_d}{Q_{x2}\left(\omega_{x2}^2-\omega_d^2\right)}\right)$, $F_c = \frac{-2\Omega_z\omega_d F_d}{m_x\sqrt{\left(\omega_{x2}^2-\omega_d^2\right)^2+\omega_{x2}^2\omega_d^2/Q_{x2}^2}}$, $\varphi_{y1,2} = -tg^{-1}\left(\frac{\omega_{y1,2}\omega_d}{Q_{y1,2}\left(\omega_{y1,2}^2-\omega_d^2\right)}\right)$.

The abovementioned equations indicate that the movement of drive and sensing modes are the compound motion of stable vibration and attenuation vibration. Since drive mode closed loop is employed, drive mode is stimulated with stable amplitude ($\omega_d = \omega_{x2}$). Then (2) can be simplified as [20]:

$$x(t) = \frac{F_d Q_{x2}}{m_x \omega_d^2} \cos(\omega_d t) = A_x \cos(\omega_d t) \tag{4}$$

The sensing in-phase and anti-phase modes movement equation can be got from (3):

$$y_{1,2}(t) = \frac{-2\Omega_z F_d Q_{x2} \sin(\omega_d t)}{m_x \omega_d \sqrt{\left(\omega_{y1,2}^2 - \omega_d^2\right)^2 + \omega_{y1,2}^2 \omega_d^2 / Q_{y1,2}^2}} = A_{y1,2} \sin(\omega_d t) \tag{5}$$

Then, the mechanical sensitivity can be expressed as:

$$S_{me} = \frac{A_{y1} + A_{y2}}{\Omega_z} \approx \frac{-F_d Q_x}{m_x \omega_d^2} \left(\frac{1}{\omega_{y1} - \omega_{x2}} + \frac{1}{\omega_{y2} - \omega_{x2}} \right) = -A_x \left(\frac{1}{\Delta\omega_1} + \frac{1}{\Delta\omega_2} \right) \tag{6}$$

The mechanical sensitivity of dual-mass sensing mode coupled structure is determined by vibration amplitude of drive mode and frequency differences between drive working mode and sensing modes (including the second and third modes). Furthermore, the one near the fourth mode is the dominant element. Therefore, the sensing anti-phase mode determines the gyro structure mechanical sensitivity. The schematic diagram of gyro sensing mode is shown in **Figure 4**. $G_{inphase}$ and $G_{anphase}$ are transform functions of in-phase and anti-phase sensing modes; K_{inyv} and K_{anyv} are displace-voltage transform parameters of sensing in-phase and anti-phase modes; K_{pre} is the preamplifier; V_{stotal} is sensing mode output and $G_{sV/F}$ can be expressed as:

$$G_{sV/F} = \left(G_{inphase} K_{inyv} + G_{anphase} K_{anyv}\right) K_{pre} \tag{7}$$

where $G_{inphase} = \frac{1}{m_y} \frac{1}{s^2 + \frac{\omega_{y1}}{Q_{y1}} s + \omega_{y1}^2}$ and $G_{anphase} = \frac{1}{m_y} \frac{1}{s^2 + \frac{\omega_{y2}}{Q_{y2}} s + \omega_{y2}^2}$ are the sensing in-phase and anti-phase mode transform functions, respectively.

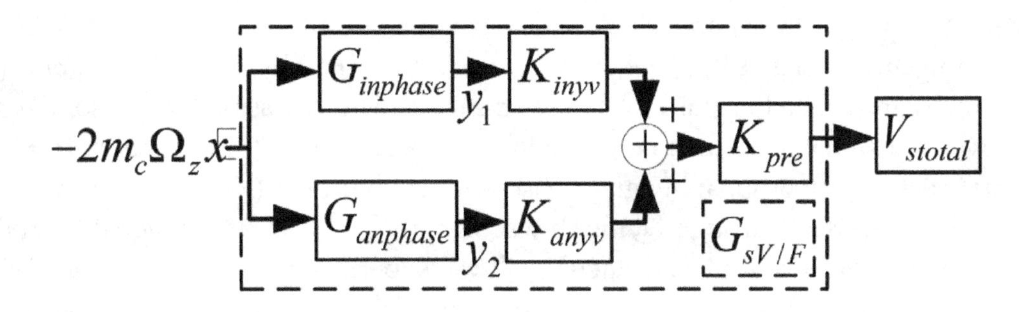

Figure 4. Schematic diagram of sensing mode real working mode.

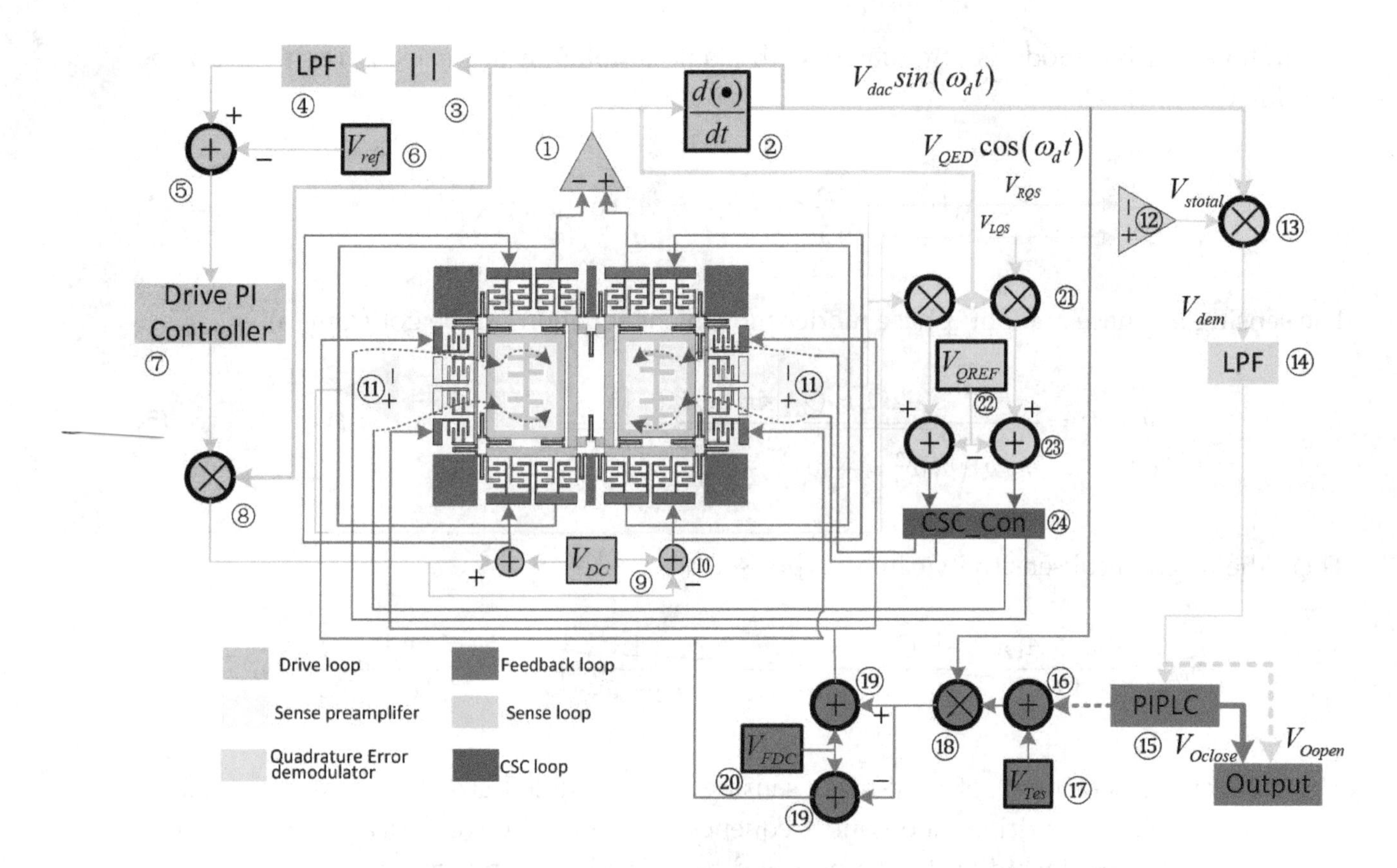

Figure 5. Dual-mass gyroscope monitoring system schematic.

2.3. Dual-mass MEMS gyroscope monitoring system

The gyro control and detection system is shown in **Figure 5**. In drive loop, the drive frame displacement *x(t)* is detected by drive sensing combs and picked up by differential amplifier ①. Then, the signal phase is delayed by 90° (through ②) to satisfy the phase requirement of AC drive signal $V_{dac}Sin(\omega_d t)$. After that, $V_{dac}Sin(\omega_d t)$ is processed by a full-wave rectifier ③ and a low pass filter ④. Afterwards, V_{dac} is compared (in ⑤) with the reference voltage V_{ref} ⑥. Next, drive PI controller ⑦ generates the control signal, which is modulated by $V_{dac}Sin(\omega_d t)$, and then the signal is superposed (through ⑩) by V_{DC} ⑨ to stimulation drive mode. The sensing system employs a closed loop, which utilizes the same interface as drive circuit. First, the left and right masses' sensing signals are detected separately with differential detection amplifier⑪. And the output signals V_{RQS} (from right mass) and V_{LQS} (from left mass) are processed by second differential amplifier⑫ to generate signal V_{stotal}. Then, V_{stotal} is demodulated by signal $V_{dac}Sin(\omega_d t)$ (in ⑬). After that, the demodulated signal V_{dem} passes through the low pass filter ⑭; so, the sensing mode's movement signal V_{Oopen} can be got. For sense closed loop ("pink" section in **Figure 5**), V_{Oopen} is first sent in proportional-integral-phase-leading controller (PIPLC) ⑮ to calculate the control signal superposed (through ⑯) with test signal V_{Tes} ⑰. Then the signal is modulated with $V_{dac}Sin(\omega_d t)$ (in ⑱). Finally, DC voltage V_{FDC} ⑳ is superposed with the modulated signal in ⑲ to generate the feedback signal. In quadrature error compensation system, V_{RQS} and V_{LQS} are demodulated by drive frame displacement signal $V_{QED}Cos(\omega_d t)$ separately in ㉑. The quadrature error compensation reference is provided in ㉒ and is compared with quadrature error amplitude in ㉓; the results are sent to coupling stiffness compensation (CSC) controller ㉔.

3. Dual-mass MEMS gyroscope quadrature error compensation

3.1. Dual-mass MEMS gyroscope structure quadrature error model

The quadrature error is caused by coupling stiffness, which is generated in structure processing stage, and the stiffness elements in Eq. (1) can be calculated by:

$$\begin{cases} k_{xx} = k_x \cos^2\beta_{Qx} + k_y \sin^2\beta_{Qy} \\ k_{xy} = k_{yx} = k_x \sin\beta_{Qx} \cos\beta_{Qx} - k_y \cos\beta_{Qy} \sin\beta_{Qy} \\ k_{yy} = k_x \sin^2\beta_{Qx} + k_y \cos^2\beta_{Qy} \end{cases} \tag{8}$$

where k_x and k_y are the design stiffness along designed axis x and y; β_{Qx} and β_{Qy} are the quadrature error angle and they are the angles between practical axis and designed axis [11, 35], and usually it is assumed that $\beta_{Qx} = \beta_{Qy}$.

The equivalent stiffness and masses system and structure motion of dual-mass gyro structure is shown in **Figure 6**. The design drive and sense stiffness axis are x and y (gray); the real drive and sense axis of left and right masses after manufacture are x_l', x_r' (with light blue) and y_l', y_r' (with light yellow), respectively; the stiffness of drive and sense modes of left and right masses after manufacture are k_{lx}, k_{rx} (with dark blue) and k_{ly}, k_{ry} (with dark yellow), respectively; the projections of k_{lx} on $-x$- and y-axis are k_{lxx} and k_{lxy}; the projections of k_{rx} on x- and y-axis are k_{rxx} and k_{rxy}; the projections of k_{ly} on $-x$- and y-axis are k_{lyx} and k_{lyy}; the projections of k_{ry} on x- and y-axis are k_{ryx} and k_{ryy}; the quadrature error angular of left and right masses are β_{ly} and β_{ry}. In design stage, $|k_{lx}| = |k_{rx}|$, $|k_{ly}| = |k_{ry}|$, and $\beta_{ly} = \beta_{ry} = 0$, but after the manufacture process, the parameters change and they do not meet the equal equations, so the coupling stiffness of two masses are different.

3.2. Dual-mass MEMS gyroscope structure coupling stiffness compensation

The CSC method utilizes quadrature error correction combs to generate negative electrostatic stiffness and correct quadrature error coupling stiffness. This special comb is unequal gap style, and is introduced in [11]; its stiffness is expressed as:

$$k_{qxy} = k_{qyx} = k_{qcoup} V_{qD} V_{qc} = -\frac{4 n_q \varepsilon_0 h}{{y_{q0}}^2}\left(1 - \frac{1}{\lambda^2}\right) V_{qD} V_{qc} \tag{9}$$

where k_{qxy} and k_{qyx} are the quadrature error correction comb stiffness along designed axes x and y; V_{qD} and V_{qc} are correction fixed voltage and controlling voltage; n_q is the number of comb; ε_0 is the permittivity of vacuum; h is the thickness of the comb; y_{q0} and x_0 are the parallel capacitance's gap and overlap length, respectively; λ is a constant. **Figure 7** shows the right mass CSC system which is same with left mass CSC system, $G_{RsV/F}$ is the transform function of

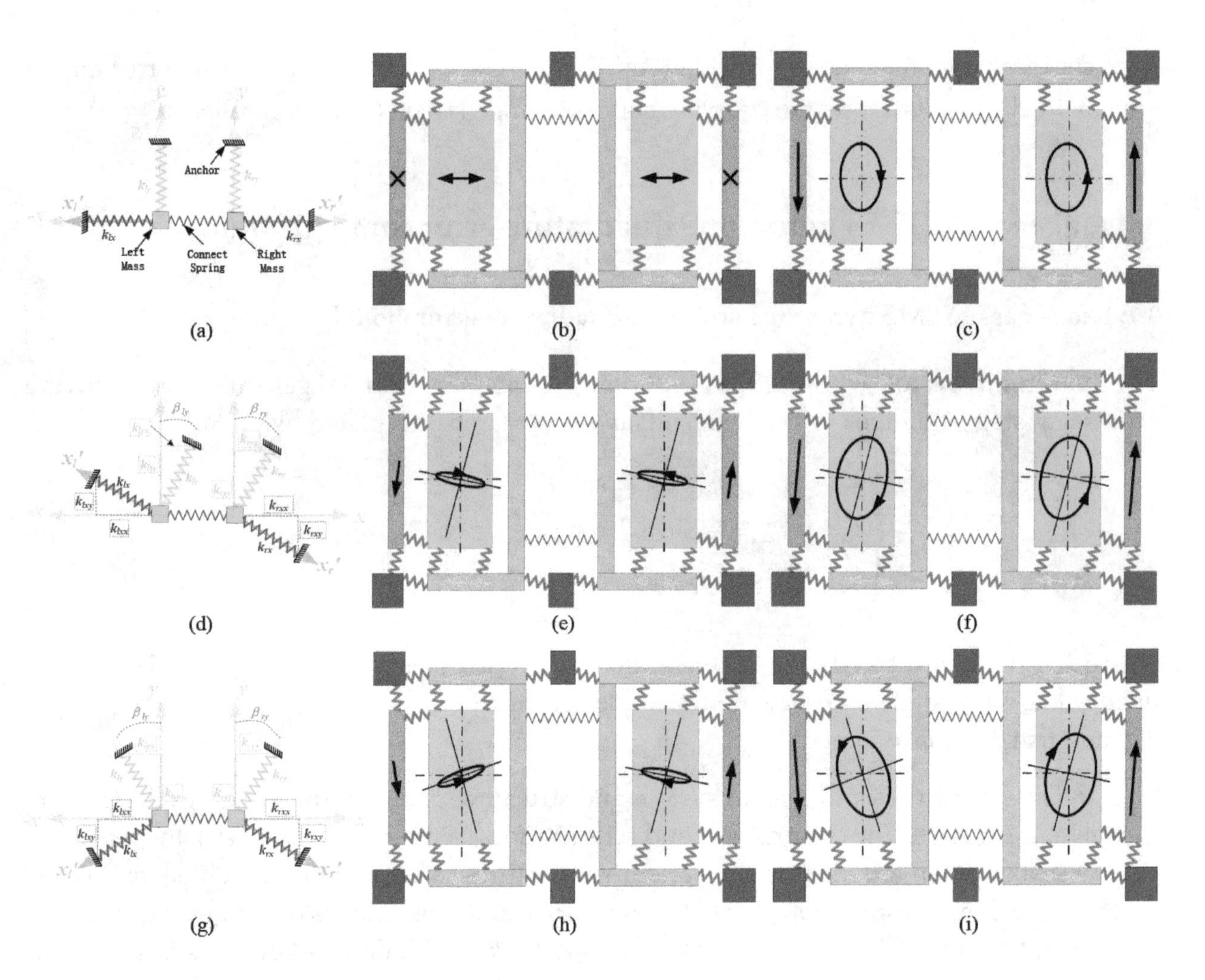

Figure 6. (a) Stiffness system of ideal dual-mass gyro structure; (b) ideal structure movement without angular rate input; (c) ideal structure movement with steady angular rate input; (d) stiffness system with in-phase quadrature error angular; (e) the movement of in-phase quadrature error angular structure without angular rate input; (f) the movement of in-phase quadrature error angular structure with steady angular rate input; (g) stiffness system with anti-phase quadrature error angular; (h) the movement of anti-phase quadrature error angular structure without angular rate input; (i) the movement of anti-phase quadrature error angular structure with steady angular rate input.

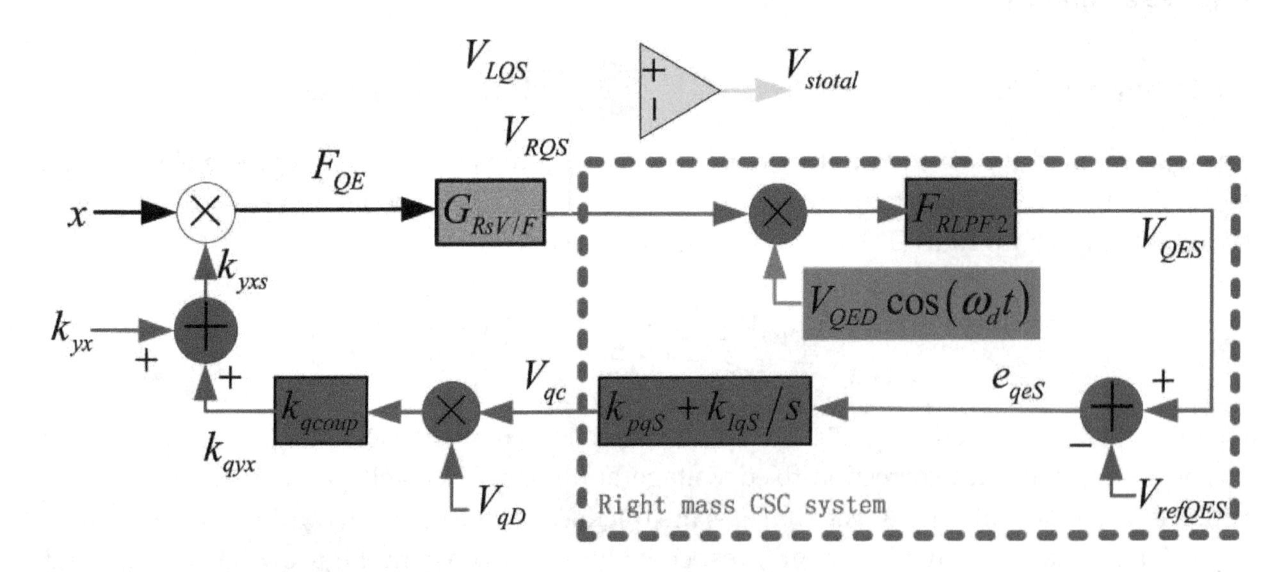

Figure 7. Right mass SCS system.

right mass. In **Figure 7**, the coupling stiffness k_{yx} is modulated by drive-mode movement, and the controller employs PI controlling technology. The correction stiffness does not need to be modulated by quadrature error in-phase signal, which is better for circuit simplification and power consumption. We can get the following equations:

$$V_{RQS} = F_{QE}G_{RsV/F} = x\left(k_{yx} + k_{qyx}\right)G_{RsV/F} \tag{10}$$

$$V_{qc} = \left(V_{QES} - V_{refQES}\right)\left(k_{pqS} + \frac{k_{IqS}}{s}\right) \tag{11}$$

We have $x = A_x cos(\omega_d t)$ and $V_{refQES} = 0$, after the low pass filter, V_{QES} can be expressed as:

$$V_{QES} = \frac{1}{2}A_x\left(k_{yx} + k_{qyx}\right)G_{RsV/F}V_{QED} \tag{12}$$

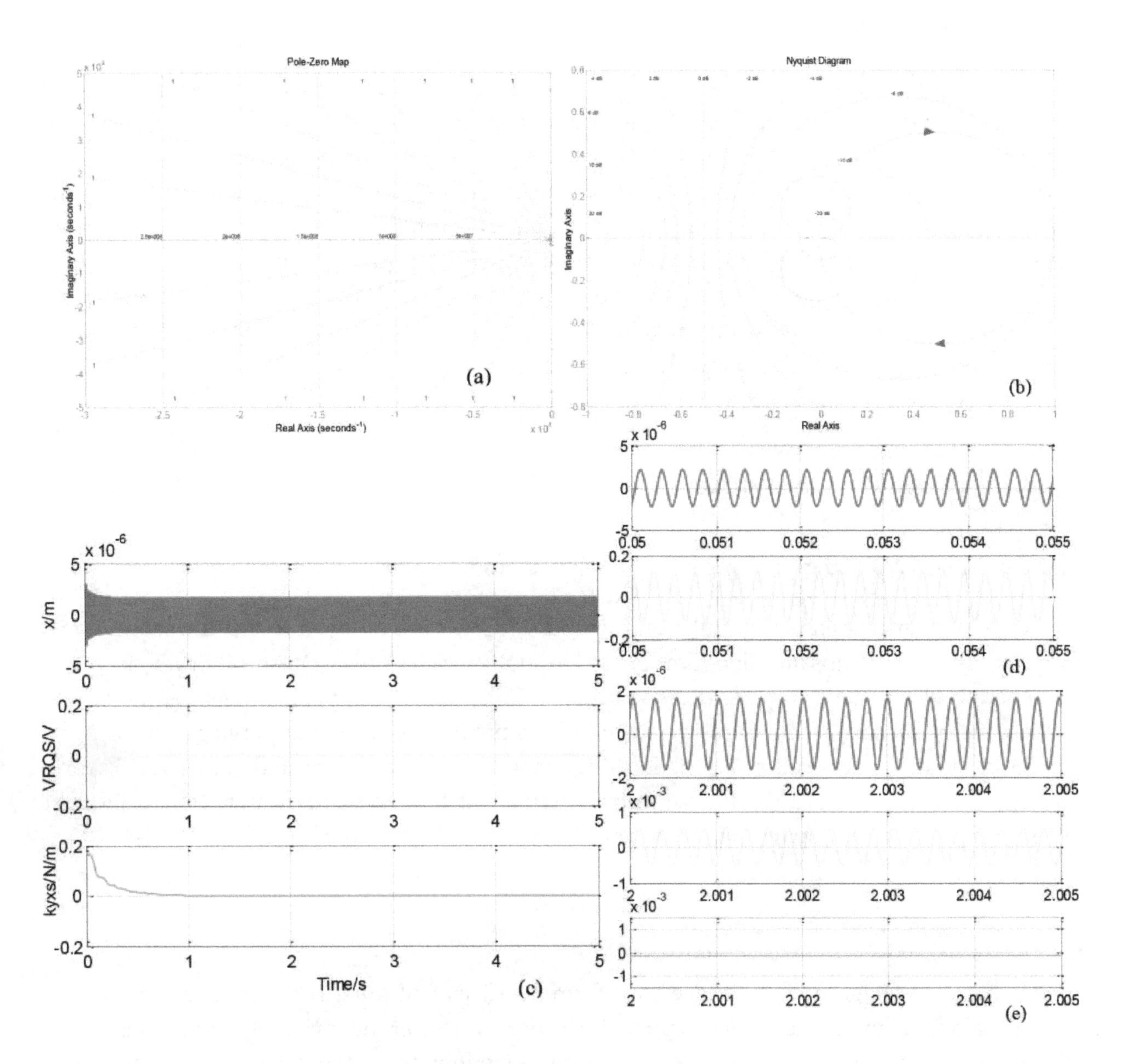

Figure 8. (a) CSC system pole-zero map; (b) CSC system Nyquist diagram; (c) CSC system simulation curves; (d) CSC system start-up stage enlarge curves; (e) CSC system stable stage enlarge curves.

Then, combining (9), (11) and (12), we have:

$$\frac{k_{qyx}(s)}{k_{yx}(s)} = \frac{\frac{A_x}{2} k_{qcoup} V_{qD} V_{QED} F_{RLPF2}(s)\left[G_{RsV/F}(s+j\omega_d) + G_{RsV/F}(s-j\omega_d)\right]\left(k_{pqS} + \frac{k_{IqS}}{s}\right)}{1 - \frac{A_x}{2} k_{qcoup} V_{qD} V_{QED} F_{RLPF2}(s)\left[G_{RsV/F}(s+j\omega_d) + G_{RsV/F}(s-j\omega_d)\right]\left(k_{pqS} + \frac{k_{IqS}}{s}\right)} \tag{13}$$

When the system is under stable state, s = 0, and the above equation has:

$$1 = \frac{A_x}{2} k_{qcoup} V_{qD} V_{QED} F_{LPF2}(s)\left[G_{RsV/F}(s+j\omega_d) + G_{RsV/F}(s-j\omega_d)\right]\left(k_{pqS} + \frac{k_{IqS}}{s}\right)\Big|_{s=0} \tag{14}$$

Then:

$$k_{qyx} \approx -k_{yx} \tag{15}$$

The coupling stiffness is corrected. The CSC system is simulated and the curves are shown in **Figure 8**. The Pole-Zero Map is shown in **Figure 8(a)**, no pole is in the positive real axis and **Figure 8(b)** is the Nyquist Diagram, the curve does not contain (−1,0j) point, which proves the system's stability. The time-domain simulation curves are shown in **Figure 8(c)–(e)**, and the curves indicate that the CSC system is under stable state after about 0.7 s. It is obvious that in start-up stage, the sense channel signal mainly consists of quadrature error signal. But, in stable state, the dominate element is Coriolis in-phase signal. Furthermore, the overall coupling stiffness k_{yxs} is suppressed from original value (about 0.18 N/m) to −0.00016 N/m, and k_{yx} is basically corrected which proves (15).

4. Dual-mass MEMS gyroscope bandwidth expansion

4.1. Dual-mass MEMS gyroscope structure sense mode model

The force rebalancing combs stimulation method (FRCSM) is employed to test the sense mode. Force rebalancing combs are slide-film form (does not vary ω_{y1} and ω_{y2}). And they are arranged to generate electrostatic force and rebalance the Coriolis force applied on sense frame. The movement of sensing frame is restricted, which improves the dynamic performance of gyro. Therefore, Coriolis simulation signal is designed to be produced through force rebalancing combs, and this method can be further applied to scale factor and bandwidth tests. The FRCSM schematic diagram is shown in **Figure 9**, and the equivalent input angular rate of V_{Tes} is:

$$\Omega_{VTes} = \frac{K_{FBy} V_{Tes} V_{dac}}{2 m_c A_x \omega_d} \tag{16}$$

where K_{FBy} is voltage-force interface transform coefficient of force rebalances combs; V_{dac} is modulate signal amplitude, which is picked up after 90° shifter in drive loop and is considered to be constant when gyro works. K_{FBy} can be got either from the calculation of structure parameter or turntable test, and it is determined by force rebalancing combs number,

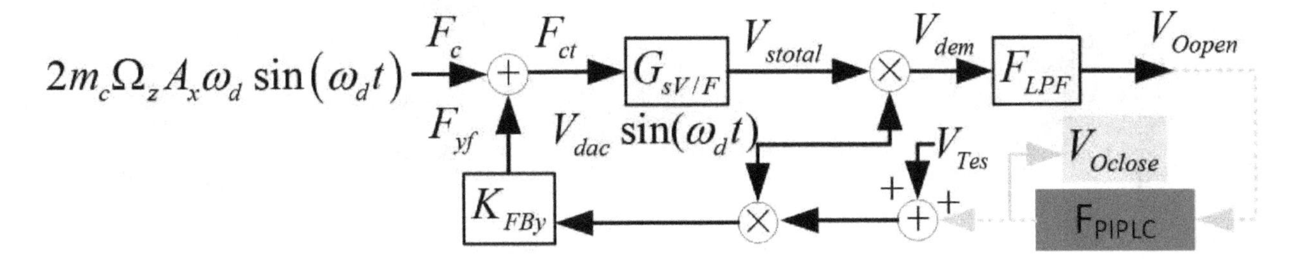

Figure 9. FRCSM schematic in sensing loop.

parameters and feedback direct voltage V_{FDC}. The system diagram of sense open loop is shown in **Figure 9**, and the equations below can be derived from **Figure 9**:

$$\begin{cases} F_{ct}(t) = 2\Omega_z(t)m_yA_x\omega_d\sin(\omega_d t) + K_{FBy}V_{Tes}(t)V_{dac}\sin(\omega_d t) \\ V_{Oopen} = F_{ct}(t)G_{sV/F}V_{dac}\sin(\omega_d t)F_{LPF} \end{cases} \tag{17}$$

After Laplace transformation, we can get [35]:

$$V_{Oopen}(s) = K_{pre}\left[\frac{A_x\omega_d V_{dac}\Omega_z(s)F_{LPF}(s)}{2} + \frac{K_{FBy}V_{dac}^2 V_{Tes}(s)F_{LPF}(s)}{4m_y}\right]G_{sV/F}(s) \tag{18}$$

where, $G_{sV/F}(s) = \left[\frac{K_{inyv}\left(s^2+\frac{\omega_{y1}}{Q_{y1}}s+\omega_{y1}^2-\omega_d^2\right)}{\left(s^2+\frac{\omega_{y1}}{Q_{y1}}s+\omega_{y1}^2-\omega_d^2\right)^2+\left(2s\omega_d+\frac{\omega_{y1}}{Q_{y1}}\omega_d\right)^2} + \frac{K_{anyv}\left(s^2+\frac{\omega_{y2}}{Q_{y2}}s+\omega_{y2}^2-\omega_d^2\right)}{\left(s^2+\frac{\omega_{y2}}{Q_{y2}}s+\omega_{y2}^2-\omega_d^2\right)^2+\left(2s\omega_d+\frac{\omega_{y2}}{Q_{y2}}\omega_d\right)^2}\right]$, and ω_d = 3488.9*2π rad/s, ω_{y1} = 3360.1*2π rad/s, ω_{y2} = 3464.1*2π rad/s, Q_{y1} = 1051 and Q_{y2} = 1224, then the Bode Diagram map is shown in **Figure 10**, and the bandwidth of the mechanical system is [20, 35]:

$$\omega_b = 0.54\Delta\omega_2 \tag{19}$$

In **Figure 10**, point A (frequency is $\Delta\omega_2$) and C (frequency is $\Delta\omega_1$) peak points are generated by two conjugate poles, respectively; B valley point ($\Delta\omega_B$) is caused by conjugate zeros shown in Eq. (18). The bandwidth obtained from the simulation curve is 12.9 Hz and from FRCSM test curve is 13 Hz, which verifies the theory calculation result in Eq. (19). In addition, the FRCSM test curve matches simulation curve well and proves the theory analysis conclusion proposed in Section 2.2.

4.2. Dual-mass MEMS gyroscope bandwidth expanding

Sensing closed-loop control method provides electrostatic force to rebalance the Coriolis force applied on sensing mode, which is one of the most effective ways to improve gyro dynamic performance. When sensing closed-loop works, the sensing frame's displacement is restricted and the Coriolis force is transformed into electronic signal directly. And it also avoids the nonlinearity of sensing mode displacement. Meanwhile, closed loop provides better anti-vibration and anti-shock characteristics to the gyroscope. The sensing closed-loop schematic is shown in **Figure 9**, where the real input angular rate Ω_z = 0°/s and the simulation angular

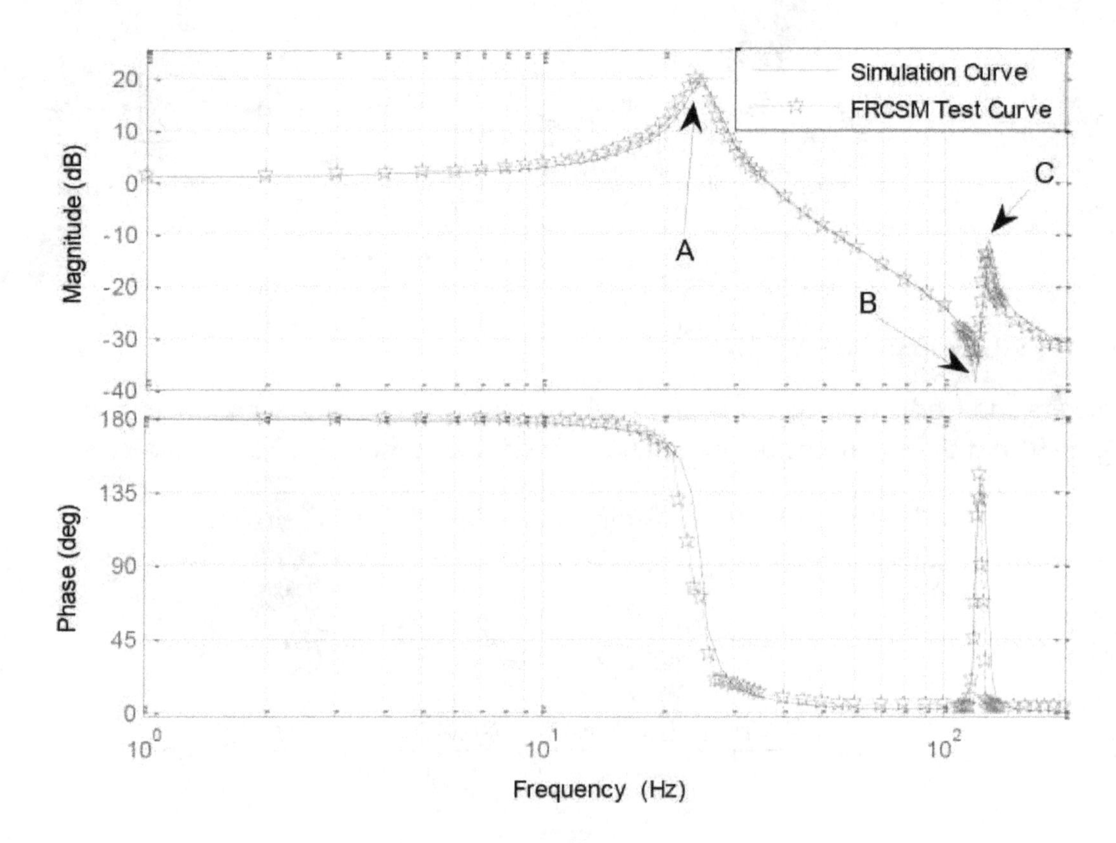

Figure 10. Simulation and FRCSM test results of dual-mass gyroscope structure bode diagram.

rate V_{Tes} is employed to test the closed system frequency response. From **Figure 8**, the equations can be derived:

$$F_{yf}(t) = K_{FBy}[V_{Oclose}(t) + V_{Tes}(t)]V_{dac}\sin(\omega_d t) \tag{20}$$

$$V_{Oclose}(s) = V_{dem}(s)F_{LPF}(s)F_{PIPLC}(s) \tag{21}$$

Combining Eq. (20) with (21), we have:

$$\left|\frac{V_{Oclose}(\omega_{open})}{\Omega_{VTes}(\omega_{open})}\right| = \frac{2m_c A_x \omega_d K_{pre} G_{se} V_{dac} F_{LPF}(s) F_{PIPLC}(s)}{4m_y - K_{pre} K_{FBy} V_{dac}^2 G_{se} F_{LPF}(s) F_{PIPLC}(s)} \tag{22}$$

Because $4m_y \ll K_{pre}K_{FBy}V_{dac}^2 G_{se}F_{LPF}(s)F_{PIPLC}(s)$within bandwidth range ($G_{se}$ reduces a lot at point B, so bandwidth range is considered before point B), Eq. (22) can be expressed as:

$$\left|\frac{V_{Oclose}}{\Omega_{VTes}}\right| = \frac{2m_c A_x \omega_d}{K_{FBy} V_{dac}} \tag{23}$$

The above equation means that the scale factor in closed loop is constant value and is not restricted by resonant peak (A point).

Generally speaking, the open-loop Bode diagram of sensing closed-loop is expected to have the following characteristics: (a) In low frequency range, a first-order pure integral element is configured to achieve enough gain and reduce the steady state error of the system, (b) In middle frequency range, the slope of magnitude line is designed to be −20 dB/dec at 0 dB crossing frequency point. The cut off frequency is ω_{cut}, which provides enough phase margin (more than 30°), (c) At $\Delta\omega_2$ frequency point (A point), the phase lags 180° acutely, so two first-order differential elements are utilized to compensate the system phase before $\Delta\omega_2$. After the phase compensation, the −180° crossing frequency is improved and optimizes the magnitude margin (expected more than 5 dB), (d) Since a pair of conjugate zero is at $\Delta\omega_B$, which provides +20 dB/dec magnitude curve, cut off frequency ω_{cut} is arranged before $\Delta\omega_B$, (e) In high frequency range, the magnitude curve is designed to be −60 dB/dec to reduce magnitude rapidly, which restrains high frequency noise and white noise effectively. F_{LPF} element is second-order type and generates −40 dB/dec, and a pair of conjugate pole is at $\Delta\omega_1$, providing −20 dB/dec slope, (f) An inertial element is required to match the phase compensation elements and its frequency should be arranged outside the bandwidth. Meanwhile, the inertial element also provides another −20 dB/dec slope in high frequency range, which is better to restrain high frequency noise and (g) Temperature compensation module is required to ensure the wide-temperature range characteristic of PIPLC. So, the PIPLC is expressed as:

$$F_{PIPLC}(s) = k_{pi}(t_{tem})\frac{s+\omega_{pi1}}{s}\frac{s+\omega_{pi1}}{s+\omega_{pi2}(t_{tem})} \tag{24}$$

According to the analysis, we make k_{pi} = 32 (experience value, over high value brings instability), ω_{pi1} = 10π rad/s and ω_{pi2} = 400π rad/s. In addition, the k_{pi} and ω_{pi2} can be adjusted by temperature t_{tem}. The PIPLC circuit diagram is shown in **Figure 11**. The transform function can be written as following:

The open-loop Bode Diagram of the sensing closed-loop is simulated in Simulink software and shown in **Figure 12(a)**. **Figure 12** indicates that the minimum phase margin of the loop is 34.6° and the magnitude margin is 7.21 dB, which satisfy the design requests. The Pole-Zero Map and Nyquist Map of sensing closed-loop are shown in **Figure 12(b)** and **(c)**. The poles distribute in negative side of real axis and the Nyquist curve does not contain (−1, 0j) point. These two criterions both illustrate the closed system is pretty stable. The Bode Diagram of sensing closed-loop simulation is shown in **Figure 12(d)**, whose curves indicate that the bandwidth of the gyro is 100 Hz, the lowest point within the bandwidth range is −13.8 dB, the DC magnitude is −12.3 dB and the highest point is −10.4 dB. The resonant peak point A (shown in **Figure 10**) is compensated, and the new bandwidth bottleneck point is valley B. Therefore, one of the best methods to expand the bandwidth under this condition is to enlarge the frequency

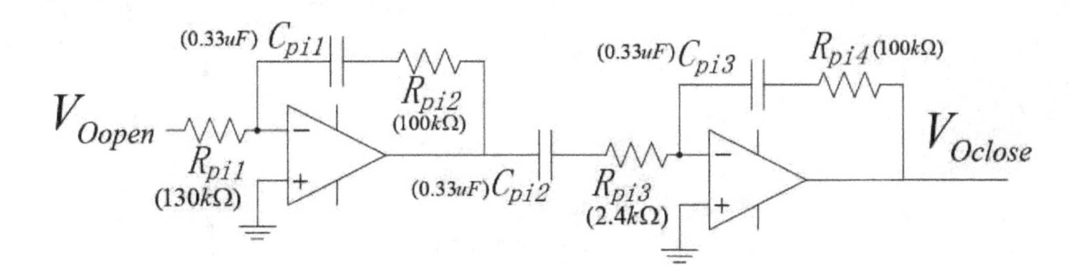

Figure 11. PIPLC circuit diagram.

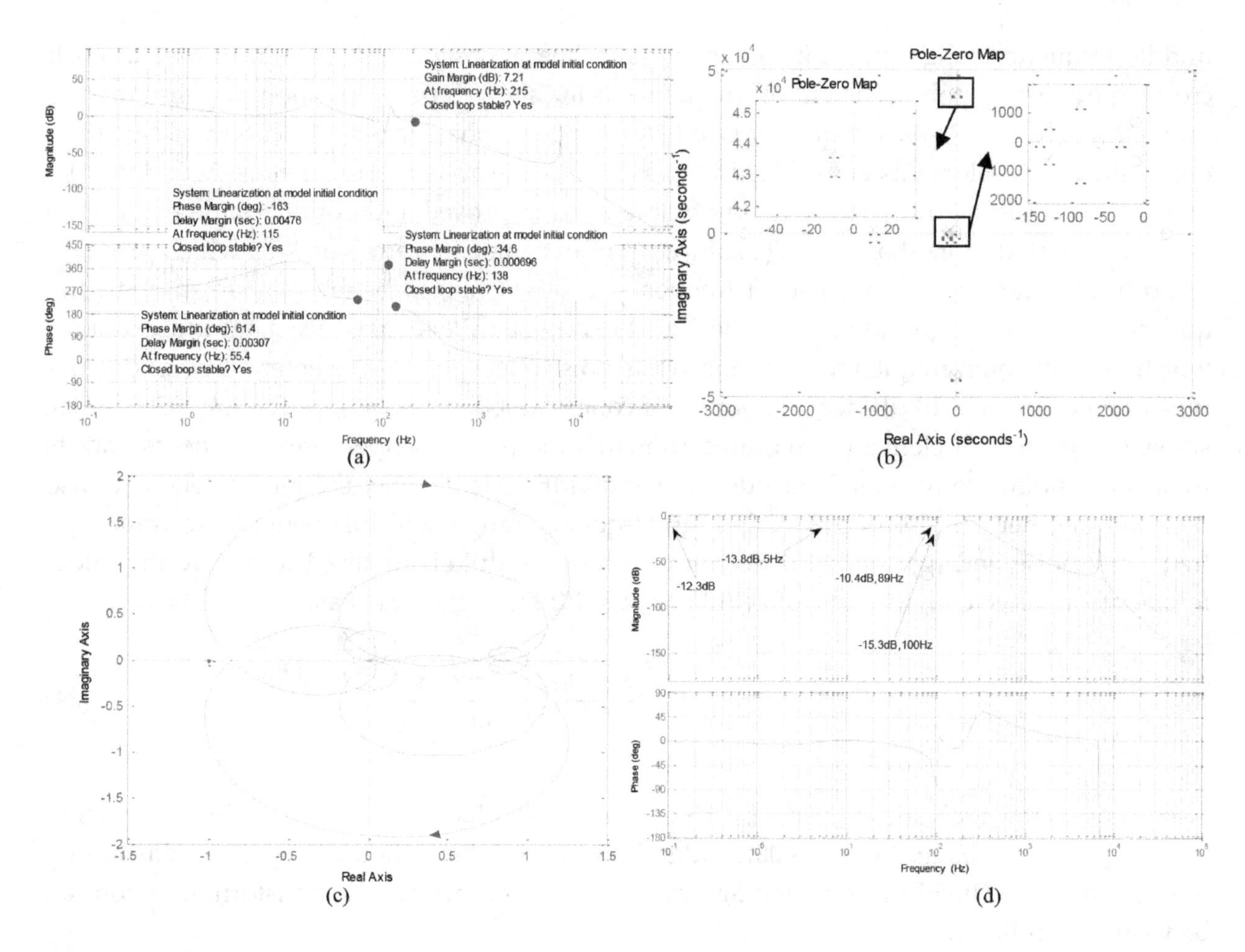

Figure 12. (a) Open-loop bode diagram of sensing close loop; (b) pole-zero map of sensing close loop; (c) Nyquist map of sense close loop; (d) sense close loop simulation bode diagram.

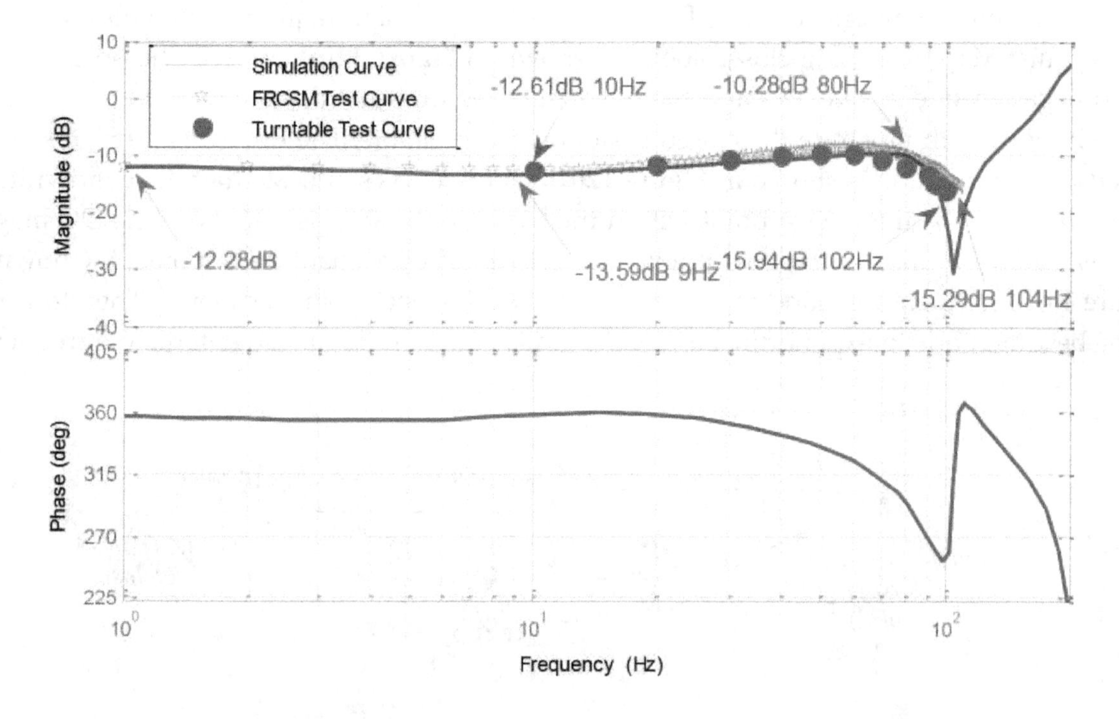

Figure 13. Dual-mass gyroscope bandwidth test.

difference between ω_{y1} and ω_{y2}. The gyro is fixed on the turntable and tested under room temperature (about 20°C) with FRCSM. More specifically, the employed method is a step-by-step test and step length is 1 Hz. First, the result is shown as **Figure 13** with red stars, where the bandwidth is about 104 Hz (as the simulation value is 102 Hz in **Figure 13**) [23]. Since the scale factor can be adjusted with the output level amplifier, the −3 dB point is the key, not the absolute value of the magnitude. After the turntable test is done, since the turntable's swing frequency value is hard to get accurately, we mainly focus on the bandwidth, not each frequency point's value. Within the bandwidth range, the step-by-step test (step length is 10 Hz) is carried out, and the bandwidth is found to be 102 Hz, which verifies the FRCSM result.

5. Conclusion

In this chapter, the recent achievements in our research group for dual-mass MEMS gyroscope are proposed. Three main parts are discussed: dual-mass gyroscope structure design, quadrature error electrostatic compensation and bandwidth expansion. First, dual-mass MEMS gyroscope structure is designed and simulated in ANSYS soft and dual-mass structure movement function is derived. Second, quadrature error is traced to the source, and the coupling stiffness electrostatic compensation method is employed to reduce the quadrature error. Finally, proportional-integral-phase-leading controller is presented in sensing close loop to expand the bandwidth from 13 to 104 Hz.

Acknowledgements

This work was supported by National Natural Science Foundation of China No.51705477. The research was also supported by Research Project Supported by Shanxi Scholarship Council of China No. 2016-083, Fund of North University of China, Science and Technology on Electronic Test and Measurement Laboratory No. ZDSYSJ2015004, and The Open Fund of State Key Laboratory of Deep Buried Target Damage No. DXMBJJ2017-15. The authors deliver their special gratitude to Prof. Hongsheng Li (Southeast University, Nanjing, China) and his group for the guidance, discussions and help.

Author details

Huiliang Cao[1]* and Jianhua Li[2]

*Address all correspondence to: caohuiliang1986@126.com

1 Science and Technology on Electronic Test and Measurement Laboratory, North University of China, Tai Yuan, China

2 National Key Laboratory of Science and Technology on Electromechanical Dynamic Control, Beijing Institute of Technology, Beijing, China

References

[1] Zaman M, Sharma A, Hao Z, et al. A mode-matched silicon-yaw tuning-fork gyroscope with subdegree-per-hour Allan deviation bias instability. Journal of Microelectromechanical Systems. 2008;(6):1526-1536. DOI: 10.1109/jmems.2008.2004794

[2] Xu Y, Chen XY, Wang Y. Two-mode navigation method for low-cost inertial measurement unit-based indoor pedestrian navigation. Journal of Chemical Information & Computer Sciences. 2016;**44**(5):1840-1848. DOI: 10.1177/0037549716655220

[3] Huang H, Chen X, Zhang B, et al. High accuracy navigation information estimation for inertial system using the multi-model EKF fusing Adams explicit formula applied to underwater gliders. ISA TRANSACTIONS. 2017;**66**:414-424. DOI: 10.1016/j.isatra.2016.10.020

[4] Antonello R, Oboe R. Exploring the potential of MEMS gyroscopes. IEEE Industrial Electronics Magazine. 2012;**3**:14-24. DOI: 10.1109/mie.2012.2182832

[5] Xia DZ, Yu C, Kong L. The development of micromachined gyroscope structure and circuitry technology. Sensors. 2014;**14**:1394-1473. DOI: 10.3390/s140101394

[6] Cao HL, Li HS. Investigation of a vacuum packaged MEMS gyroscope structure's temperature robustness. International Journal of Applied Electromagnetics & Mechanics. 2013;**4**: 495-506. DOI: 10.3233/jae-131668

[7] Tatar E, Alper S, Akin T. Quadrature- error compensation and corresponding effects on the performance of fully decoupled MEMS gyroscopes. Journal of Microelectromechanical Systems. 2012;(3):656-667. DOI: 10.1109/jmems.2012.2189356

[8] Chaumet B, Leverrier B, Rougeot C, et al. A new silicon tuning fork gyroscope for aerospace applications. In: Proceedings of Symposium Gyro Technology. 2009, 1.1-1.13

[9] Saukski M, Aaltonen L, Halonen K. Zero-rate output and quadrature compensation in vibratory MEMS gyroscopes. IEEE Sensors Journal. 2007;(12):1639-1652. DOI: 10.1109/jsen.2007.908921

[10] Tally C, Waters R, Swanson P. Simulation of a MEMS Coriolis gyroscope with closed-loop control for arbitrary inertial force, angular rate, and quadrature inputs. Proceedings of IEEE Sensors. 2011:1681-1684

[11] Li HS, Cao HL, Ni YF. Electrostatic stiffness correction for quadrature error in decoupled dual-mass MEMS gyroscope. Journal of Micro-Nanolithography Mems And Moems. 2014;**13**(3): 033003. DOI: 10.1117/1.jmm.13.3.033003

[12] Walther A, Blanc CL, Delorme N, Deimerly Y, Anciant R, Willemin J. Bias contributions in a MEMS tuning fork gyroscope. Journal of Microelectromechanical Systems. 2013;**22**(2): 303-308. DOI: 10.1109/jmems.2012.2221158

[13] Cao HL, Li HS, Liu J et al. An improved interface and noise analysis of a turning fork microgyroscope structure. Mechanical Systems and Signal Processing. 2016;**s 70–71**:1209-1220. DOI: 10.1016/j.ymssp.2015.08.002

[14] Su J, Xiao D, Wu X, et al. Improvement of bias stability for a micromachined gyroscope based on dynamic electrical balancing of coupling stiffness. Journal of Micro-Nanolithography Mems And Moems. 2013;**12**(3):033008. DOI: 10.1117/1.jmm.12.3.033008

[15] Ni YF, Li HS, Huang LB. Design and application of quadrature compensation patterns in bulk silicon micro-gyroscopes. Sensors. 2014;**14**:20419-20438. DOI: 10.3390/s141120419

[16] Seeger J, Rastegar A, Tormey T. Method and apparatus for electronic cancellation of quadrature error : United States Patent. No. 7290435B2; 2007

[17] Maurer M, Northemann T, Manoli Y. Quadrature compensation for gyroscopes in electromechanical bandpass ΣΔ-modulators beyond full-scale limits using pattern recognition. Procedia Engineering. 2011;**25**:1589-1592. DOI: 10.1016/j.proeng.2011.12.393

[18] Antonello R, Oboe R, Prandi L, et al. Open loop compensation of the quadrature error in MEMS vibrating gyroscopes. In: Proceedings of the IEEE Industrial Electronics Society Conference; 2009. pp. 4034-4039

[19] Cao HL, Li HS, Sheng X, et al. A novel temperature compensation method for a MEMS gyroscope oriented on a periphery circuit. International Journal of Advanced Robotic System. 2013;**10**(5):1-10. DOI: 10.5772/56759

[20] M H Bao. Handbook of sensors and actuators. 1st ed. Amsterdam; The Netherlands, Elsevier; 2000. DOI: 10.1016/9780444505583

[21] Ezekwe CD, Boser BE. A mode-matching ΣΔ closed-loop vibratory gyroscope readout interface with a 0.004°/s/√Hz noise floor over a 50Hz band. IEEE Journal of Solid-State Circuits. 2008;**43**(12):3039-3048. DOI: 10.1109/jssc.2008.2006465

[22] Sonmezoglu S, Alper SE, Akin T. An automatically mode-matched MEMS gyroscope with wide and tunable bandwidth. Journal of Microelectromechanical Systems. 2014;**23**(2): 284-297. DOI: 10.1109/jmems.2014.2299234

[23] Sung WT, Sung S, Lee JG, Kang T. Design and performance test of a MEMS vibratory gyroscope with a novel AGC force rebalance control. Journal of Micromechanics & Microengineering. 2007;**17**(10):1939. DOI: 10.1088/0960-1317/17/10/003

[24] Cui J, Guo Z, Zhao Q, et al. Force rebalance controller synthesis for a micromachined vibratory gyroscope based on sensitivity margin specifications. Journal of Microelectromechanical. 2011;**20**(6):1382-1394. DOI: 10.1109/jmems.2011.2167663

[25] He C, Zhao Q, Liu Y, et al. Closed loop control design for the sensing mode of micromachined vibratory gyroscope. Science China-Technological Sciences. 2013;**56**(5): 1112-1118. DOI: 10.1007/s11431-013-5201-x

[26] Alshehri A, Kraft M, Gardonio P. Two-mass MEMS velocity sensor: Internal feedback loop design. IEEE Sensors Journal. 2013;**13**(3):1003-1011. DOI: 10.1109/jsen.2012.2228849

[27] Trusov AA, Schofield AR, Shkel AM. Study of substrate energy dissipation mechanism in in-phase and anti-phase micromachined vibratory gyroscope. In: IEEE Sensors Conference; 2008. pp. 168-171

[28] Y Ni, H Li, L Huang, et al. On bandwidth characteristics of tuning fork micro-gyroscope with mechanically coupled sensing mode. Sensors 2014;**14**(7):13024-13045. DOI: 10.3390/s140713024

[29] Si C, Han G, Ning J, Yang F. Bandwidth optimization design of a multi degree of freedom MEMS gyroscope. Sensors. 2013;**13**(8):10550-10560. DOI: 10.3390/s130810550

[30] Acar C, Shkel AM. An approach for increasing drive-mode bandwidth of MEMS vibratory gyroscopes. Journal of Microelectromechanical Systems. 2005;**14**(3):520-528. DOI: 10.1109/jmems.2005.844801

[31] Feng ZC, Fan M, Chellaboina V. Adaptive input estimation methods for improving the bandwidth of microgyroscopes. IEEE Sensors Journal. 2007;**7**(4):562-567. DOI: 10.1109/jsen.2007.891992

[32] Cui J, He C, Yang Z, et al. Virtual rate-table method for characterization of microgyroscopes. IEEE Sensors Journal. 2012;**12**(6):2192-2198. DOI: 10.1109/jsen.2012.2185489

[33] Y Yin, S Wang, C Wang, et al. Structure-decoupled dual-mass MEMS gyroscope with self-adaptive closed-loop. In: Proceedings of the 2010 5th IEEE International Conference on Nano/Micro Engineered and Molecular Systems; Xiamen. China: 2010. p. 624-627

[34] Cao HL, Li HS, Shao XL, Liu ZY, Kou ZW, Shan YH, Shi YB, Shen C, Liu J. Sensing mode coupling analysis for dual-mass MEMS gyroscope and bandwidth expansion within wide-temperature range. Mechanical Systems & Signal Processing. 2018;**98**:448-464. DOI: 10.1016/j.ymssp.2017.05.003

[35] Cao HL, Li HS, Kou ZW, Shi YB, Tang J, Ma ZM, Shen C, Liu J. Optimization and experiment of dual-mass MEMS gyroscope quadrature error correction methods. Sensors. 2016;**16**(1):71. DOI: 10.3390/s16010071

7

Integrated Power Supply for MEMS Sensor

Hai-peng Liu, Lei Jin, Shi-qiao Gao and
Shao-hua Niu

Abstract

The recent expansion of wireless sensor networks and the rapid development of low-power consumption devices and MEMS devices have been driving research on harvester converting ambient energy into electricity to replace batteries that require costly maintenance. Harvesting energy from ambient environment vibration becomes an ideal power supply mode. The power supply module can be integrated with the MEMS sensor. There are many ways to convert ambient energy into electrical energy, such as photocells, thermocouples, vibration, and wind and so on. Among these energy-converting ways, the ambient vibration energy harvesting is more attractive because the vibration is everywhere in our daily environment. Based on the analysis of the basic theory of the electret electrostatic harvester, the basic equations and equivalent analysis model of electret electrostatic harvester are established. The experimental tests for the output performance of electret electrostatic harvester are completed. For the electret material, the material itself can also provide a constant voltage to avoid the use of additional power, which provides an effective way for electrostatic harvesting. Therefore, the electret electrostatic harvesting structure is a kind of ideal energy harvesting method using ambient vibration and can be easily integrated with the MEMS system.

Keywords: integrated, MEMS, vibration, energy harvesting, electromagnetic harvester

1. Introduction

With the development and progress of science and technology, information technology has already entered the era of micro-nano, and the emergence and development of large-scale integrated circuits have made great and rapid progress in computer technology, information technology, and control technology [1–4]. With the development of micro-nano-technology, the volume of the sensor becomes smaller and smaller, and the signal processing circuit becomes smaller and smaller. The overall system has become smaller, but the power supply has not become smaller [5, 6].

With the system becoming smaller, the energy power required is also getting smaller, but the power supply must be wired. Independent battery energy has limited energy, and its own energy is decaying, requiring periodic replacement or periodic charging, which is an extremely difficult or almost impossible task for an information system that requires long-term work or working in a confined space things. Therefore, an energy harvesting system that takes advantage of ubiquitous energy in the environment is imperative [4–6].

According to the application characteristics of MEMS wireless sensors, harvesting energy from ambient environment vibration becomes an ideal power supply mode. The power supply module can be integrated with the MEMS sensor. There are many ways to convert ambient energy into electrical energy, such as photocells, thermocouples, vibration, and wind and so on [7, 8]. Among these energy converting ways, the ambient vibration energy harvesting is more attractive because the vibration is everywhere in our daily environment [9, 10]. So it has broad application to convert energy directly from the ambient vibration.

Compared with the traditional chemical batteries, micro-vibration energy harvester has the following advantages:

1. Long-term storage. Micro-energy harvester can directly convert the mechanical energy into electrical energy. This will overcome the shortcoming of electrical energy release due to long-term storage of the traditional battery.

2. Small size and high energy density. It is much smaller than the traditional battery size and has higher energy density.

3. Easy to integrate. The manufacturing technology of the micro-harvester structure is compatible with the manufacturing process of MEMS so as to realize the integration of the overall system (including the micro-power supply, the micro-mechanical structure, and the circuit system).

4. Low cost. The harvesting structures are easy to integrate with the MEMS structures for mass production. Therefore, the production and maintenance cost can be greatly reduced.

According to the analysis, the vibration energy harvester possesses many advantages that other micro-energy sources do not have. Therefore, the research on the vibration harvester has become a hotspot in the field of micro-energy [11–13]. Recently, the energy harvesting mechanism of vibration energy harvesting mainly includes four types: piezoelectric, electromagnetic, electrostatic, and magnetostrictive [12, 14]. Among these mechanisms, the electrostatic harvesting is suitable for MEMS device power supply because the electrostatic materials are easy to integrate with the micro-mechanical structures.

2. Electrostatic vibration energy harvesting

Self-powered microsystems are implemented as MEMS technology enables miniaturization of devices and MEMS monoliths can be monolithically integrated with other MEMS devices. For the current MEMS technology, compatible with MEMS technology to achieve self-powered MEMS system service technology is mainly electrostatic harvesting technology.

Electrostatic vibration energy harvesting is a kind of energy harvesting method through variable capacitance to generate electrical energy. When the capacitance of the variable capacitor is changed due to the vibration of the external environment, the amount of charge stored in the capacitor is changed. And a charge flow formed in the circuit due to the change of capacitor will provide electrical power to the load. The electrostatic harvester has high output voltage and is easy to integrate with MEMS structure. The coupling coefficient of electrostatic harvester is easy to adjust, and the capacitance can be adjusted by adjusting the capacitor size [15].

Generally, electrostatic harvesting system consists of two-unit modules, the vibration unit and the electrostatic harvesting unit. The vibration unit can be regarded as a second-order vibration system composed of a mass spring, which can convert the vibration excitation in the external environment into the kinetic energy of the vibration mass. Electrostatic harvesting unit usually consists of a variable capacitor, a load resistor, and an external power source (such as electret, charging capacitor, or other power source) that converts the kinetic energy of the mass into electrical energy. There are two main types of electrostatic harvester: one is electret-free electrostatic harvester, and the other is based on electret-based electrostatic harvester. The operation mode of electret-free electrostatic harvester is generally divided into constant charge and constant voltage [16–18]. Before the vibration energy harvesting structure starts to output electric energy, an initial voltage needs to be applied to the variable capacitor, and the upper and lower electrodes of the variable capacitor are charged and discharged during energy harvesting. The control circuit of electret-free electrostatic harvester is relatively complicated and difficult to achieve. While the electret of electret-based electrostatic harvester can be regarded as one pole of a variable capacitor, and the vibration energy can be directly converted into electric energy without charging or discharging the variable capacitor [19, 20].

2.1. Variable capacitance model

The output characteristics of electrostatic harvesting system depend on the kinetic energy obtained by vibrating unit and on the energy conversion of the harvesting unit, while the power conversion depends on the size of capacitance and the change rate of capacitance. In order to design a relatively large capacitance in a limited space, people take a variety of ways. A variety of structures have been proposed at the micrometer scale, of which the comb structure is relatively common. Most of these variable capacitance structures originated from the structures used in micro-accelerometers and micro-gyroscopes. There are four major capacitive models, which are:

a. The comb capacitor model based on pitch tuning: The comb capacitors overlap each other and spacing is variable.

b. The comb capacitor model based on area tuning: The comb capacitors overlap each other and the overlap area is variable.

c. The plane capacitance model based on pitch tuning: The two-plate electrodes are parallel to each other and spacing is variable.

d. The plane capacitance model based on area tuning: The two-plate electrodes are parallel to each other and the overlap area is variable.

Since the spacing between combs is relatively small and the area of the combs is relatively large, the capacitance can also be approximated calculated by an infinite plate capacitance calculation model. According to the general formula of infinite panel capacitance calculation model, we can give the calculation of above four models. The capacity of the plate capacitor is not only related to the size, shape, and spacing of the two electrodes but also related to the dielectric between the two plates. When the electret is added between the two plates of the plate capacitor, the capacitance calculation formula of the plate capacitor will change. In terms of the electrostatic harvesting structure, since the variable capacitance between the two electrodes is generally attached to electret, the corresponding calculation formula of the plate capacitance needs to be adjusted. Capacitance values calculated below include the electrets.

The structure of the plate capacitor with the electret is shown in **Figure 1**. The plane capacitor consists of the upper and lower electrodes and an electret. The electret has an air gap with the upper electrode, and the electret attaches to the lower electrode surface. C_1 is the capacitance between electret and the upper electrode; C_2 is the capacitance between electret and the lower electrode. The total capacitance C is

$$C = \frac{C_1 C_2}{C_1 + C_2} = \frac{\varepsilon_0 S}{g_0 + d/\varepsilon_r} \tag{1}$$

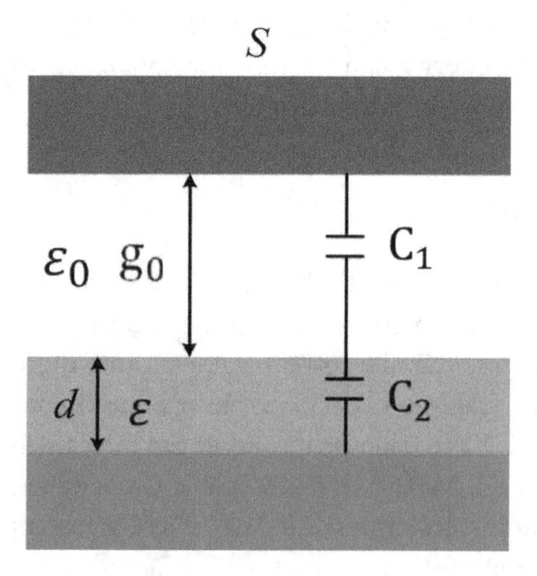

Figure 1. Plate capacitance model.

where S is the overlap area of the upper electrode and electret, g_0 is the distance between the upper electrode and electret, d is the thickness of electret, and $\varepsilon_r = \varepsilon/\varepsilon_0$ is the relative dielectric constant of the electret.

a. The comb capacitor model based on pitch tuning (**Figure 2**)

The comb capacitor model based on pitch tuning refers to the spacing between the movable comb and the fixed comb that is variable (shown in **Figure 2**), according to the equation $C_{p1}(x) = \frac{\varepsilon_0 S}{g_0 + d/\varepsilon_r - x}$ and $C_{p2}(x) = \frac{\varepsilon_0 S}{g_0 + d/\varepsilon_r + x}$. The total capacitance $C(x)$ is

$$C(x) = N \times (C_{p1}(x) + C_{p2}(x)) = \frac{2N\varepsilon_0 S(g_0 + d/\varepsilon_r)}{(g_0 + d/\varepsilon_r)^2 - x^2} \quad (2)$$

where N is the number of the overlapping comb, S is the overlap area between the movable comb and the fixed comb, g_0 is the initial distance between the movable comb and the fixed comb, d is the thickness of the electret attached to the fixed comb, $\varepsilon_r = \varepsilon/\varepsilon_0$ is the relative dielectric constant of the electrets, and x is the displacement of the movable comb relative to the fixed comb.

b. The comb capacitor model based on area tuning (**Figure 3**)

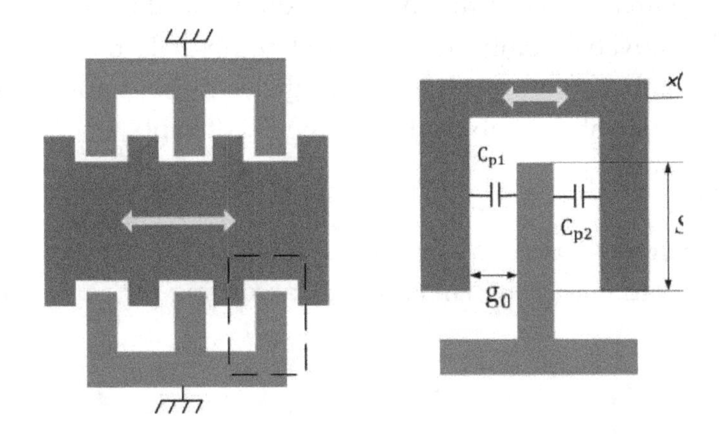

Figure 2. The comb capacitor model based on pitch tuning.

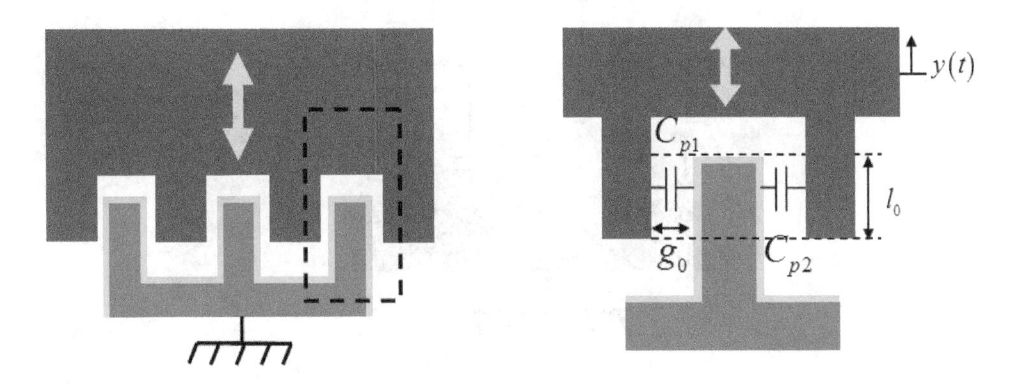

Figure 3. The comb capacitor model based on area tuning.

The comb capacitor model based on area tuning refers to the overlap area between the movable comb and the fixed comb that is variable (shown in **Figure 3**), according to the Eq. (1):

$$C(y) = \frac{2N\varepsilon_0 w}{g_0 + d/\varepsilon_r}(l_0 - y) \tag{3}$$

where N is the number of the overlapping comb, l_0 is the overlap length when the movable comb C_{p1} and the fixed comb C_{p2} are in the initial position and $C_{p1} = C_{p2}$, w is the width (depth) of the comb, g_0 is the initial distance between the movable comb and the fixed comb, d is the thickness of the electret attached to the fixed comb, $\varepsilon_r = \varepsilon/\varepsilon_0$ is the relative dielectric constant of the electrets, and y is the displacement of the movable comb relative to the fixed comb.

c. The plane capacitance model based on pitch tuning (**Figure 4**)

The plane capacitance model based on pitch tuning refers to the distance between the two plane electrodes that can be changed (shown in **Figure 4**); according to Eq. (1), the capacitance based on pitch tuning is

$$C(y) = \frac{\varepsilon_0 S}{g_0 + d/\varepsilon_r - y} \tag{4}$$

where S is the overlap area between the movable comb and the fixed comb, g_0 is the initial distance between the movable comb and the fixed comb, d is the thickness of the electret attached to the fixed comb, $\varepsilon_r = \varepsilon/\varepsilon_0$ is the relative dielectric constant of the electrets, and y is the displacement of the movable comb relative to the fixed comb.

d. The plane capacitance model based on area tuning (**Figure 5**)

The plane capacitance model based on area tuning refers to the overlap area between the two plane electrodes that is variable (shown in **Figure 5**); according to Eq. (1), the capacitance based on area tuning is

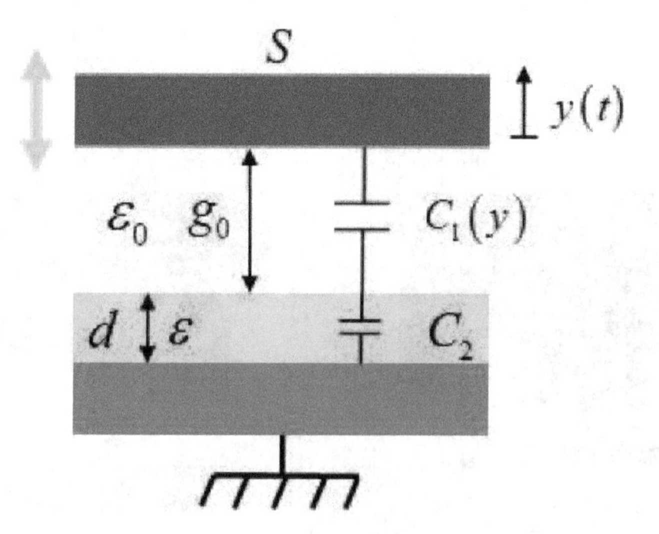

Figure 4. The plane capacitance model based on pitch tuning.

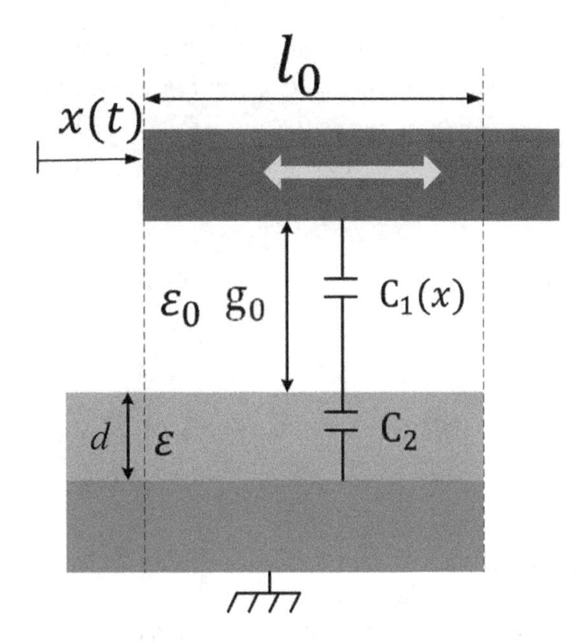

Figure 5. The plane capacitance model based on area tuning.

$$C(x) = \frac{\varepsilon_0 w(l_0 - x)}{g_0 + d/\varepsilon_r} \tag{5}$$

where w is the width (depth) of the comb, l_0 is the overlap length when the movable comb and the fixed comb are at the initial time, g_0 is the initial distance between the movable comb and the fixed comb, d is the thickness of the electret attached to the fixed comb, $\varepsilon_r = \varepsilon/\varepsilon_0$ is the relative dielectric constant of the electrets, and x is the displacement of the movable comb relative to the fixed comb.

2.2. Capacitive electrostatic force

Actually, the energy stored between any two conductors is the capacitance energy W; it can be expressed as

$$W = \frac{1}{2} C U^2 \tag{6}$$

where C is the capacitance between two conductors and U is the voltage between the conductors.

If the charge Q between conductors is constant, the conductor should be an isolated system that is not connected to the external power source. The work of the field force can only come from the decrease of the electric field energy, that is:

$$F_n = -\frac{\partial W}{\partial n} = -\frac{1}{2}\frac{\partial}{\partial n}\left(\frac{1}{C}\right)Q^2 = \frac{1}{2}\frac{\partial C}{\partial n}\frac{Q^2}{C^2} = \frac{1}{2}\frac{\partial C}{\partial n}U^2 \tag{7}$$

where n is the relative motion direction coordinates between two conductors; if the voltage U between the conductor is constant, the external power source must be added. When there is an external power source, according to the previous analysis, the work done by the electric field force should be equal to the increase of the electric field energy, that is:

$$F_n = \frac{\partial W}{\partial n} = \frac{1}{2}\frac{\partial C}{\partial n}U^2 \tag{8}$$

Although it looks like a symbol is different, the final form of the formula is the same. For a more general case, when the voltage or charge is variable, the electrostatic force should be expressed as

$$F_n = \frac{\partial}{\partial n}\left(\frac{1}{2}CU^2\right) \text{ or } F_n = \frac{\partial}{\partial n}\left(\frac{1}{2}\frac{Q^2}{C}\right) \tag{9}$$

According to Eq. (1), the capacitance of the variable capacitor structure shown in **Figure 6** is

$$\mathrm{C(t)} = \frac{C_1(t)C_2}{C_1(t) + C_2} = \frac{\varepsilon_0 w l(t)}{g(t) + d/\varepsilon_r} \tag{10}$$

where $\varepsilon_r = \varepsilon/\varepsilon_0$ is the relative dielectric constant of the electrets, w is the width (depth) of the comb, $l(t) = l_0 - x(t)$ is the overlap length between the movable comb and the fixed comb, and $g(t) = g_0 + y(t)$ is the initial distance between the movable comb and the fixed comb. According to the principle of energy derivation, the change of capacitance energy along a certain direction is the electrostatic force in this direction. When the overlap length $l(t)$ between the upper electrode and the electret changes, the electrostatic force between the upper electrode and the electret can be expressed as

$$F_x = \frac{d}{dx}(W) = \frac{d}{dx}\left(\frac{1}{2}\frac{Q_C^2(x)}{C(x)}\right) \text{ or } F_x = \frac{d}{dx}(W) = \frac{d}{dx}\left(\frac{1}{2}C(x)U_C(x)^2\right) \tag{11}$$

When the gap $g(t)$ between the upper electrode and the electret changes, the electrostatic force between the upper electrode and the electret can be expressed as

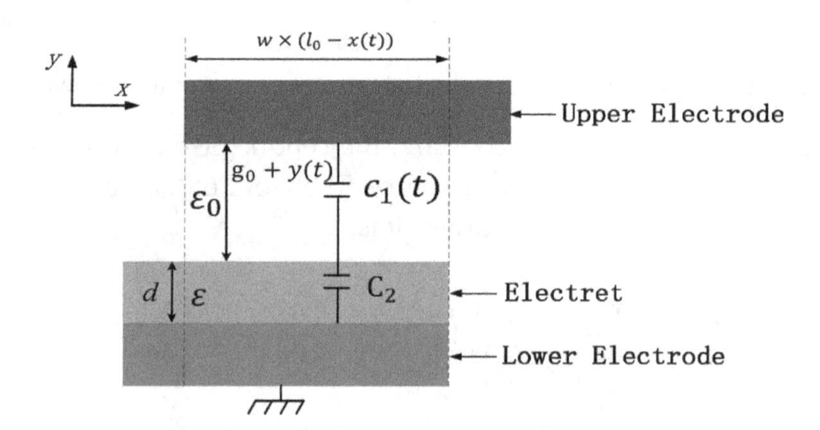

Figure 6. Variable capacitance model.

$$F_y = \frac{d}{dy}(W) = \frac{d}{dy}\left(\frac{1}{2}\frac{Q_C^2(y)}{C(y)}\right) \text{ or } F_y = \frac{d}{dy}(W) = \frac{d}{dy}\left(\frac{1}{2}C(y)U_C(y)^2\right) \quad (12)$$

where $U_C(n)$ is the voltage between the upper and lower electrodes of the capacitor, $Q_C(n)$ is the amount of charge stored on the upper electrode of the capacitor, and n is the movement direction of the upper electrode of the capacitor relative to the electret or the lower electrode. Electrostatic force changes along the direction of the overlapping electrodes when n is x direction, and electrostatic force changes along the direction of the pitch when n is y direction.

2.3. Electrostatic harvesting mechanism

Although the electrostatic harvesting is based on the environmental forces to change the capacitance and then converts into charge or voltage changes to achieve energy harvesting, there are some mechanism differences for the constant charge mode or constant voltage mode.

2.3.1. *Electret-free electrostatic harvester*

A. Constant charge work mode

The constant charge work mode is shown in a light-colored (blue) area of **Figure 7**, which indicates that charges are continually injected into the variable capacitor using an external power source (battery, charged capacitor, etc.) until the charge on both ends of the variable capacitor reaches the maximum Q_{cst} when the capacitance of the variable capacitor reaches the maximum C_{max}, and the voltage of the variable capacitor is U_{min} (Apoint), and then the electromechanical transformation will start. During the energy conversion stage, the charge Q_{cst} remains unchanged, and the variable capacitance changes under the action of external force and gradually decreases from the maximum value until the variable capacitance reaches the minimum value C_{min}. At the same time, the voltage U increases until the voltage reaches the maximum value U_{max} (Bpoint). After that, through the circuit control, the charge in the variable capacitor is transferred to the external charge storage element until the voltage is zero and the charge is cleared; returning to the O point, an energy conversion is completed and enters the next harvesting cycle. The output electrical energy converted from mechanical energy at a harvesting cycle is

$$E = \frac{1}{2}Q_{cst}^2\left(\frac{1}{C_{min}} - \frac{1}{C_{max}}\right) \quad (13)$$

In order to keep the charge of the external power supply (battery, charged capacitor, etc.) from loss, it is usually controlled by the circuit switch to transfer the charge in the charge storage element back to the external power supply at the same time as the variable capacitance changes.

b. Constant voltage work mode

Constant voltage work mode is shown in **Figure 7** dark (red) area, where it also indicates that the electrostatic harvester starts to harvest when variable capacitance reaches the maximum C_{max}.

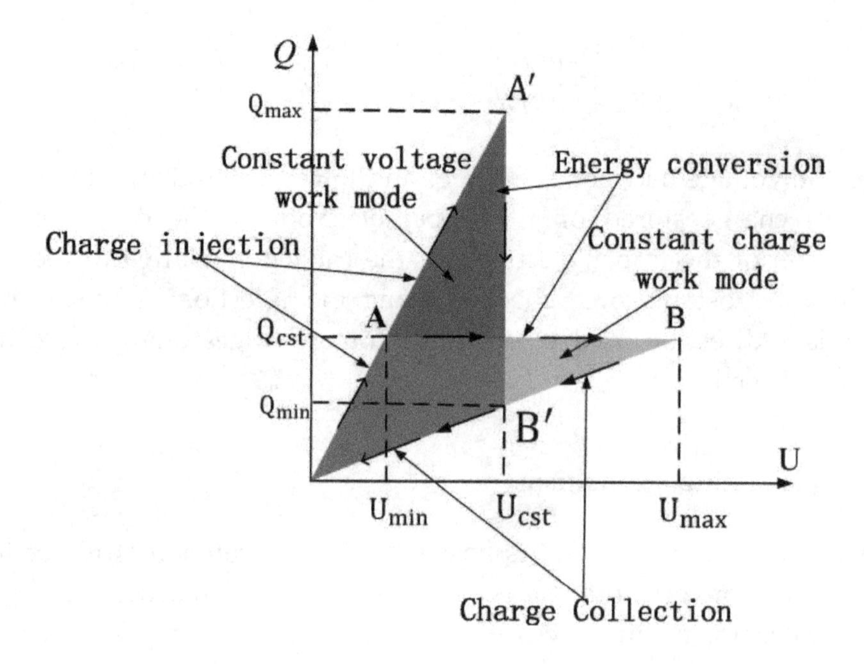

Figure 7. Basic energy conversion cycle of the electret-free electrostatic harvester.

The variable capacitor is charged up to a certain voltage U_{cst} (A' point) by an external power supply (battery, charged capacitor, etc.), the amount of charge stored in the capacitor reaches its maximum Q_{max}, and thereafter the voltage U_{cst} remains constant throughout the electromechanical conversion. As the voltage across the variable capacitor is constant, when the variable capacitance decreases with the external force, the amount of charge stored in the variable capacitor decreases, and the reduced amount of charge generates current through the external circuit loop. When the variable capacitor reaches its minimum value C_{min} (B' point), the residual charge Q_{min} continues to be transferred to the external circuit in a reducing voltage manner until the voltage and charge are both equal to zero. At this point, one energy conversion process is completed. The electrical energy converted from mechanical energy in a work cycle is

$$E = \frac{1}{2} U_{cst}^2 (C_{max} - C_{min}) \tag{14}$$

For electret-free electrostatic harvester, whether it works on the constant charge mode or constant voltage mode, an external power supply (battery, charged capacitor) is required to supply charge for variable capacitance either before the start of electromechanical conversion or at the first energy cycle. Whether in constant charge mode or constant voltage mode of operation, variable capacitance provides charge. Obviously, this is a drawback of the electret-free electrostatic harvester.

2.3.2. *The electrostatic harvester based on the electret*

The electrostatic harvester based on the electret can directly convert the vibration energy into electric energy without requiring external power to charge and discharge the variable capacitor, thus simplifying the control circuit.

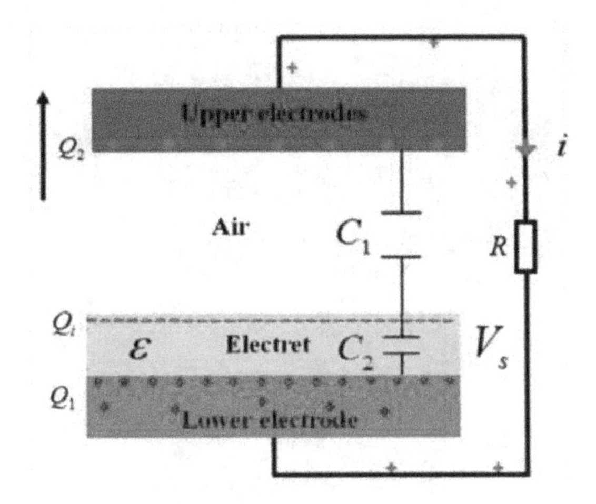

Figure 8. Energy conversion model of electret electrostatic harvester.

In the electret electrostatic harvester, mechanical energy is also converted into electrical energy by the change of the capacitance structure (**Figure 8**). The electret is attached on the surface of the lower electrode. The electrets are separated from the upper electrode by air, and the upper and lower electrodes are connected together by a resistor. According to the electrostatic induction and the charge conservation, the upper and lower electrodes will induct the same polarity charge as the charge polarity of the electret. Q_1 indicates the amount of charge induced by the lower electrode, Q_2 indicates the amount of charge induced by the upper electrode, Q_i indicates the amount of charge in electret, and it is equal to the sum of Q_1 and Q_2. When the upper electrode moves relative to the electret and the lower electrode under the action of external force, the capacitance between the upper electrode and the electret changes, resulting in the redistribution of the charges carried by the upper electrode and the lower electrode. When the upper electrode is close to the electret under the action of external force, the capacitance formed between the upper electrode and the electret will increase. The amount of charge in the upper plate increases, and the amount of charge in the lower plate decreases (**Figure 9a**). When the upper electrode is far away from the electret under the action of external force, the capacitance formed between the upper electrode and the electret will be reduced. The amount of charge in the upper plate decreases, and the amount of charge in the lower plate increases (**Figure 9b**). When the charge flows in the circuit, the current is produced; the voltage across the resistor is generated, so the mechanical energy is converted into electric energy.

2.3.3. *Equivalent circuit model*

The electrostatic harvesting unit of the electrostatic harvester based on the electret can be equivalent to a series connection of a voltage source and a variable capacitor. **Figure 10** shows the circuit formed by the electrostatic harvesting unit and the load resistor.

When the electrostatic harvesting unit is connected to the load resistor to form a loop, according to Kirchhoff's law, there is $iR = R\frac{dQ_2}{dt} = U = V_s - U_c = V_s - \frac{Q_2}{C(t)}$, then

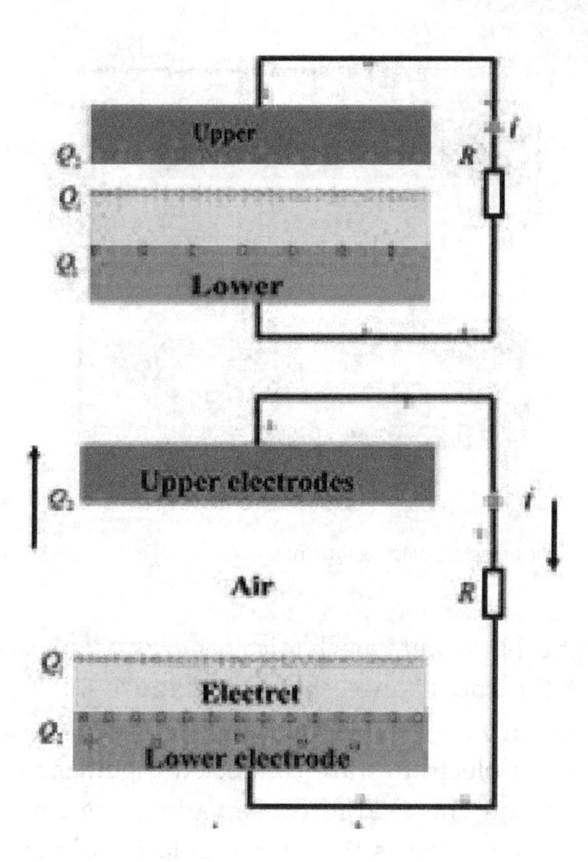

Figure 9. Charge cycle model.

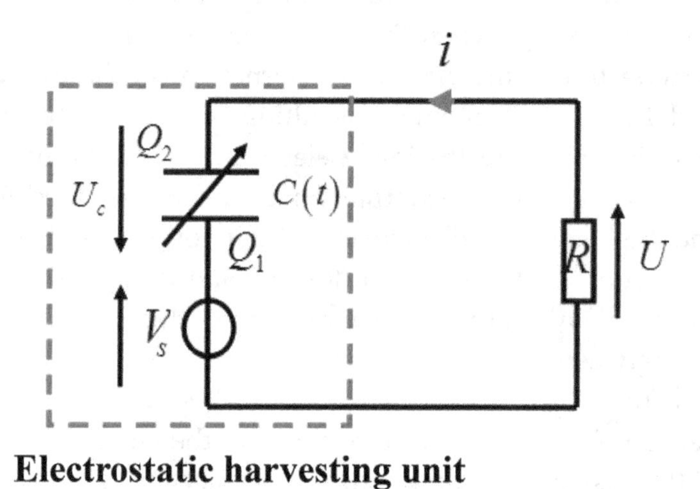

Figure 10. The circuit formed by the electrostatic harvesting unit and the load resistance.

$$\frac{dQ_2}{dt} = \frac{V_s}{R} - \frac{Q_2}{RC(t)} \tag{15}$$

where V_s is the surface potential of the electrets, $C(t)$ represents the capacitance value of the variable capacitor, R represents the load resistance, U represents the voltage across the load resistance, U_c represents the voltage across the capacitor, and Q_2 represents the charge carried by the upper electrode of the variable capacitor.

2.4. The cantilever electrostatic harvester based on electret

In 2011, Boisseau proposed a cantilever electret electrostatic harvester structure based on a - pitch-tuning planar capacitive structure (**Figure 11**). This structure better illustrates the working mechanism of the electret electrostatic harvester and gives its theoretical model. The electret electrostatic harvester consists of a vibrating unit and an electrostatic harvesting unit. The vibration unit is a second-order vibration system consisting of a cantilever beam and a mass. The electrostatic harvesting unit consists of a variable capacitor and a load resistor. The upper surface of the free end of the cantilever is fixedly connected with the mass, and the lower surface of the free end of the cantilever is covered with a layer of electrode as upper electrode of the variable capacitor. The structure of the pitch-tuning planar capacitor is shown in **Figure 8**, Q_i is the amount of charge carried by the electret, Q_1 is the amount of charge induced by the lower electrode, Q_2 is the amount of charge induced by the upper electrode, and Q_i is equal to the sum of Q_1 and Q_2. Under the action of external acceleration, the space between the free end of the cantilever and the electret will change. As a result, the capacitance of the variable capacitor will change, and the charge between the upper and lower electrodes will be redistributed. There will be a charge flow in the circuit and the electric current formed. So, a part of the mechanical energy excited by the vibration of the external environment is converted into electric energy. The equivalent circuit model is shown in **Figure 10**.

The kinetic equation and electrical equation of a cantilever electret electrostatic harvester are as follows (**Figure 12**):

$$\begin{cases} m\ddot{x}+c\dot{x}+kx-F_{elec}-mg=m\ddot{y} \\ \dfrac{dQ_2}{dt}=\dfrac{V_s}{R}-\dfrac{Q_2}{RC(t)} \end{cases} \tag{16}$$

where the capacitance of the variable capacitor is

$$C(t)=\frac{\varepsilon_0 w\lambda}{g_0+{}^{d}/_{\varepsilon_r}-x(t)} \tag{17}$$

where $\ddot{y}$ represents the external excitation acceleration, m represents the mass of the mass, c represents the damping coefficient of the vibrating unit, k represents the stiffness of the

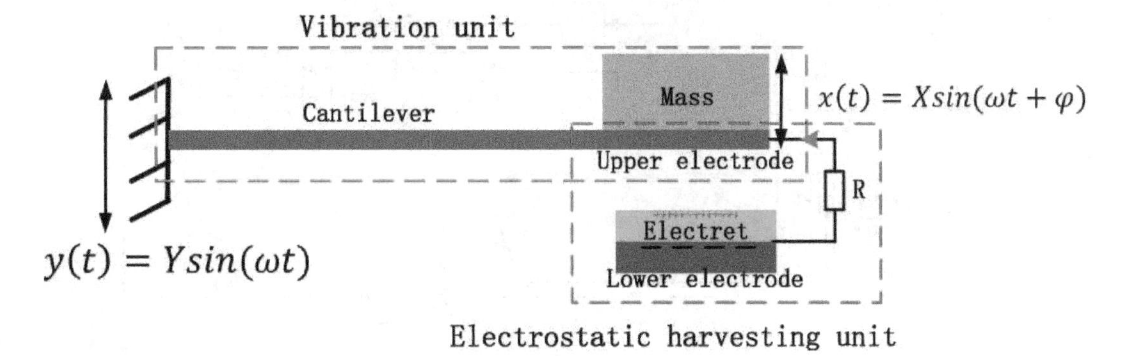

Figure 11. The schematic of cantilever electrostatic harvester based on electret.

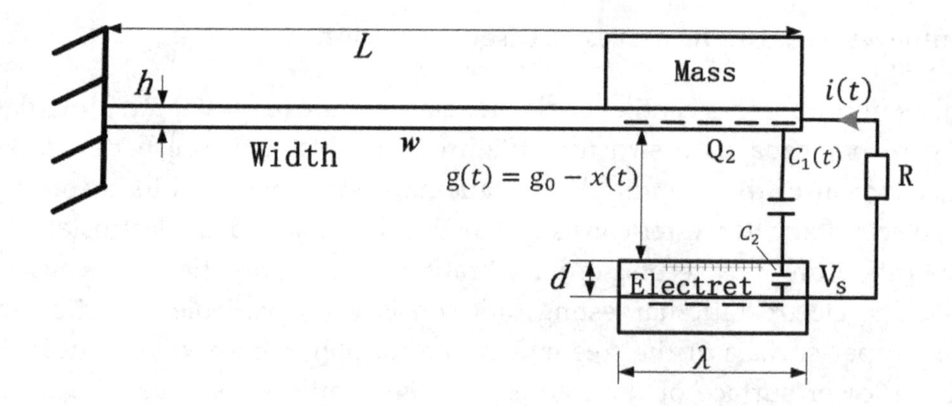

Figure 12. The parameter schematic of cantilever electrostatic harvester based on electret.

vibrating unit, Q_2 represents the amount of charge induced on the upper electrode of variable capacitor, V_s represents the electret surface potential (it keep constant during the entire harvesting process), R represents the load resistance, w represents the width of the electrode, λ represents the length of the electrode, g_0 represents the initial spacing between the upper electrode and the electret, d represents the thickness of the electret, ε_r represents the relative permittivity of the electret, and $x(\mathrm{t})$ is displacement of free end of cantilever under the action of external vibration. The instantaneous power at both ends of the load resistance is given by

$$P(t) = \frac{U^2}{R} = \frac{1}{R}\left(V_s - \frac{Q_2}{C(t)}\right)^2 \tag{18}$$

From Eq. (18), it can be seen that the electrical output of the electrostatic harvesting unit will affect the vibration response of the vibrating unit due to the electromechanical coupling characteristics of system, which in turn changes the electrical output characteristics of the harvesting unit. Since it is difficult to obtain the solution of the system of Eq. (18) by analytical method, it can usually be analyzed by simulation software such as Simulink. The Simulink simulation model is shown in **Figure 13.**

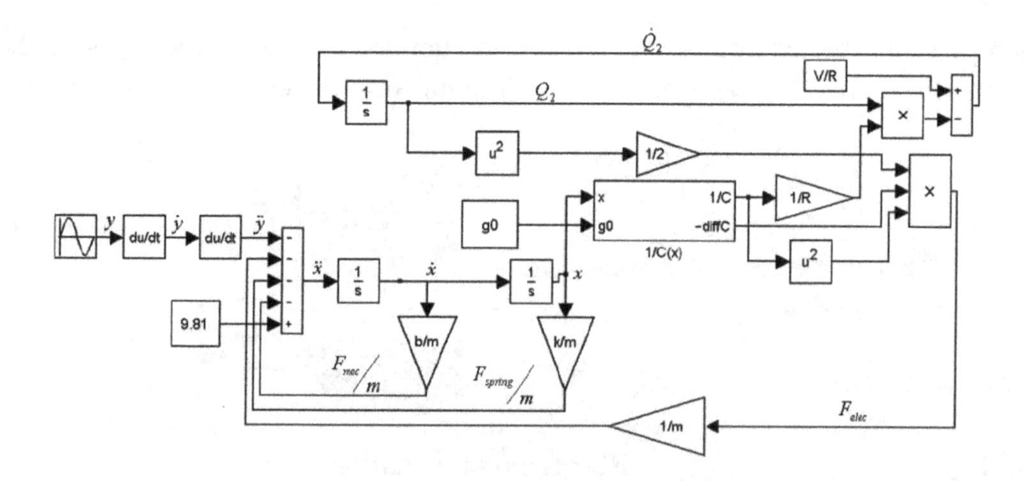

Figure 13. The Simulink model of cantilever electrostatic harvester based on electret.

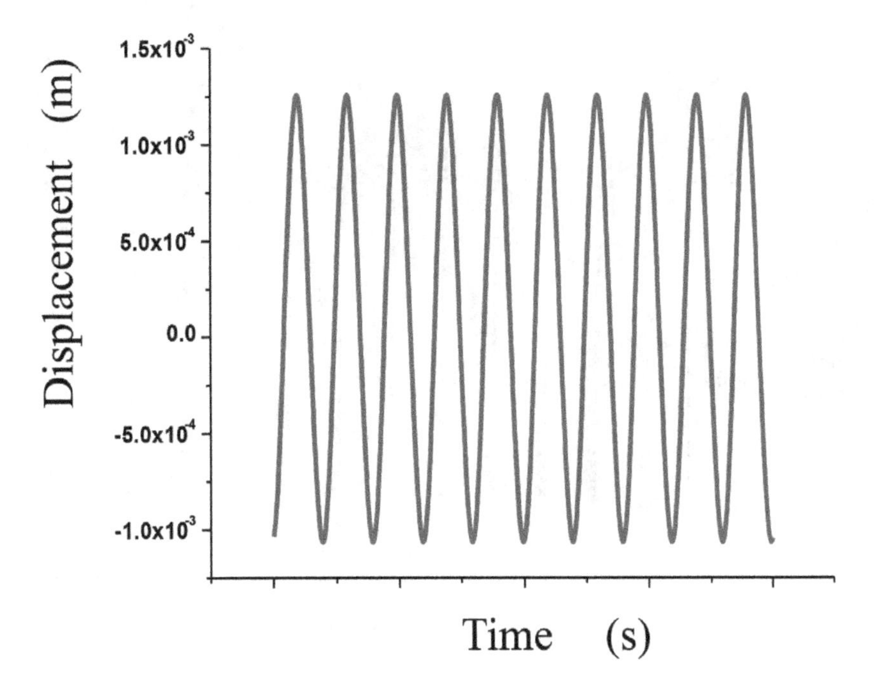

Figure 14. Vibration displacement of cantilever free end.

When an external sinusoidal excitation in which the frequency is 50 Hz and the amplitude is 1.5 m/s^2 is applied to the vibrating unit, the vibration displacement at the free end of the cantilever is shown in **Figure 14**, and **Figure 15** shows the current varies with time in the circuit. When the capacitance reaches the minimum and maximum values, current direction changes. **Figure 16** shows the voltage across the load resistance varies with time, and the peak voltage can be as high as a few hundred volts. It is found through analysis that the surface potential V_s of the electret, the air gap g_0 of the capacitor, the length λ of the electret, and the load resistance R all affect the output power of the cantilever electret electrostatic harvester.

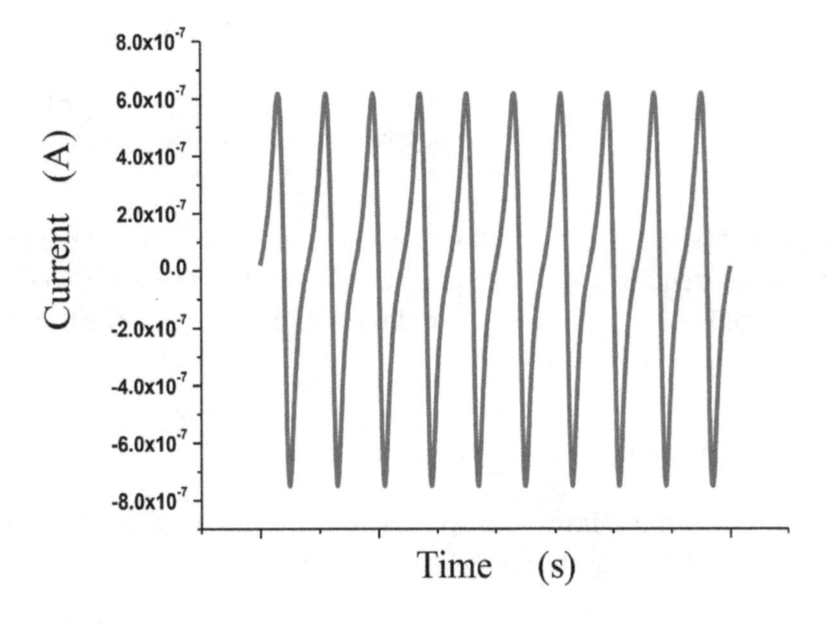

Figure 15. Current varies with time in the circuit.

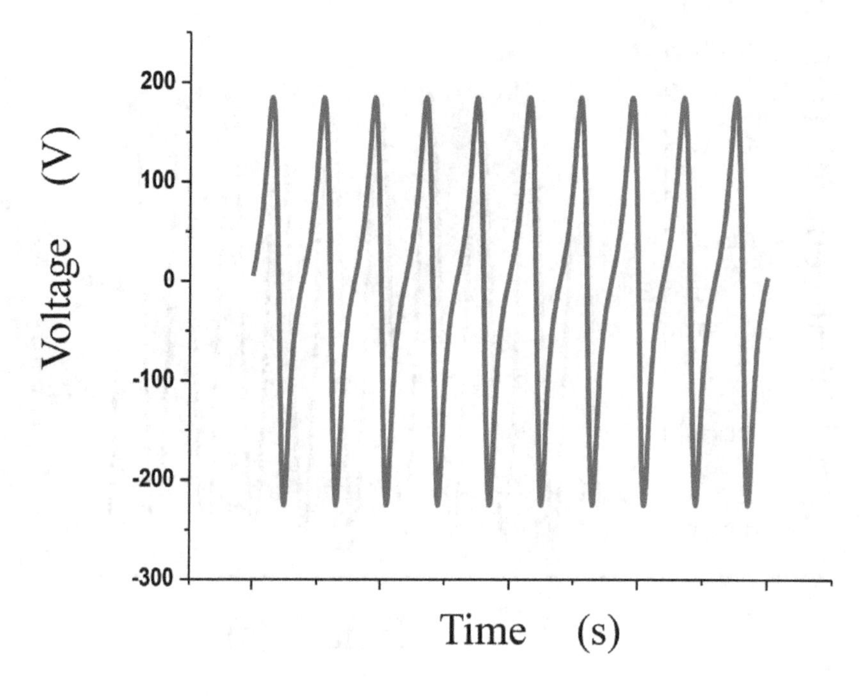

Figure 16. Voltage across the load resistance varies with time.

3. Experimental test on the electret electrostatic harvester output performance

In order to further study the performance of electret electrostatic harvester, a prototype of electrostatic harvester with double-ended fixed-beam electret was fabricated and tested experimentally.

The effects of excitation frequency, air gap, load resistance, and other factors on the output characteristics of the electret electrostatic harvester are tested by the experimental method.

3.1. Effect of excitation frequency on the output of electrostatic harvester

The output of the harvester is related to the excitation frequency. In the acceleration peak of 1.5 harmonic excitation, the output voltage and output power with frequency curve are shown in **Figures 17** and **18.**

The experimental results show that when the initial air gap is 0.2 mm and the acceleration peak is 1.5 m/s^2, the resonant frequency of the electrostatic harvester is 96.2 Hz, the maximum peak-to-peak output voltage is 63.6 V, the corresponding half-power bandwidth is 3.2 Hz, and the maximum output power is 0.054 mW.

3.2. Effect of air gap on the output of electrostatic harvester

The air gap is one of the most important parameters of the electret electrostatic harvester, which plays a key role in the output of the electrostatic harvester.

When the external excitation acceleration peak and the electret surface potential are constant, as shown in **Figure 19**, the output voltage decreases with the increase of air gap. When the air

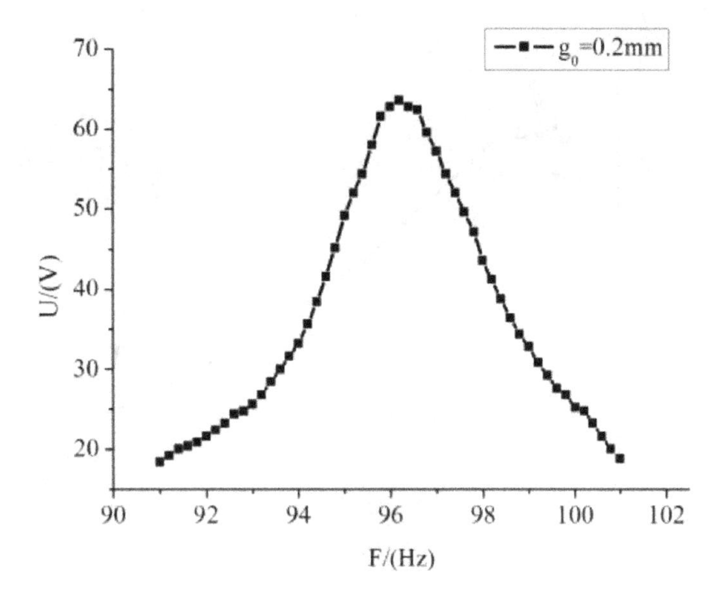

Figure 17. Output voltage with frequency curve.

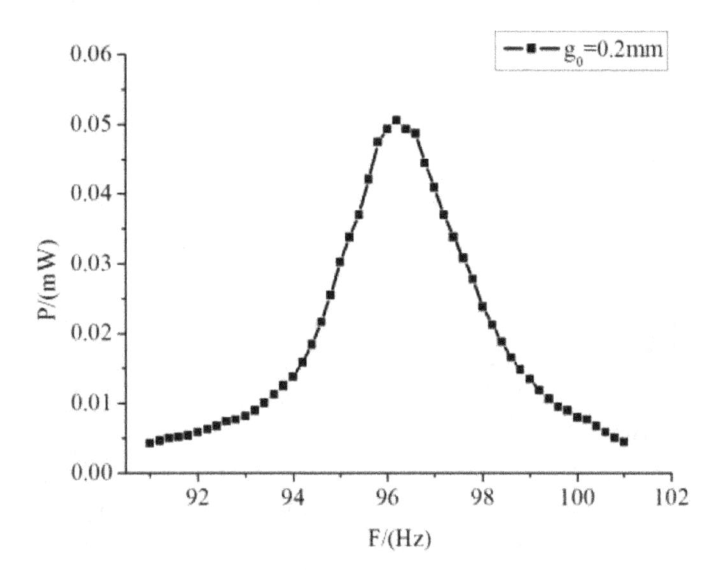

Figure 18. Output power with frequency curve.

gap is 0.15 mm, the peak-to-peak output voltage is 66 V; when the air gap is reduced to 0.5 mm, the output voltage is reduced to 20.8 V.

As shown in **Figure 20**, as the air gap increases, the output power first increases and then decreases. When the gap is 0.2 mm (optimal air gap), the output power reaches a maximum of 0.08 mW. When the air gap increases to 0.5 mm, the output power is reduced to 0.015 mW.

As shown in **Figure 21**, as the air gap decreases, the electrostatic force between the upper and lower plates gradually increases, the soft spring effect increases, and the stiffness coefficient of the spring decreases. As a result, the resonant frequency shifts from 96.8 to 94.8 Hz.

As shown in **Figure 22**, as the air gap decreases, the half-power bandwidth gradually increases. When the air gap is 0.5 mm, the bandwidth is 1.8 Hz, and when the air gap is reduced to 0.15 mm, the bandwidth reaches a maximum of 6 Hz.

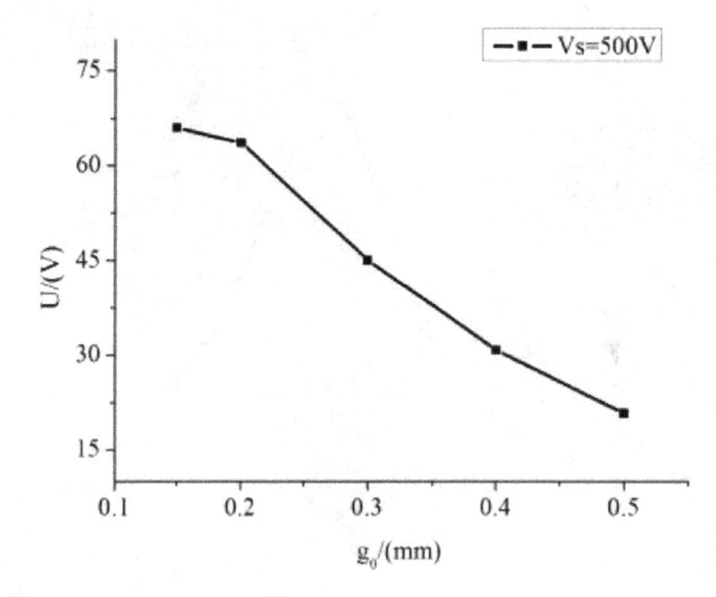

Figure 19. Output voltage with air gap curve.

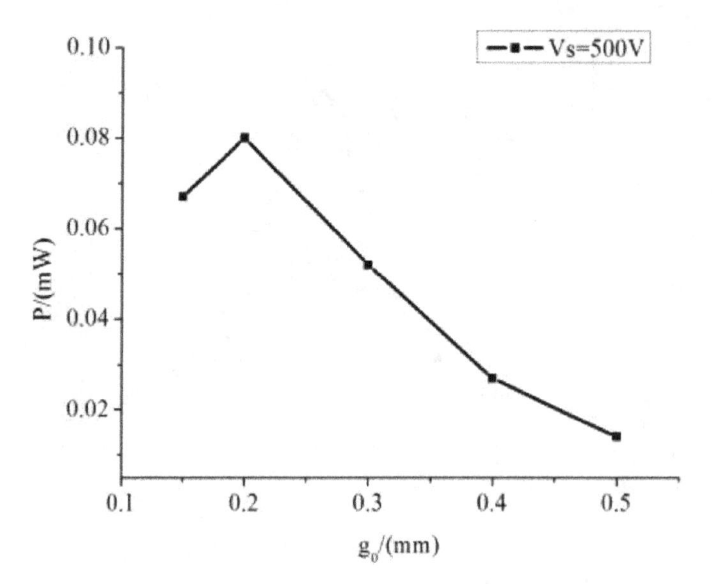

Figure 20. Output power with air gap curve.

3.3. Effect of load on electrostatic harvester output

In addition to the excitation frequency and air gap, the output power of the electrostatic harvester is also related to the external load resistance. The best load test method is as follows:

1. Keeping the acceleration peak of 1.5 m/s^2, the air gap is 0.2 mm.
2. The external load resistance must be connected in series with the oscilloscope during the test.
3. Measuring the output voltage corresponding to different resistances sequentially at the resonance point and plotting the recorded data with Matla (shown in **Figure 23**).

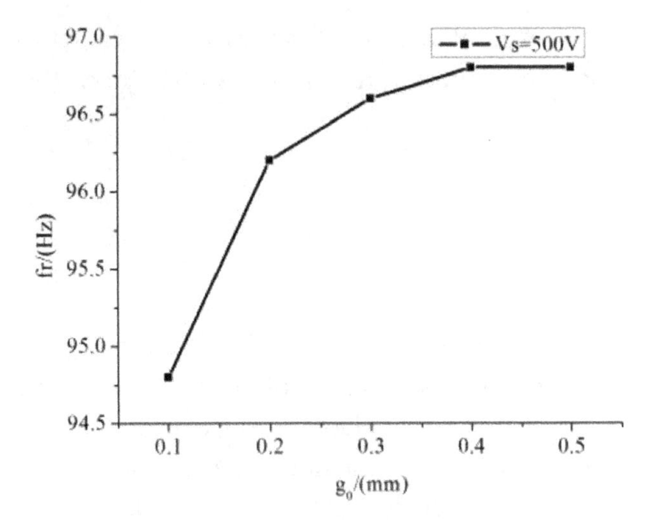

Figure 21. Output power with frequency curve.

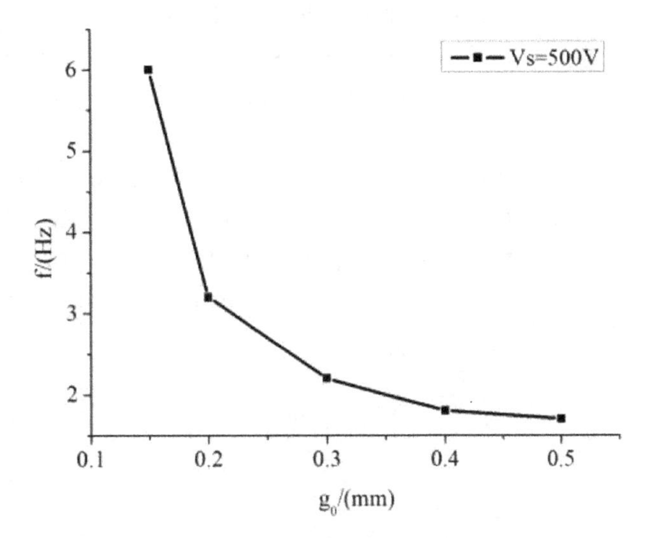

Figure 22. Half power bandwidth with air gap curve.

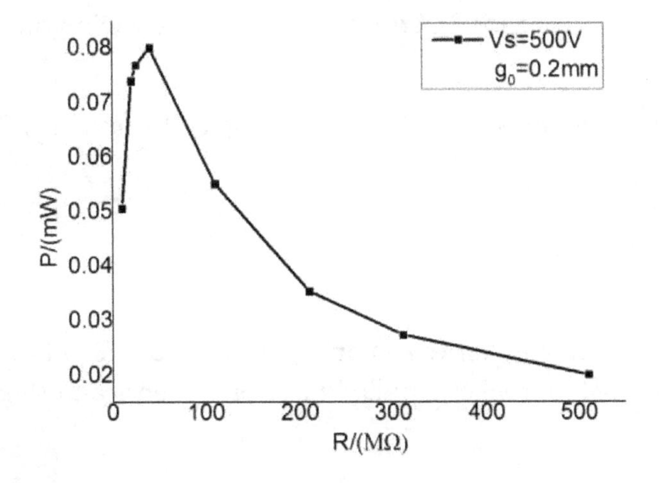

Figure 23. The relationship between load resistance and output power.

As shown in **Figure 23**, as the external load resistance increases, the output power first increases and then decreases. There is a maximum output power of 0.08 mW; the corresponding optimal load is 40 MΩ.

4. Conclusion

Based on the analysis of the basic theory of the electret electrostatic harvester, the basic equations and equivalent analysis model of electret electrostatic harvester are established. Through the experimental tests of electrets electrostatic harvester, it can be seen that the output performance of electret electrostatic harvester is influenced by the parameters, such as excitation frequency, air gap between electrets, and electrode and load resistance and so on. Thus, when designing the electret electrostatic harvester, the influence of some important parameters on the output performance of electrostatic harvester should be considered as much as possible. On the other hand, some microscale effect also should be considered when the MEMS device and the electrostatic energy harvester are integrated design and fabrication. Electrostatic harvesting requires a constant voltage or constant charge condition; it seems that a separate power supply is needed. And it seems contrary to the idea of energy harvesting. However, with the electret material, the material itself can also provide a constant voltage to avoid the use of additional power, which provides an effective way for electrostatic harvesting. Therefore, the electret electrostatic harvesting structure is a kind of ideal energy harvesting method using ambient vibration and can be easily integrated with the MEMS system because of its compatibility with MEMS technology.

Author details

Hai-peng Liu[1]*, Lei Jin[2], Shi-qiao Gao[1] and Shao-hua Niu[2]

*Address all correspondence to: lhp@bit.edu.cn

1 State Key Laboratory of Explosion Science and Technology, Beijing Institute of Technology, Beijing, China

2 School of Mechatronical Engineering, Beijing Institute of Technology, Beijing, China

References

[1] Li P, Gao S, Cai H. Design, fabrication and performances of MEMS piezoelectric energy harvester. International Journal of Applied Electromagnetics and Mechanics [J]. 2015;**47**(1): 125-139

[2] Li P, Gao S, Shi Y, Liu J. Effects of package on performance of MEMS piezoresistive accelerometers [J]. Microsystem Technologies. 2013;**19**:1137-1144

[3] Priya S, Inman DJ. Energy Harvesting Technologies. USA, New York: Springer; 2009

[4] Erturk A, Inman DJ. Piezoelectric Energy Harvesting. U.K.: Wiley; 2011

[5] Ling CS, Dan H, Steve GB. Technological challenges of developing wireless health and usage monitoring systems. Proceedings of SPIE. 2013;**8695**:86950K-869501

[6] Karami MA, Inman DJ. Powering pacemakers from heartbeat vibrations using linear and nonlinear energy harvesters. Applied Physics Letters. 2012;**100**:042901-042904

[7] Marin A, Turner J, Ha DS. Broadband electromagnetic vibration energy harvesting system for powering wireless sensor nodes. Smart Materials and Structures. 2013;**075008**(13pp):22

[8] Farid k. Vibration-based Electromagnetic Energy Harvesters for MEMS Application [R]. CA: The University of British Columbia; 2011:4

[9] Priya DYM. Sand Hills Grams. Energy Harvesting Technology. Nanjing: Southeast University Press; 2011. (in Chinese)

[10] Renwen C. New Ambient Energy Collection Technology. Beijing: National Defense Industry Press; 2011. (in Chinese)

[11] Erturk A, Inman JD. A distributed parameter electromechanical model for cantilevered piezoelectric energy harvesters. Journal of Vibration and Acoustics. 2008;**130**:1257-1261

[12] Roundy S. On the effectiveness of vibration-based energy harvesting. Journal of Intelligent Material Systems and Structures. 2005;**16**:809-823

[13] Sodano HA, Inman DJ, Park G. Comparison of piezoelectric energy harvesting devices for recharging batteries. Journal of Intelligent Material Systems and Structures. 2005;**16**: 799-807

[14] Mateu L, Moll E. Review of energy harvesting techniques and applications for microelectronics. Proceeding of SPIE: Conference on VLSI Circuits and Systems. 2005;**II**: 359-373

[15] Boisseau S, Despesse G, Ricart T, Defay E, Sylvestre A. Cantilever-based electret energy harvesters. IOP Smart Materials and Structures. 2011;**20**:105013

[16] Mitcheson PD, Green TC, Yeatman EM. Power processing circuits for electromagnetic, electrostatic and piezoelectric inertial energy scavengers. Microsystem Technologies. 2007;**13**:1629-1635

[17] Cheng S, Wang N, Arnold DP. Modeling of magnetic vibrational energy harvesters using equivalent circuit representations. Journal of Micromechanics and Microengineering. 2007;**17**:2328-2335

[18] Maurath D, Becker PF, Spreemann D. Efficient energy harvesting with electromagnetic energy transducers using active low-voltage rectification and maximum power point tracking. IEEE Journal of Solid-State Circuits. 2012;**47**(6):1369-1380

[19] Yang Y, Tang L. Equivalent circuit Modeling of piezoelectric energy harvesters. Journal of Intelligent Material Systems and Structures. 2009;**20**:124-136

[20] Chia-Che W, Chen C-S. An electromechanical model for a clamped–clamped beam type piezoelectric transformer. Microsystem Technologies. 2011;**18**:75-80

8

Enzyme Biosensors for Point-of-Care Testing

Chunxiu Liu, Chenghua Xu, Ning Xue, Jian Hai Sun,
Haoyuan Cai, Tong Li, Yuanyuan Liu and Jun Wang

Abstract

Biosensors are devices that integrate a variety of technologies, containing biology, electronics, chemistry, physics, medicine, informatics, and correlated technology. Biosensors act as transducer with a biorecognition element and transform a biochemical reaction on the transducer surface directly into a measurable signal. The biosensors have the advantages of rapid analysis, low cost, and high precision, which are widely used in many fields, such as medical care, disease diagnosis, food detection, environmental monitoring, and fermentation industry. The enzyme biosensors show excellent application value owing to the development of fixed technology and the characteristics of specific identification, which can be combined with point-of-care testing (POCT) technology. POCT technology is attracting more and more attention as a very effective method of clinic detection. We outline the recent advances of biosensors in this chapter, focusing on the principle and classification of enzyme biosensor, immobilization method of biorecognition layers, and fabrication of amperometric biosensors, as well as the applications of POCT. A summary of glucose biosensor development and integrated setups is included. The latest applications of enzyme biosensors in diagnostic applications focusing on POCT of biomarkers in real samples were described.

Keywords: enzyme biosensors, point-of-care testing, classification, immobilization, low cost, fabrication of biosensor

1. Introduction

Biosensor is a delicate analytical device that combines a biologically derived sensitive element with a physical transducer that produces a tip or a continuous digital electrical signal (**Figure 1**) that is proportional to the analyte [1].

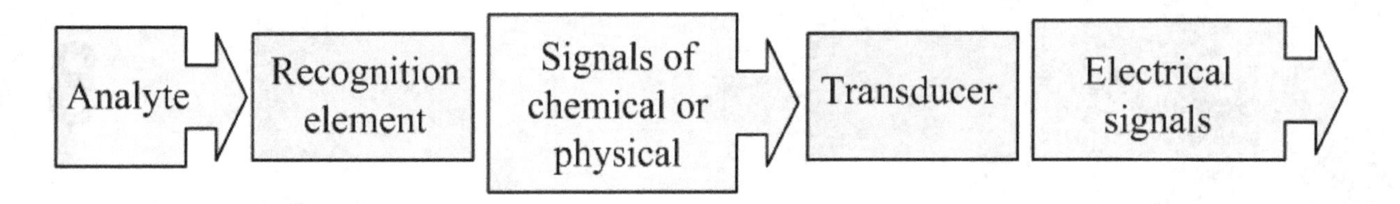

Figure 1. The principle of biosensor.

Biosensor is a new detection technology developed in recent decades. It is an advanced detection method and monitoring method necessary for the development of biological technology. It is also a rapid and micro effective analysis method at the level of material molecules.

Biosensors have become an active research field and demonstrated a very promising prospect, showing important practical value in environmental monitoring, clinical inspection, food and drug analysis, and biochemical analysis, such as rapid and sensitive detection of pathogens and pesticide residue in food and beverage, remote detection of air pollution or heavy metal ion pollution, real-time detection of human blood components and pathogens, long-term monitoring of the health of the human body, as well as the rapid detection of the battlefield weapons, which can be extended to molecular device development, neural network simulation, bionic intelligent devices, and basic research of biological computer [2].

The biosensor has several advantages as the following: (1) the biosensor has good selectivity owing to its high-sensitive molecular recognition element. The biosensor is highly integrated, and the detected components in the sample can be detected directly without the sample pretreatment, and no additional reagents are needed in the determination. (2) The biosensor with small size can be used for continuous monitoring. (3) The biosensor has rapid response and requires only a small amount of sample; as the sensitive material is fixed, it can be used repeatedly. (4) The cost of the sensor and the matched measuring instrument is lower than that of the large analytical instrument, which is very helpful for point-of-care testing (POCT) detection.

Biosensors are devices that are sensitive to biological substances and convert their concentrations to electrical signals. Biosensors act as analytical tools including biologically sensitive material immobilized as recognition element (**Figure 2A**) (including enzyme, antibody, antigen, microorganism, cell, tissue, nucleic acid, and other biologically active substances), physicochemical transducer (**Figure 2B**) (such as the electrochemical electrode, photodiodes, FET, piezoelectric, etc.), and signal amplifying device [1–3]. It has been widely used in biological medicine, drug development and testing, environmental quality testing, and other fields.

The development of enzyme electrode is the most representative and the most studied in biosensor field. Electrochemical biosensors are the largest group of biosensors, which have the earliest development and the most abundant research content. A large number of research achievements have been achieved, and some of them have been widely used [4, 5].

Electrochemical enzyme biosensors have unique advantages in improving selectivity and sensitivity. The microstructure on the surface of the electrochemical biosensor can provide a variety of potential fields that can effectively separate and enrich the analytes, which can

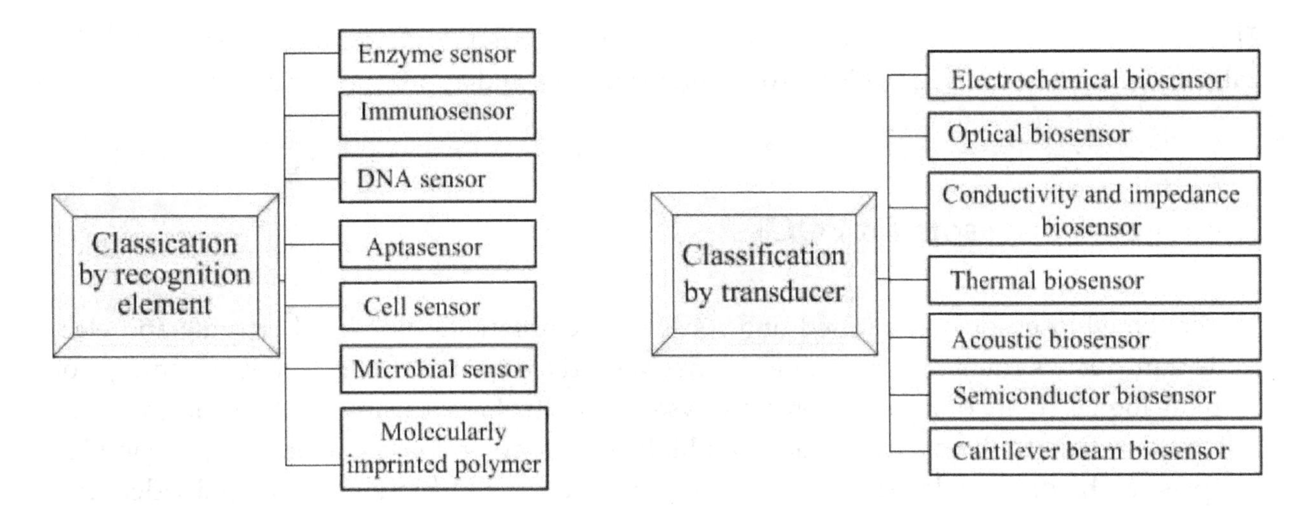

Figure 2. Classification of biosensors.

control the electrode potential, further improve the selectivity, and also combine the sensitivity of the determination method with the selectivity of the chemical reaction of the surface material. The electrochemical enzyme sensor is an ideal system for the integration of separation, enrichment, and selective determination. Among them, the development of amperometric enzyme biosensor is the most representative and the most studied in biosensor field [6, 7]. Biosensors are developing to the following aspects in recent years. (1) Biosensor with low cost, high sensitivity, high stability, and long life can meet the needs of field or on-site sampling and measurement. With the development of biosensor technology, it is necessary to reduce the cost of products and improve the sensitivity, stability, and life. The improvement of these characteristics will accelerate the commercialization and commercialization of biosensors. (2) With the development of microfabrication technology and nanotechnology, biosensors based on MEMS technology has been received a rapid development [7], and biosensors have been further miniaturized. The emergence and application of various portable biosensors will enable people to diagnose diseases at home, so they can directly detect food in the market, which meet the requirements of testing in vivo intracellular detection, online detection, and so on.

POCT technology has emerged and progressed with the development of computers, biosensors, electronics, and medicine. As a new method of clinical detection, POCT is getting more and more attention. POCT test equipment is portable, easy to operate, and suitable for non-laboratory professionals. Therefore, POCT is fast, sensitive, and free from site conditions. It plays an increasingly important role in medical field, environmental monitoring, and safety monitoring. The advantage of POCT is the immediate determination of the whole blood specimen, without the anticoagulant steps, without specimen preparation, a small amount of specimens, the short specimen turnaround time (TAT), the miniaturized instrument, and the patient's bedside test. The central laboratory test model of multistep and multi-staff was avoided. At present, POCT can include blood gas, electrolytes, blood sugar, renal function, liver function, hemoglobin, and cardiac markers, such as creatine kinase (CK), myoglobin (MYO), troponin (TNI, TNT), and so on. The POCT test project has already met some needs of emergency patients. And, the new POCT detection program is coming out. This shows that the real-time detection result of POCT has opened up a new prospect for clinical testing.

This chapter discusses several topics related to enzyme-based biosensors, including immobilization of recognition element, biosensor miniaturization, and application of biosensors.

2. Enzyme biosensors for POCT

The enzyme biosensor is composed of a sensitive membrane-immobilized enzyme and electrode transducer system, which combined enzyme and electrode together. It has advantages of both the high stability of insoluble enzyme system and the high sensitivity of the electrochemical system. Due to the specificity and the high selectivity of enzyme reaction, the complex samples can be directly determined by the sensor. The enzyme biosensor can be divided into two kinds of electric potential sensor and amperometric sensor. The development of the amperometric enzyme electrode is more widely [6]. The amperometric enzyme electrode is defined as an output of a corresponding current signal was generated by the targets reacting or redox reaction on the electrode, which has a linear relationship with the concentration of the tested substance under certain conditions. The basic biosensor mainly uses oxygen electrode and hydrogen peroxide electrode.

2.1. The enzyme immobilization method

2.1.1. Definition and characteristics of biosensor immobilization technology

Enzyme immobilization technology is of great importance in the development of biosensors. The main purpose of immobilization is to restrict the enzyme and other biologically sensitive elements to a certain space, but it does not interfere with the free diffusion of the analytes. It is one of the key factors that affect the stability (or life span), the sensitivity, and selectivity of the biosensor. The immobilization method can be divided into two kinds: direct method and indirect method. The direct method is the surface of the probe which is directly fixed to the transducer by physical and chemical modification. The direct method is the surface of the probe which is directly immobilized by the biomolecules on the transducer using the physical and chemical modification. This method is to integrate sensitive materials and probes, which is helpful to improve response performance. It is the main research direction of the commercialization of the portable enzyme sensors. The indirect rule is to fix the biological component on a carrier first and then install it on the probe of the sensor. As the sensitive part is independent of the converter, it is conducive to prolonging the service life of the converter and is suitable for long time monitoring. Indirect immobilization has become the main direction of the commercialization of enzyme sensors in the process of online analysis and process control. Compared with the free phase of biological materials, solid biological material has a series of advantages, for example, the thermal stability, the repeated use, no need to be separated and reactive substances in catalytic material after reaction, and the determination of the film life according to the known half-life. The immobilization method plays a decisive role in the performance and usage of the biosensors.

2.1.2. *Classification of biosensor immobilization methods*

The commonly used immobilization methods of biosensors include adsorption, embedding, covalent bonding, cross-linking, and electrochemical polymerization (**Figure 3**).

(1) Physical adsorption

The biosensor element is immobilized by the physical adsorption or ion binding of non-water-soluble carrier, which is defined as the adsorption method.

These binding forces may be hydrogen bonds, van Edward forces, or ionic bonds and may also play a role in a variety of bonding forms. There are a wide variety of carriers, such as active carbon, hydroxyl limestone, aluminum powder, gold, chitosan, cellulose, and ion exchangers. The firmness of the adsorption is related to the pH, ionic strength, temperature, properties and types of solvent, and the concentration of the enzyme. The enzyme is directly adsorbed on the surface of the electrode. The method is simple, and the enzyme activity is seldom degraded. However, the stability of the enzyme is not good, and this technology is not widely used at present. The more distinctive work is the co-deposition of enzyme and platinum/palladium; therefore, the enzyme is adsorbed during the growth of platinum black or palladium black particles. Platinum black holes have larger surface area, and their affinity for protein can keep the stability of the enzyme [8].

In addition, the adsorption method combined with nanomaterials can also help to stabilize the enzyme activity and improve the life span. The adsorption process is mainly used to prepare enzymes and immune membranes. The adsorption process usually does not require chemical reagents, and it has little effect on the activity of protein molecules. Because the protein molecules are easy to fall off, especially when environmental conditions changed, so they are often used by combining with other immobilization methods.

(2) Embedding method

Embedding enzyme molecules or antibody and immobilizing them in the three-dimensional spatial grid matrix of polymers are defined as the embedding method. The process is generally not necessary for binding reaction with the residues of biological material and rarely changing

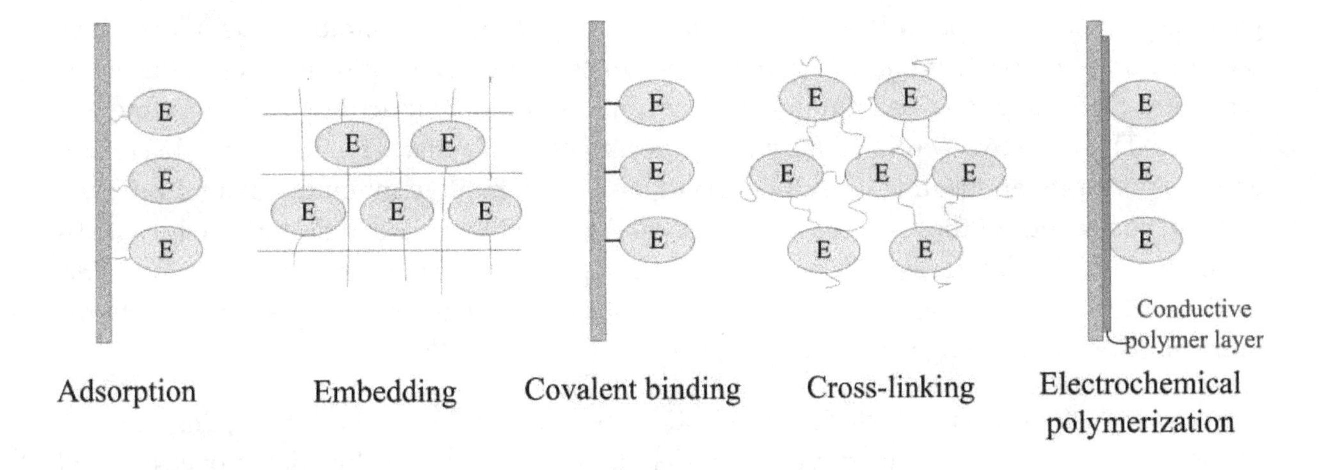

Figure 3. Schematic diagram of enzyme immobilization.

advanced structure of biological active substances; therefore, the loss of biological activity is very small, the pore diameter can be controlled arbitrarily, the encapsulated material is not easy to leak, and the substrate molecules can diffuse freely in the film [9, 10]. The traditional method of encapsulation is to bury the enzyme in the gel or membrane and then immobilize it to the electrode. The main gels are polyacrylic acid amine, polyvinyl chloride (PVC), polyvinyl alcohol (PVA), photosensitive resin, polycarbonate, nylon, acetic acid fiber, and other synthetic polymers and alginate, gelatin, collagen, and other colloidal polymers. This method is generally suitable for the bioactive substances acting on small molecular substrates and products, while the bioactive substances acting on macromolecular substrates and products are too large for mass transfer resistance. The enzyme is easy to leak from the sensing layer, but this method is more suitable for the encapsulation of organelles, lipids, and microbes.

Synthetic polymers, such as Nafion, can be directly embedded on the surface of the enzyme electrode. These films can embed biological molecules, anti-interference and anti-poison and are not easy to leak out, which is suitable for acting as encapsulation and immobilization material. Carbon paste embedding is a method of immobilization of enzyme directly with electrode material (carbon paste and carbon epoxy resin). It is a very common method to prepare biosensors. The carbon paste electrode can not only immobilize enzymes but also immobilize cofactors NAD, media, stabilizers, cells, and tissues. In order to eliminate interference and increase sensitivity, it can also immobilize two enzymes or enzymes in carbon paste.

Photopolymerization has recently been introduced into the gel of immobilized enzyme. The enzyme is dissolved in the mixture solution of polymer monomer and the photoinducer. It can form redox polymer hydrogel-immobilized enzyme when irradiated under the ultraviolet light conditions [10]. These light-induced polymers include polyethylene glycol (PEG), acrylic acid (acrylic acid), vinyl ferrocene, and other monomers.

(3) Covalent bond method

The method of binding the bioactive molecule through the covalent bond to the insoluble carrier is defined as the covalent bond.

The carrier includes the inorganic carrier and the organic carrier. Organic carriers are cellulose and its derivatives, dextran, agar powder, and so on; the inorganic carriers are mainly porous glass, graphite, and so on. It is particularly important to protect the amino acids of the active center during the covalent immobilization of enzymes. If the amino acid residues of the active center are chemically modified during covalent immobilization, the activity of the enzyme will decrease. There are two ways to effectively protect the active center. The first method is to add enzyme substrate or substrate structure analogues or competitive inhibitors before the covalent reaction to protect the active center. The second method is the use of bifunctional reagents, which have a chemoselectivity and a biologically specific source of covalent reactions. This dual selectivity can not only protect the active sites of protein but also realize directional immobilization of the enzyme molecules [11, 12].

The covalent bonding forms include diazotization method, peptide method, alkylating method, amino reaction, etc. The characteristics of covalent bond are strong combination, not easy to fall off, not easy to be biodegraded, and long life for service. The disadvantage of covalent bond is that the operation procedure is tedious, the enzyme activity may be reduced by chemical modification, and it is difficult to prepare high active immobilized enzyme. Covalent bonding is mostly used for production of enzyme membrane and immune molecular membrane. It usually requires operation at low temperature (0 degrees), low ionic strength, and physiological pH.

(4) Chemical cross-linking method

Cross-linking is a method of using bifunctional or multifunctional reagents, such as glutaraldehyde, to bind the bioactive substances to the transducer directly. The functional groups involved in the coupling of bioactive substances are -NH_3, -COOH, -SH, -OH, imidazolyl, phenol base, etc., but these groups cannot be active center groups. Most transducers have to be pretreated before cross-linking to produce their surfaces for coupling groups. The pretreatment can be treated by electrochemical method and plasma treatment. The cross-linking and immobilization method combined with supramolecular self-assembled monolayers is to assemble thiols on the surface of gold electrodes and then immobilize enzymes on self-assembled membranes with bifunctional reagents [13, 14].

This method gives the film a specific function of highly ordered and complete structure, good stability through molecular design and is the highest form of chemically modified electrodes. At present, a fast and sensitive biosensor can be prepared. Cross-linking is widely used in the preparation of enzyme membrane and immune molecular membrane. The operation is simple and the combination is firm. When the enzyme source is more difficult, it is often necessary to add inert protein which is times the enzyme as substrate. The problem of this method is that pH must be strictly controlled during immobilization. It is usually operated near the isoelectric point of protein. The concentration of cross-linking agent should also be carefully adjusted, such as glutaraldehyde itself can cause protein poisoning, usually under 2.5% (volume fraction) concentration. In the cross-linking reaction, the enzyme molecules are inevitably partially inactivated.

(5) Electrochemical polymerization

Electrochemical polymerization is based on electrostatic interaction. In the process of electropolymerization, enzymes act as anions and interact with positively charged polymer skeletons, and enzyme is added to the polymer membrane and directly immobilized on the electrode surface [15].

The interaction between polymer, metal, and carbon conductor lays a foundation for immobilization of enzyme on the electrode surface, thereby improving the communication efficiency between oxidation–reduction center and electrode. For conductive polymer membrane, this method can not only control the thickness of the membrane but also control the density of the membrane. But the amount of enzyme is relatively small, the background current is large, and a large amount of enzymes are wasted in the solution of electropolymerization. It is also useful for nonconducting polymer membrane-immobilized enzymes, which are limited in thickness. The oxidoreductase can be doped into polypyrrole, polyaniline, and polyphenol membrane [16, 17].

2.2. Amperometric biosensor for glucose detection using POCT

The amperometric enzyme biosensor is the most representative biosensor. The development of the amperometric enzyme electrode mainly consists of three stages [18, 19]. The classic enzyme electrode is represented by the Clark enzyme electrode as the first-generation electrochemical sensor (**Figure 4A**). The mediated enzyme electrode is the second-generation electrochemical sensor (**Figure 4B**), which solves the interference of oxygen and the interference of electrode active substances and overcomes the problem of high working potential of electrodes, which are often used for POCT [20–23]. The direct electrochemical enzyme electrode is the third-generation electrochemical sensor (**Figure 4C**), which is mainly used to solve the problem of low-efficiency transmission between enzymes and other biometric elements and electrodes.

2.2.1. The first-generation enzyme electrode (Clark enzyme electrode)

The first generation of enzyme electrodes was first proposed by Clark in 1962, and it was based on oxygen reduction and used glucose oxidase (GOD) as an example to catalyze glucose [22, 24].

The oxidation–reduction reaction occurs on the enzyme layer:

$$\text{GOD (ox)} + \text{glucose} \rightarrow \text{gluconolactone} + \text{GOD (red)} \tag{1}$$

$$\text{GOD (red)} + O_2 \rightarrow \text{GOD (ox)} + H_2O_2 \tag{2}$$

The measurements of peroxide are carried out on the electrode at a moderate anodic potential of +0.6 V (vs. Ag/AgCl):

$$H_2O_2 \rightarrow O_2 + 2H_2 + 2e \tag{3}$$

The most used oxygen electrodes are electrolyzed Clark oxygen electrodes, which are made up of platinum cathode, Ag/AgCl anode, KCl electrolyte, and air permeable membrane. The overpotential of H_2O_2 on both metal and carbon electrodes is higher, generally from +0.6 to +0.8 V (vs. Ag/AgCl). The high detection potential can interfere with the electroactive substances such as antiblood acid and uric acid in samples. Various selective osmotic membranes can be used to remove interfering substances or chemically modified electrodes to reduce the overpotential, but their sensitivity is always limited by the concentration of dissolved oxygen in the system.

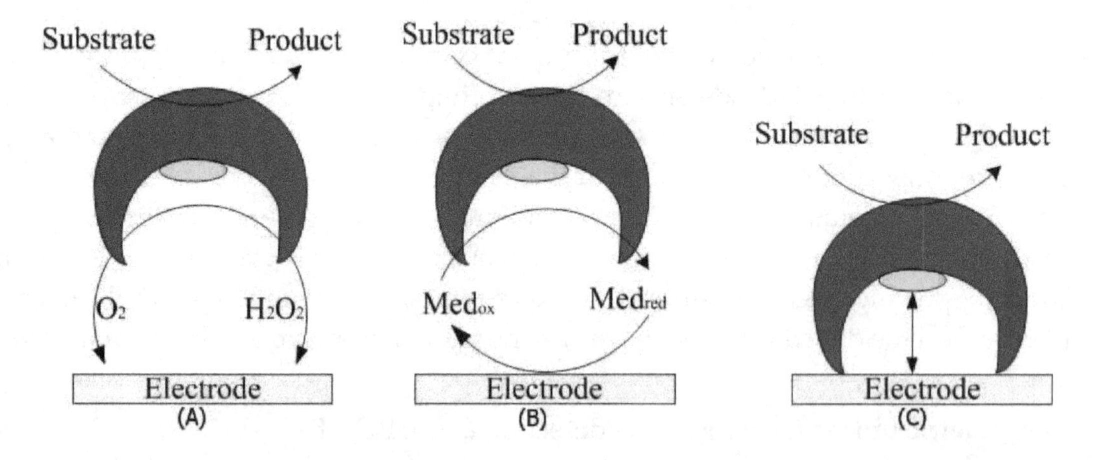

Figure 4. Electron transfer comparison diagram of the three generation amperometric enzyme electrodes, which are (A) first-generation enzyme electrode (Clark enzyme electrode), (B) second-generation enzyme electrode (mediated enzyme electrode), and (C) third-generation enzyme electrode (direct enzyme electrode).

2.2.2. *The second-generation enzyme electrode (mediated enzyme electrode)*

In order to overcome the shortcomings of the first generation of enzyme electrodes, the second generation of enzyme electrodes (mediated enzyme electrode) is widely used nowadays, which is the use of artificial electronic media to solve the problem of transferring electrons. The chemical modification layer was added at the second generation of enzyme electrodes. The chemically modified layers on the electrode are used to expand the range of chemical substances and also to improve the sensitivity of the determination. The modified substrate electrode can be regarded as an improved signal converter, and the modifier becomes an electron transfer medium. The role of electronic media is to promote the electron transfer process, reduce the working potential, and eliminate the interference of other electroactive substances.

In 1984, Cass established a mediated enzyme electrode method [23]. The chemical mediated ferrocene substituted the molecular oxygen as the electron acceptor for the enzymatic reaction and immobilized on the graphite electrode with the glucose oxidase (GOD). This research contributed to the successful development of a printed enzyme electrode in the US MediSense company in 1987. However, due to the difference between the reduced state of the medium and the oxidation state, the leakage occurred, which leads to the gradual decrease of the response activity of the electrode with time.

Dong used Nafion (a cationic resin) to fix ferrocene and then fixed the GOD to the Nafion-ferrocene electrode. Because of the hydrophilic structure and hydrophobic structure of Nafion, Fc/Fc + is retained in the membrane. Therefore, the service life of electrodes is greatly prolonged, and the stability of the mediated enzyme electrode is improved [25, 26]. In commercial products, in order to avoid the problems of dielectric loss and electrode calibration, many enzyme electrodes have been designed to be disposable.

Electron mediator refers to the transfer of electrons generated from the enzyme reaction process to the surface of the electrode, so that the electrodes can generate corresponding molecular electric conductors. The most basic requirement for mediator is low redox potential and high electrochemical reaction rate, and the reaction is reversible [27]. Cyclic voltammetry is usually used to investigate the electrochemical properties of the mediator.

The electron transfer mechanism of glucose oxidase catalyzed by glucose is acted as an example and described as the following:

$$\mathrm{GOD(ox) + glucose \rightarrow gluconolactone + GOD\ (red)} \tag{4}$$

$$\mathrm{GOD\ (red) + M\ (ox) \rightarrow GOD(ox) + M\ (red)} \tag{5}$$

$$\mathrm{M\ (red) \rightarrow M\ (ox) + nH^{+} + ne^{-}} \tag{6}$$

Media has been developing rapidly in the past 10 years, the species is also increasing, and at present, the electron transfer mediator according to the fixed method is divided into three types [28, 29] (1) the simple electron transfer medium, (2) the electron transfer mediator covalent bond with enzyme, and (3) the electron transfer medium covalent bond with the polymer.

(1) Simple electron transfer medium

A simple electron transfer medium refers to the existence of an electron transfer medium in the form of a monomer in the application, including the oxidation–reduction pair and the organic conductor salt. Ferrocene and its derivatives [30–33], and quinone and its derivatives [34], are widely used as electron transfer mediators. The glucose biosensor was constructed by Lange [30], using a glassy carbon electrode as the substrate, and immobilized on the electrode with glucose oxidase, ferrocene, and polyamide gel. The sensor response principle is that GOD (FAD) catalyzes the oxidation of glucose and itself is reduced to GOD($FADH_2$), the GOD ($FADH_2$) was oxidized by ferrate into GOD(FAD), and reduced ferrocene was oxidized into high iron state on electrode. The oxidation current produced by the oxidation of two ferrocenes on the electrode is related to the concentration of glucose. Two ferrocenes here have the electron transfer between the GOD($FADH_2$) and the electrode.

However, the simple electron transfer mediator acts as effective electron transfer between enzyme redox active center and electrode, overcomes the limitations of O_2 or H_2O electrode in direct electrochemical reduction or oxidation, and greatly improves the biosensor performance in response to speed, detection sensitivity, and anti-interference ability. However, simple electron transfer mediators have encountered some difficulties in practical applications, such as dissolution or partial dissolution of mediators, diffusion of mediators away from electrode surface, and so on. Because of the loss of mediators, the stability and service life of biosensors are affected, which limits the clinical application of such biosensors as subcutaneous probes and for long time online analysis. However, the simple electron transfer medium has an irreplaceable advantage; that is, the electron transfer speed is fast.

(2) Electron transfer mediator covalent bond with enzyme

In order to prevent the loss of the electron transfer medium from the electrode surface due to dissolving and other reasons, the electron transfer medium can be chemically bonded to the enzyme. Bonding can not only better fix the electron transfer mediator but also make the redox active center of the enzyme more closely contact with the electron transfer mediator, which is more conducive to the electron transfer [35–40]. For example, Degani and Heller [41, 42] have successfully made glucose biosensors, which are chemically bonded to glucose oxidase by electron transfer mediator ferrocene derivative and ammonia ruthenium. The electron transfer mediator is chemically bonded with enzyme, which successfully prevents the loss of electron transport mediator, improves the stability of biosensor, and prolongs the service life. However, the catalytic activity of the enzyme is reduced after chemical modification, which will also reduce the sensitivity of the biosensor [43].

(3) The electron transfer medium covalent bond with the polymer

In recent years, great progress has been made in the research of bonding of electron transfer mediator and suitable polymer in order to prevent the loss of electron mediator and overcome the shortcomings of reducing enzyme activity after mediator and enzyme binding. A more typical example is the bonding of ferrocene and its derivatives to a siloxane polymer. A new type of ferrocene-modified siloxane polymer was synthesized by Hale [44]. The glutamate biosensor was prepared by combining this polymer with glutamate oxidase. Niwa [45] et al.

made a complex sensor of horseradish peroxidase and glutamate oxidase based on the complexes of polypyridine and osmium. The sensor has the characteristics of fast response, strong anti-interference ability, large current response, and high stability. The electrochemical behavior of the cations, such as Os(bpy) $_2$PVPCl, is also proposed by Han [46]. In addition to the electron transfer mediator bonded to the siloxane polymer with ethylene oxide, silicon can also be bonded to the polymer polypyrrole and polypyridine. The electron transfer mediator covalently bonded to a polymer medium, insoluble in water; it is not easy to spread out the surface of the electrode. The biosensor made of it avoids the pollution of the sample solution and improves the stability of biosensor by prolonging the life of biosensor.

2.2.3. The third-generation enzyme electrode (direct enzyme electrode)

The biosensor performance of the third-generation enzyme electrode is realized by the direct electrocatalysis of the enzyme on the electrode. It is named as a non-reagent sensor. It takes advantage of the direct electron transfer between enzyme and electrode and does not need to add other media reagents to reduce operation steps. It is a real reagent-free biosensor. Glucose oxidase is used as an example to illustrate the reaction mechanism on enzyme film and electrode separately:

$$\text{GOD(ox)} + \text{glucose} \rightarrow \text{gluconolactone} + \text{GOD (red)} \tag{7}$$

$$\text{GOD (red)} \rightarrow \text{GOD (ox)} + ne \tag{8}$$

The implementation of the enzyme's direct electrochemistry is of great significance for the development of a non-media enzyme sensor. Because the enzyme molecule belongs to protein and has large molecular weight, the active center of enzyme molecule is deeply embedded in the interior of the molecule, and it is easy to deform after adsorption on the electrode surface. Therefore, it is difficult to directly transfer electrons between enzyme and electrode. Theoretically, the direct electron transfer process between enzyme and electrode is closer to the original model of biological redox system, which lays the foundation for revealing the mechanism of biological redox process. Therefore, although the study of this direction is very difficult, it is still the direction of development in this field. In addition, the combination of nano-modification technology, biological simulation, and molecular imprinting technology can provide power for the development of direct electrochemical measurement [47].

2.3. Fabrication of amperometric enzyme biosensor and POCT detection

POCT detector is usually combined with biosensors. A biosensor coupled to a specific biological detector (such as enzyme, antibody, or nucleic acid probe) to a transducer is used for direct determination of target analytes without separation from matrix. It embodies the combination of enzyme chemistry, immunochemistry, electrochemistry, and computer technology. It can be used to carry out ultramicroanalysis of the analyte in the organism's fluid.

At present, more and more factories in the world are developing POCT devices [48]. The main products are hematology analyzer, electrolysis analyzer, and blood gas analyzer, and the analytical performance of some selected commercial POCT devices was shown in **Table 1**.

Manufacturer	System	Target analytes	Sample volume (μL)
Bayer	RAPIDLab 800	pO_2, pCO_2, pH, electrolytes, glucose, lactate	140–175
Nova Biomedical	Xpress, Nova 16	Blood gases, electrolytes, glucose, lactate, urea nitrogen, creatinine, hematocrit	85–190
Radiometer	ABL 725 ABL 77	pO_2, pCO_2, pH, Na^+, K^+, Ca^{2+}, glucose, lactate	80–135
Instrumentation Laboratory	Synthesis 1745 GEM Premier 3000	pO_2, pCO_2, pH, Na^+, K^+, Ca^{2+}, Hct, glucose, lactate	135
Abbott Diagnostics	i-STAT	pO_2, pCO_2, pH, electrolytes, Hct, urea nitrogen, glucose, lactate, creatinine	65–95
Agilent Technologies	IRMA	pO_2, pCO_2, pH, electrolytes, Hct, glucose, lactate	125
Roche	OMNI 9	pO_2, pCO_2, pH, electrolytes, Hct, Hb, urea nitrogen, glucose, lactate, creatinine	40–161
Yellow Springs Instruments	2300 Stat Plus	Glucose, lactate	120

Electrolytes: Na^+, K^+, Ca^{2+}, Cl^-.

Table 1. Electrochemical sensors in commercial systems for critical care and POCT.

There are two kinds of handheld and portable products. Handheld products (**Figures 5, 6**) are small in size, such as mobile phones, automatic calibration, and less blood collection. Analysis items can be preset, records can be stored, maintenance is free, and quality control procedures are completed in 1 min. The practical significance of such products is that they can be simplified and routinely used for blood analysis, such as body temperature and blood pressure, under the premise of clinical diagnosis and treatment.

Blood glucose monitoring is the most common application for enzyme biosensor. The design and structure of a disposable electrochemical enzyme sensor, as well as the use of the media, are determined by the concentration and sampling method of the analytes. Cost and convenience must be taken into account in structural design. Therefore, the most commonly used single parameter amperometric enzyme sensor is a double-electrode system or a tri-electrode system.

2.3.1. *The fabrication of disposable biosensor with dual electrodes modified by ferrocene*

The electrode arrays (**Figure 6**) were fabricated with the vacuum sputtering technology through sputtering gold on polycarbonate (PC) plastic substrate material. The two-electrode system was made up of a working electrode (1×2 mm) and a pseudo-reference/counter electrode (1×1 mm). The channel region where reaction took place and bonding pads were defined by two-piece insulating plastic layer. The volume of capillary-fill region channel was 3 μL. The gold working and counter electrodes were deposited to form nanoporous platinum layer. The electrodes were then activated by the method of plasma cleaning, and significant improvements of hydrophilicity and electroactivity were achieved by the simple treating. 3 μL of the reagent (containing enzyme, potassium ferricyanide, BSA, and 0.1% triton) was added

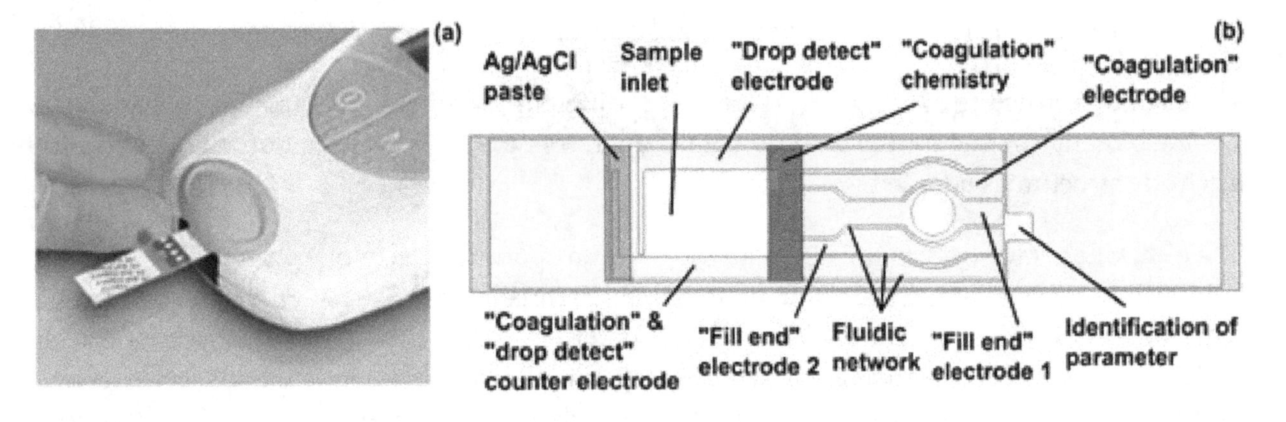

Figure 5. The biosensor and handheld POCT device for blood coagulation test. (a) Handheld POCT meter. (b) the disposable biosensor for blood coagulation with the design of the microfluidic structure and four electrodes by electrochemical detection (from [49, 50]).

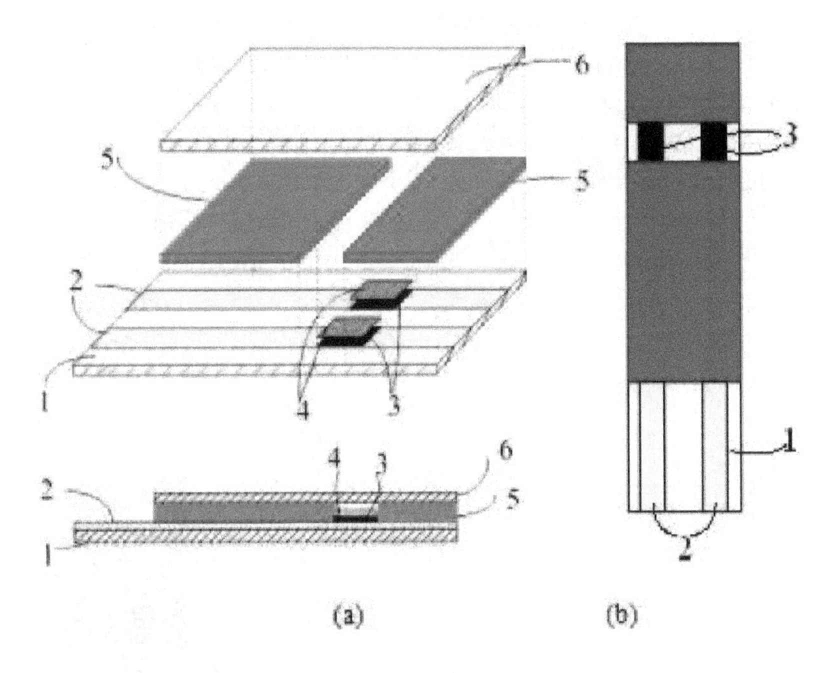

Figure 6. Schematic diagram of disposable biosensor with dual electrodes. Three-dimensional structure schematic diagram. (b) Planar graph of the biosensor. (1) The plastic substrate, (2) lead, (3) the reaction region modified by Pt-black, (4) dry reagent, (5) insulating double-sided adhesive, and (6) the cover layer (from [51]).

on the channel region of the electrodes and dried at 25°C. After the reagent was immobilized on the capillary-fill channel, a covering sheet was coated on the insulation layer to prevent the electrode from staining. The dry strip/biosensor was stored at room temperature. During the experiments, a potential of 0.2 V vs. reference/counter electrode was employed. 3 μL serum doses were added on the biosensor through capillary effect. The whole blood samples were determinated with biosensors combined with handheld test meter.

The two-electrode system biosensor is suitable for detection of analytes with mM concentration, which is simple in preparation, simple in driving system, and low in cost. Many of the blood glucose and lactate biosensors in commercial systems adopt the two-electrode design.

2.3.2. The fabrication of biosensor with tri-electrodes modified by Os polymer mediator

The electrode structure of the sensor (**Figure 7**) is different for different parameters and analytes. For the biosensors used for the detection of the analytes with low concentration, the classical tri-electrode structure is adopted.

The electrode arrays were fabricated with the vacuum sputtering technology through sputtering gold on polycarbonate (PC) substrate material. Each three-electrode system was made up of a working electrode with surface of 3 mm^2, a counter electrode, and a reference electrode with surface of 6 mm^2. The channel region where reaction took place and bonding pads were defined by two-piece insulating plastic layer. Nanoporous platinum layer was deposited on the gold working and counter electrodes. The gold electrodes were first cleaned with plasma cleaner and then were electroplated in a solution containing H_2PtCl_6 (hexachloroplatinate); then, the electrodes were washed with distilled water and dried in desiccator. The Ag|AgCl layer was electrochemically formed on the reference electrode. On the surface of the freshly prepared silver electrode, silver chloride coating was formed by anodizing for 30 s in 0.1 M HCl solution using 10 mA/cm^2 current density. The electrodes were then activated by the method of plasma cleaning, and significant improvements of hydrophilicity and electroactivity were achieved. Then 1 μL PVP-Os-HRP redox polymer was added on the surface of the working electrodes, and the polymer was allowed to dry overnight under ambient conditions. 3 μL of the enzyme reagent was added on the channel region of the electrodes and dried at 25°C. After the reagent was immobilized on the capillary-fill channel, a covering sheet was coated on the insulation layer to prevent the biosensors from staining. Finally, the electrode arrays were cut in pieces, and dry biosensors were stored at 4°C.

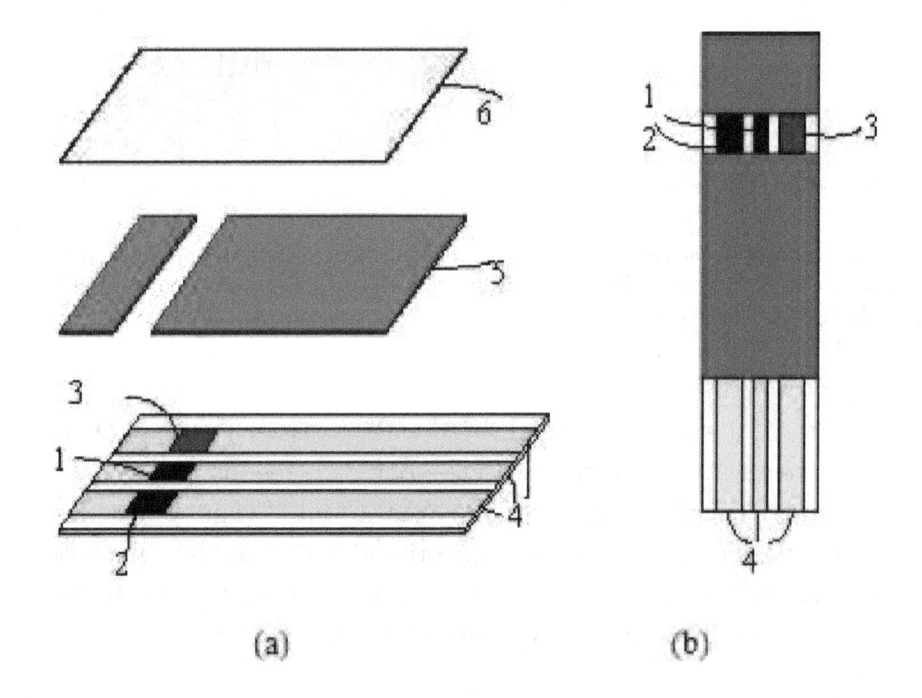

Figure 7. Schematic diagram of disposable biosensor with tri-electrodes. (a) Three-dimensional structure schematic diagram. (b) Planar graph of the biosensor. (1) Working electrode, (2) counter electrode, (3) Ag|AgCl reference electrode, (4) lead line, (5) insulating double-sided adhesive, and (6) the cover layer (from [52, 53]).

During the amperometric experiments, a potential of 0.1 V vs. integrated Ag|AgCl reference electrode was employed. The dry reagent biosensor combined with handheld meter was used to determine analytes in blood. The osmium redox polymer medium covalently combined on the electrode reduces the working potential of the sensor and reduces the interference of other active substances.

The disposable biosensors have the advantages of easy handling, quick test in several minutes, friendly to untrained users. In order to realize multiparameter detection and more accurate measurement, more complex biosensors are designed and developed. The biosensors designed by i-STAT (**Figure 8**) have prestorage of two kinds of reagents (dry and liquid), using the airbag driving mode, which further improves the precision and linearity of the detection.

2.4. Continuous real-time in vivo monitoring

Physiological diseases and various physiological activities of human are often identified as biochemical substances. So far, the determination of these identification substances is mainly in vitro. In general, in vitro determination can satisfy the diagnosis of the disease and the basic judgment of some physiological states [55]. However, the monitoring and treatment of some sudden and severe diseases are best to be monitored in real time. Therefore, the measurement

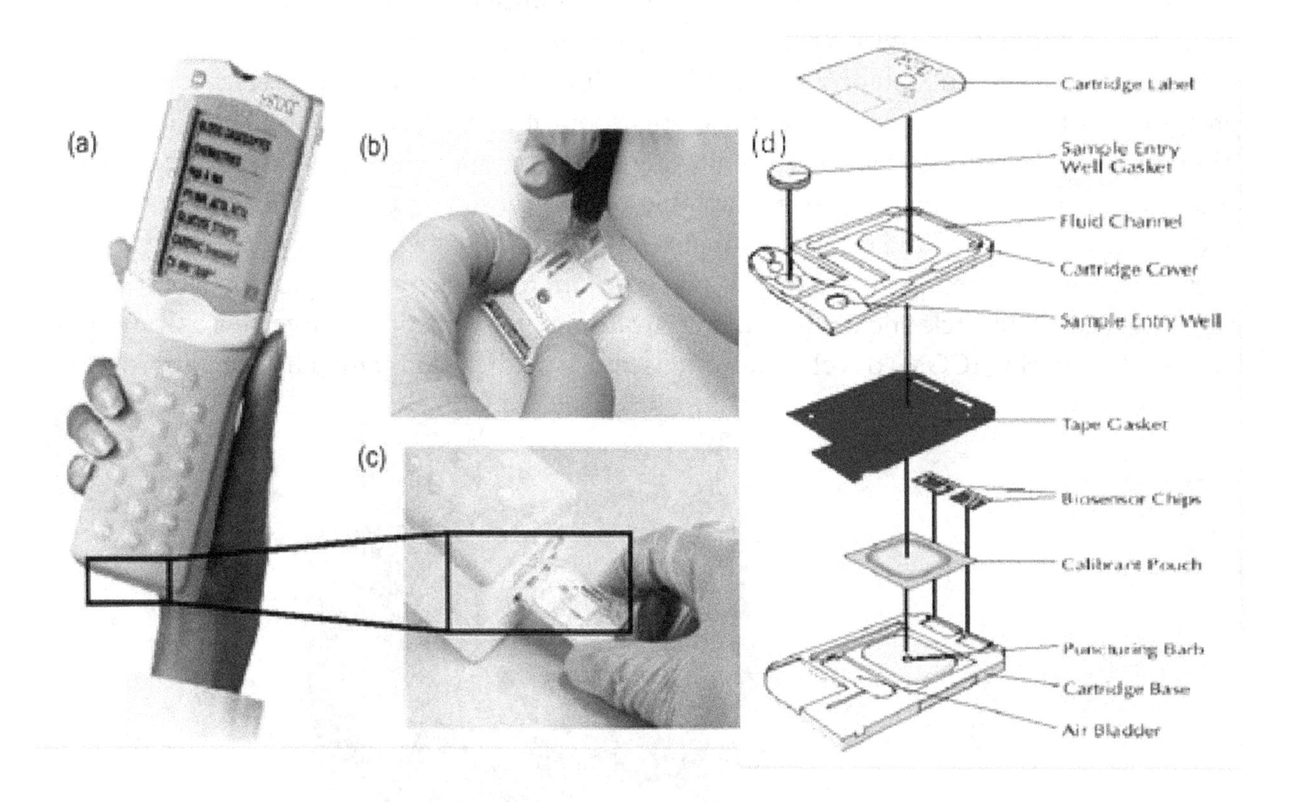

Figure 8. The integrated biosensor and handheld reader of i-STAT analyzer. (a) the portable i-STAT analyzer. (b) the disposable biosensor with integrated structure; 65–95 μL whole blood sample is filled by capillary forces and pressure force. (c) the biosensor is loaded into the meter, and the samples is pushed by pressure and capillary forces for electrochemical detection. (d) the fabrication of the biosensor (from [50, 54]).

of in vivo in the biosensor has its special significance both in the clinical monitoring of the patients and in the basic physiological research [56–58].

The internal environment is very complex, not as easy to control as in vitro environment. After implantation of biosensor, it is important not only to ensure normal performance parameters of biosensors but also to ensure that implanted sensor systems cannot pose any danger to patients or subjects. Therefore, we should not only face some special technical problems but also follow strict regulations on medical instrument management and obtain experimental licenses. Biosensor for in vivo monitoring needs to be implanted into the body (such as blood vessels, cerebral cortex, subcutaneous tissue) and faces many complex environments, including the tissue response, biocompatibility, disinfection problems, oxygen interference, and other technical issues.

Many companies and research institutes carry out the research on in vivo sensor system in recent years. Several glucose sensor systems that continuously monitor the blood glucose level under the subcutaneous have been commercially developed and applied. Nowadays, these blood glucose sensors belong to this kind of "percutaneous," which directly pierce the fine-needle sensor into the skin, and the base and data part are fixed on the skin. The CGMS probe sensor from Medtronic company has three layers containing the semipermeable membrane, glucose oxidase, and platinum electrode [59]. Among them, the platinum electrode of the Medtronic probe is the tri-electrode probe system, which can reduce the interference of the impurity to the current signal. The CGMS probe was implanted subcutaneously for 3 days in the patient's abdomen (**Figure 9**). The probe generates electrical signals by chemical reaction with glucose in the subcutaneous intercellular fluid of the patient. The electrical signal is proportional to the concentration of blood sugar, but it is delayed for about 20 minutes. The Abbott freestyle continuous glucose monitoring system contains implantable needle biosensor and wireless reading device. The implantable needle biosensor adopted the same principal with the size of length 5 and 0.4 mm diameter on a coin-sized sticker (**Figure 10**), long term for 14 days sticking on the upper arm [60]. When we want to check blood sugar, the blood glucose meter is close to the circle sticker, which is accurate and convenient. The Eversense Continuous Glucose Monitoring (CGM) developed by Senseonics medical company has been approved by

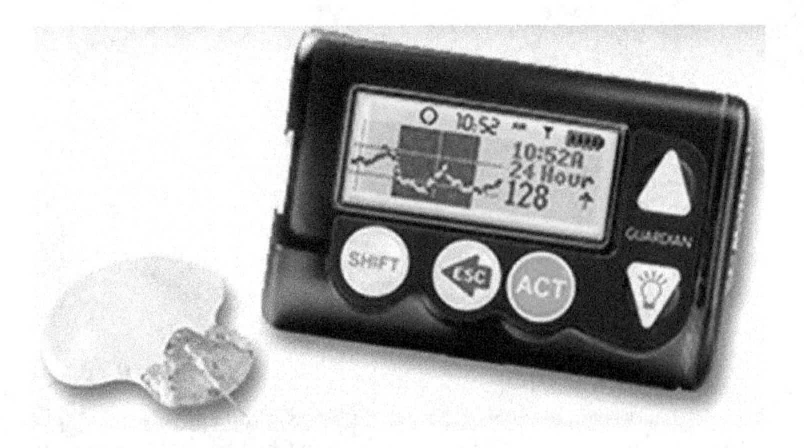

Figure 9. The Medtronic needle sensor and reader.

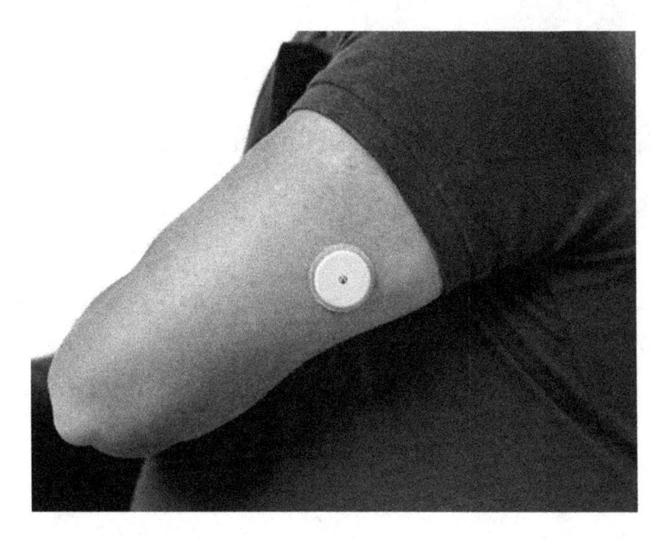

Figure 10. The Abbott continuous glucose monitoring sensor.

the EU management department for CE. The Eversense system implanted a needle-type sensor subcutaneously into the upper arm of the patient for a period of 90 days [61].

The new sensing technology improves the value of biological analysis and improves the level of diagnosis and monitoring of disease. These smaller, more easy-to-use new devices not only broaden the scope of diagnosis but also provide more effective methods for monitoring special conditions.

3. Conclusion

Biosensor has the characteristics of cross disciplinary integration. The development trends of biosensors are miniaturization, multifunction, integration, and intelligence. The introduction of some new frontier technology such as microelectronics, nanotechnology, and microelectromechanical technology (Microelectronic Mechanism) is applied to the biomedical sensor to develop a new generation of biosensor with low cost, high sensitivity, high reliability, high life and bionic function, achieving rapid and accurate test of trace components in the sample. At the same time, the biomedical sensor has developed from planar two-dimensional to three-dimensional microelectronic mechanical system sensors. With the development of computer software and hardware technology, sensor technology will also develop, and the biological system of closed-loop control will be possible. In the future, biosensors will be widely used in many fields, such as food detection, medical care, disease diagnosis, environmental monitoring, fermentation industry, and so on.

Acknowledgements

This work is supported by the NSFC (Nos. 61774157, 61271147, 61372052, 11574219, 81771388), Beijing NSFC (100063), and the major national scientific research plan (2016YFC1304302) and (2017YFF0107704).

Author details

Chunxiu Liu[1]*, Chenghua Xu[1], Ning Xue[1], Jian Hai Sun[1], Haoyuan Cai[1], Tong Li[1], Yuanyuan Liu[2] and Jun Wang[3]

*Address all correspondence to: cxliu@mail.ie.ac.cn

1 State Key Laboratory of Tranducer Technology, Institute of Electronics, Chinese Academy of Sciences, Beijing, China

2 Institute of Semiconductors, Chinese Academy of Sciences, Beijing, China

3 School of Materials Science and Engineering, Beijing Institute of Technology, Beijing, China

References

[1] Turner APF. Biosensors: Past, Present and Future. www.cranfield.ac.uk/biotech/chinap.htm; 1996

[2] Turner APF, Kaube I, Wilson GS. Biosensors Fundamentals and Applications. Oxford, UK: Oxford University Press; 1987. pp. 86-92

[3] Jaffari S, Turner APF. Novel hexacyanoferrate (III) modified-graphite disc electrodes and their application in enzyme electrodes (Part I). Biosensors and Bioelectronics. 1995;**12**:1-9

[4] Kost GJ. Guidelines for point-of-care testing. Improving patient outcomes. American Journal of Clinical Pathology. 1995;**104**:S111-S127

[5] Luppa PB et al. Point-of-care testing (POCT): Current techniques and future perspectives. Trends in Analytical Chemistry. 2011;**30**:887-898

[6] Cass AEG, Davis G, Francis GD, Hill HAO, Aston WJ, Higgins IJ, Plotkin EV, Scott LDL, Turner APF. Ferrocene-mediated enzyme electrode for amperometric determination of glucose. Analytical Chemistry. 1984;**56**:667-671

[7] Mark AE, Priscilla QV, Zhang KX, Kang DK, Ali MM, Xu C, Zhao W. Novel molecular and nanosensors for in vivo sensing. Theranostics. 2013;**3**(8):583-594

[8] Kros A, Gerritsen M, Sprakel VSI, et al. Silica-based hybrid materials as biocompatible coatings for glucose sensors. Sensors & Actutors B. 2001;**81**(1):68-75

[9] Rhemrev BMM, Korf J, Venema K, et al. A versatile biosensor device for continuous biomedical monitoring. Biosensors & Bioelectronics. 2001;**16**(9–12):839-847

[10] Rahman MA, Kumar P, Park D, Shim Y. Electrochemical sensors based on organic conjugated polymers. Sensors. 2008;**8**:118-141

[11] Ates M. A review study of (bio) sensor systems based on conducting polymers. Materials Science and Engineering: C. 2013;**33**:1853-1859

[12] Volder MFLD, Tawfick SH, Baughman RH, Hart AJ. Carbon nanotubes: Present and future commercial applications. Science. 2013;**339**:535-539

[13] Chen A, Chatterjee S. Nanomaterials based electrochemical sensors for biomedical applications. Chemical Society Reviews. 2013;**42**:5425-5438

[14] Tasis D, Tagmatarchis N, Bianco A, Prato M. Chemistry of carbon nanotubes. Chemical Reviews. 2006;**106**:1105-1136

[15] Shi W, Liu C, Song Y, Cai X. An ascorbic acid amperometric sensor using over-oxidized polypyrrole and palladium nanoparticles composites. Biosensors and Bioelectronics. 2012;**38**(1):100-106

[16] Parkin MC, Hopwood SE, Strong AJ, et al. Resolving dynamic changes in brain metabolism using biosensors and on-line microdialysis. TrAC Trends in Analytical Chemistry. 2003;**22**(9):487-497

[17] Sirkar K, Pishko MV. Amperometric biosensors based on oxidoreductases immobilized in photopolymerised poly(ethylene glycol) redox polymer hydrogels. Analytical Chemistry. 1998;**70**:2888-2894

[18] Wang J. Electrochemical glucose biosensors. Chemical Reviews. 2008;**108**:814-825

[19] Reza KD, Azadeh A. Biosensors: Functions and applications. Journal of Biology and Today's World. 2013;**2**(1):53-61

[20] Ozcan A. Mobile phones democratize and cultivate next-generation imaging, diagnostics and measurement tools. Lab on a Chip. 2014;**14**:3187-3194

[21] Boonlert W et al. Comparison of the performance of point-of-care and device analyzers to hospital laboratory instruments. Point Care. 2003;**2**:172-178

[22] Kiechle FL, Main RI. Blood glucose: Measurement in the point-of-care setting. Laboratoriums Medizin. 2000;**31**:276-282

[23] Hu J et al. Advances in paper-based point-of-care diagnostics. Biosensors & Bioelectronics. 2014;**54**:585-597

[24] Liu X, Liu O, Gupta E, et al. Quantitative measurements of NO reaction kinetics with a Clark-type electrode. Nitric Oxide. 2005;**13**(1):68-77

[25] Dong SJ, Wang BX, Liu BF. Amperometric glucose sensor with ferrocene as an electron transfer mediator. Biosensors & Bioelectronics. 1991;**7**:215

[26] Dong SJ, Lu ZL. Ferrocene–Nafion modified electrode and its catalysis for cerium(IV). Molecular Crystals and Liquid Crystals. 1990;**190**:197

[27] Ghica ME, Brett CMA. Poly(brilliant green) and poly(thionine) modified carbon nanotube coated carbon film electrodes for glucose and uric acid biosensors. Talanta. 2014;**130**:198-206

[28] APF T. Amperometric sensors based on mediator modified electrodes. Methods in Enzymology. 1988;**137**:90-103

[29] Davis G. Electrochemical techniques for the development of amperometric biosensors. Biosensor. 1985;**1**:161-178

[30] Lange MA, Chambers JQ. Amperometric determination of glucose with a ferrocene-mediated glucose oxidase/polyacrylamide gel electrode. Analytica Chimica Acta. 1985; **175**:89-97

[31] Yamamotok K, Ohgaru T, Torimura M, a1. Highly-sensitive flow injection determination of hydrogen peroxide with a peroxidase immobilized electrode and its application to clinical chemistry. Analytica Chimica Acta. 2000;**406**:201-207

[32] Kunzelmann U, Bottcher H. Biosensor properties of glucose oxidase immobilized within SiO_2 gels. Sensors and Actuators B. 1997;**38-39**:222-228

[33] Niu JJ, Lee JY. Reagentless mediated biosensors based on polyelectrolyte and sol−gel derived silica matrix. Sensors and Actuators B. 2002;**82**:250-258

[34] Ikeda T, Shibata T, Senda M. Journal of Electroanalytical Chemistry. 1989;**261**:351

[35] Hogan CF, forster RJ. Mediated electron transfer for electroanalysis: Transport and kinetics in thin films of $[Ru (bpy)_2PVP_{10}] (ClO_4)_2$. Analytica Chimica Acta. 1999;**396**:13-21

[36] Mullane APO, Macpherson JV, Unwin PR. Measurement of lateral charge propagation in [Os(bpy)2(PVP)nCl]Cl thin films: A scanning electrochemical microscopy approach. The Journal of Physical Chemistry. B. 2004;**108**:7219-7227

[37] Ohara TJ, Rajagopalan R, Heller A. "Wired" enzyme electrodes for amperometric determination of glucose or lactate in the presence of interfering substances. Analytical Chemistry. 1994;**66**:2451-2457

[38] Ju HX, Leech D. $[Os(bpy)_2(PVI)_{10}Cl]Cl$ polymer-modified carbon fiber electrodes for the electrocatalytic oxidation of NADH. Analytica Chimica Acta. 1997;**345**:51-58

[39] Trudeau F, Daigle F, Leech D. Reagentless mediated laccase electrode for the detection of enzyme modulators. Analytical Chemistry. 1997;**69**:882-886

[40] Chen L, Gorski W. Determination of glucose at the $Ru(NH_3)_6^{3+}$ based paste enzyme electrode. Electroanalysis. 2002;**14**(1):78-81

[41] Degani Y, Heller A. Direct electrical communication between chemically modified enzymes and metal electrodes. I. Electron transfer from glucose oxidase to metal electrodes via electron relays, bound covalently to the enzyme. The Journal of Physical Chemistry. 1987;**91**:1285-1289

[42] Degani Y, Heller A. Direct electrical communication between chemically modified enzymes and metal electrodes. II. Methods for bonding electron-transfer relays to glucose oxidase and D-amino-acid oxidase. Journal of the American Chemical Society. 1988;**110**: 2615-2620

[43] Barsan MM, Emilia Ghica M, Brett CMA. Electrochemical sensors and biosensors based on redox polymer/ carbon nanotube modified electrodes: A review. Analytica Chimica Acta. 2015;**881**:1-23

[44] Hale PD, Lee SH, Okamoto Y. Electrical communication between glucose oxidase and novel ferrocene containing siloxane-ethylene oxide copolymers: Biosensor Application. Analytical Letters. 1993;**26**(1):l-16

[45] Niwa O, Kurita R, Liu Z, Horiuchi T, Torimitsu K. Subnanoliter volume wall-jet cells combined with interdigitated microarray electrode and enzyme modified planar micro-electrode. Analytical Chemistry. 2000;**72**:949-955

[46] Sue H, Britta L. Electrochemistry at ultra-thin polyelectrolyte films self-assembled at planar gold electrodes. Electrochimica Acta. 1999;**45**:845-853

[47] Nikolelis DP, Petropoulou SSE. Investigation of interaction of a resorcin[4]arene receptor with bilayer lipid membranes (BLMs) for the electrochemical biosensing of mixture of dopamine and ephedrine. Biochimica et Biophysics Acta-Biomembranes. 2002;**1558**:238-245

[48] Sandeep KV, Peter BL, Leslie YY, Aydogan O, John HTL. Emerging technologies for text-generation point-of-care testing. Trends in Biotechnology. 2015;**1294**(14)

[49] Evaluation of the CoaguChek XS System, International Evaluation Workshop, Heidelberg, Germany; 2009

[50] Daniel M, Stefan H, Gunter R, Felix S, Roland Z. Microfluidic lab-on-a-chip platforms requirements, characteristics and applications. Chemical Society Reviews. 2010;**39**:1153-1182

[51] Liu C, Liu H, Yang Q, Tian Q, Cai X. Development of an amperometric lactate biosensor modified with Pt-black nanoparticles for rapid assay. Chinese Journal of Analytical Chemistry. 2009;**37**(4):624-628

[52] Liu C, Jiang L, Guo Z, Cai X. A novel disposable amperometric biosensor based on trienzyme electrode for the determination of total creatine kinase. Sensors and Actuators B. 2007;**122**(1):295-300

[53] Shi W, Lin N, Song Y, Liu C, Zhou S, Cai X. A novel method to directionally stabilize enzymes together with redox mediators by electrodeposition. Biosensors and Bioelectronics. 2014;**51**:244-248

[54] Abbott Point-of-Care, USA, www.abbottpointofcare.com. [Accessed: November 30, 2017]

[55] Wilson GS, Johnson MA. Invivo electrochemistry: What can we learn about living systems? Chemical Reviews. 2008;**108**(7):2462-2481

[56] Ronkainen NJ, Halsall HB, Heineman WR. Electrochemical biosensors. Chemical Society Reviews. 2010;**39**(5):1747-1763

[57] Vaddiraju S, Tomazos I, Burgess DJ, Jain FC, Papadimitrakopoulos F. Emerging synergy between nanotechnology and implantable biosensors: A review. Biosensors & Bioelectronics. 2010;**25**(7):1553-1565

[58] Crespi F. Wireless in vivo voltammetric measurements of neurotransmitters in freely behaving rats. Biosensors & Bioelectronics. 2010;**25**(11):2425-2430

[59] http://www.medtronicdiabetes.com.cn/. [Accessed: November 20, 2017]

[60] http://www.medgadget.com/2014/09/abbott-freestyle-libre-flash-continuous-glucose-monitoring-system.html. [Accessed: May 20, 2014]

[61] http://www.senseonics.com. [Accessed November 30, 2017]

9

MEMS Devices for Miniaturized Gas Chromatography

Imadeddine Azzouz and Khaldoun Bachari

Abstract

In the era of the Internet of Things, the need for mobile devices able to analyze accurately real samples with sometimes very small volumes is a must. Gas chromatography (GC) is a common laboratory technique widely used for analyzing semi-volatile and volatile compounds. However, this technique is not suitable to be used outside labs. The development of micro-machined processes encouraged the development of miniaturized gas chromatographs. This chapter focuses on the recent development in the field of miniaturized gas chromatography and its component up to the present in various fields of analyses.

Keywords: gas chromatography, MEMS, micro-detector micro-column, microinjector, miniaturization

1. Introduction

Gas chromatography (GC) is a common and a complex analytical technique involving the separation of different types of gas or molecules easily vaporized without decomposition. The gas molecules are carried through the column using a carrier gas, typically nitrogen or helium. Based on their affinity for the coating material inside the column which is called the "stationary phase," molecules are separated based on certain molecular characteristics such as their molecular weight, polarity, and presence of certain functional groups. At the end of the column, molecules are separated and detected by a detector.

Commercially available GC analyzers use conventionally manufactured components (~30 kg) and need power and gas sources that often limit their portability and suitability of "outside-laboratory" use. Miniaturization of GC is based on theoretical and practical considerations. This chapter describes the miniaturization of analytical system, in order to give a complete view of miniaturized chromatographic separations.

2. Need for portable analytical systems

Conventional GCs provide accurate analysis of complex mixtures but at the cost of using large, power-hungry, and relatively expensive table-top instruments. Usually, samples are collected and brought back to the laboratory for analysis. On-site analysis is becoming increasingly important, especially in the area of environmental monitoring. It reduces the risk of contamination, sample loss, and sample decomposition during transport. Furthermore, on-site monitoring also results in much shorter analysis turnaround times and thus allows for faster response to the analytical results. Lightweight devices with low maintenance are needed. In order to achieve these features, the miniaturization of the main components of GCs is performed.

Miniaturization of GC is based both on theoretical and practical considerations [1]. Theory predicts that reducing the dimensions of flow channels enhances the analytical performances. In practice, miniaturization also enhances analysis of small-volume samples and increases analysis speed. A microfabricated GC system requires a number of components to function properly: preconcentrator, micro-valves for injecting the sample into the carrier gas, micro-fabricated columns well-functionalized for the specific use, heaters and temperature sensors for controlling column temperature, and detector(s) for detecting the arrival of different types of molecules. Temperature stability is also critical for GC operation, as the adsorption/desorption processes responsible for molecular separation in the column are very sensitive to temperature. The issues of microfluidic integration are therefore critical in GC microsystems.

Despite the fact that the first work on microchip-based chromatographic system was a miniaturized gas chromatograph in 1979 [2] using microelectromechanical systems (MEMS), this development was hardly pursued afterward, probably because the analytical community was not yet ready to embrace this new technology.

3. Injectors-preconcentrators

The injector is a device used for introducing liquid or gas samples into the gas chromatograph. The sample is introduced directly into the carrier gas stream via a temperature-controlled chamber temporarily isolated from the system by gas sampling valves. Among all reported studies, several research teams have used commercial injectors (part of a convention GC) in split mode or gas sample valves to introduce samples into the micro-columns. Some other teams designed and fabricated a chip-based preconcentrator instead of an injector to increase sensitivity and selectivity when solute concentration is below detection limit of the detector [3, 4]. In both cases, the device must be capable of generating sharp injection plugs.

A six-valve MEMS-based injector with constant 250 nL of sample volume and suitable for harsh environment was introduced in 2010 emulating Valvo® six-valve injector. Each valve is made from sandwiching polyether ether ketone (PEEK) membranes between silicon substrate and glass. The six valves opened and closed by changing the pressure through their actuation holes. In sampling mode, valves A, D, and E are closed, while for injecting samples onto

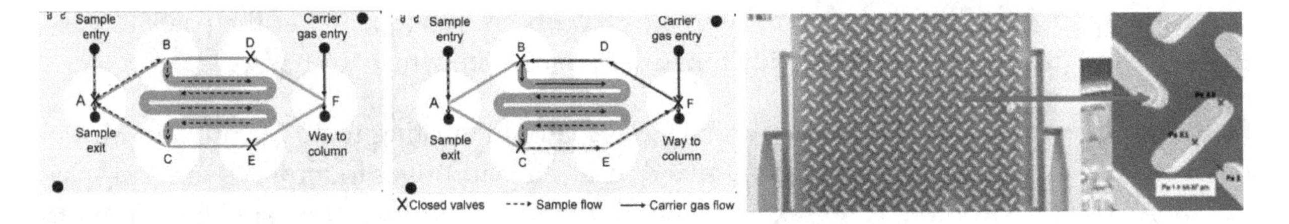

Figure 1. Left: Illustration of the MEMS injector in loading and injecting modes, adapted from [6], right: Optical image of a 3D preconcentrator with embedded pillars, adapted from [8].

separation channels, valves B, C, and F are closed (**Figure 1**) [5, 6]. Moreover, a customized volume injector (0.5–15 μL) was designed by Holland et al. [7].

As a general definition, the preconcentrator relies on an adsorbing material deposited on the active area adjacent to the heating element [8]. Ideally, the sorptive material must adsorb selectively one or more chemical species of interest over a time period necessary to concentrate the chemical compound in the adsorptive material. Then, the sorptive layer must be heated with a pulse of temperature for providing narrow desorption peaks with relatively high concentration to the connecting sensor or detector. This process must allow the analytes present in a large air volume to be purified and concentrated, so increasing the efficiency of detection. Since the first micro-machined preconcentrator designed by the ChemLab at Sandia National Laboratories in 1999 [9], many works have been carried out. In literature, different preconcentrating microstructures are now available in different designs and are combined with a wide range of adsorbents [10–12].

The optimization of the device performance (adsorption and desorption duration and flow rates, heating rates) is rather important for achieving a high preconcentration factor. A compromise must be then established between a suitable adsorbent, low power consumption, and simple fabrication technology.

4. Columns

The gas chromatographic column is considered the "heart" in a gas chromatograph. Over the last three decades, the nature and design of the column have changed considerably. Conventional GCs are equipped with conventional columns: a silica or stainless steel tube containing an immobilized or a cross-linked stationary phase bound to the inner surface. Terry et al. [2] were the first group to introduce "miniaturized GC" and "planar column" concepts by etching channels into a substrate rather by using capillaries of conventional GC technology (**Figure 1**). However, this groundbreaking work had not led to further developments of related skills or technology until the early 1990s.

4.1. Technology fabrication

Silicon is a very common substrate for microelectronics. The material is relatively inexpensive, is abundant in nature, and can be ordered with well-controlled crystal orientation, thickness, and surface roughness. A large number of processes have been developed over the past 50 years, giving the microsystem designer a wide range of options from which to choose.

Glass, via its optical and mechanical properties, is very interesting to be included in MEMS devices. Additionally, glass can be customized by adding additives to improve some properties, boron oxides, to produce Pyrex well known to its low thermal expansion or sodium to easily to bond with silicon.

The combination of glass and silicon provides the most versatile fabrication technique for producing GC columns (**Figure 2**). However, silicon and glass fabrication requires the use of a clean room, making this technology relatively expensive and not within the reach of every academic laboratory. Fabrication processes for both glass and silicon can be divided into three main steps: patterning, etching, and bonding.

Many processes involve the deposition and patterning of thin films (e.g., for heating or as a stationary phase) [13, 14]. There is a wide variety of methods for performing such depositions, from nano- to microscale, such as physical vapor deposition (PVD), sputtering [15, 16], and atomic layer deposition (ALD).

4.2. Performance of MEMS columns

Theoretical plate number N defines the efficiency of the column or sharpness of peaks. The concept of plate theory was originally proposed for the performance of distillation columns. It is proportional to the square root of the retention time and inversed proportional to the peak width following the normal distribution law. The theory assumes that the column is divided into a number of zones called "theoretical plates." Moreover, the zone thickness is considered as height equivalent to a theoretical plate (H or HEPT):

$$N = 16\frac{(t_r)^2}{w} \tag{1}$$

where t_r is the retention time of a compound, w is the width of the peak at the base.

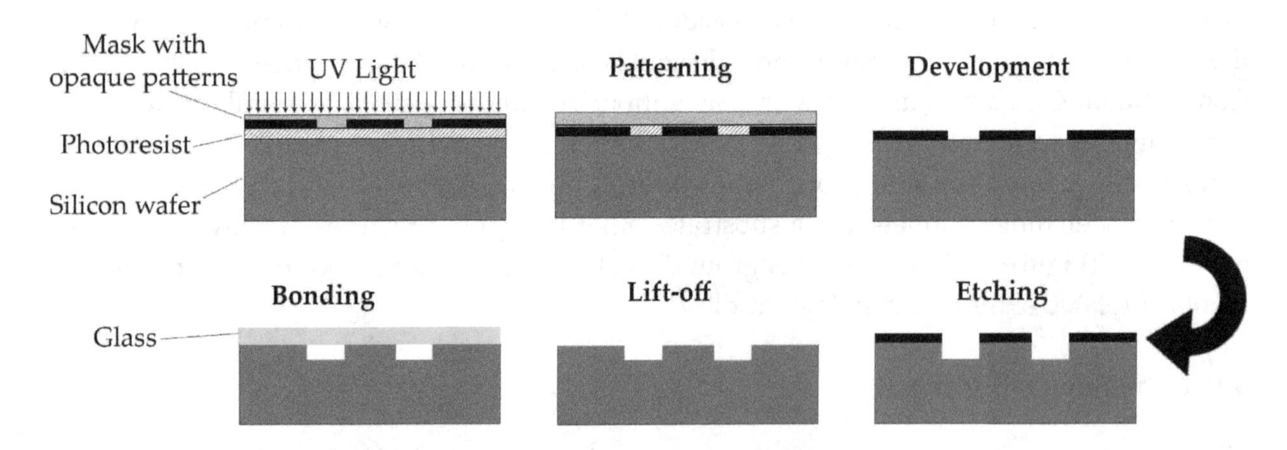

Figure 2. Illustrating steps to obtain a MEMS column.

The fundamental equation underlying the performance of a gas chromatographic column is the Van Deemter equation expressed as

$$H = A + \frac{B}{\overline{u}} + C\overline{u} \tag{2}$$

where H is the height equivalent to a theoretical plate, A is the eddy diffusion or multiple path term, B is the longitudinal diffusion contribution, C is the resistance to mass transfer term, $\overline{u}$ is the average linear velocity of the carrier gas.

Thus, equation is simplified in case of open columns. The A term is equal to zero because there is no packing. This abbreviated expression is often referred to as the *Golay Eq.* [17].

The profile of "H" versus "u" graphic goes through a minimum value of "H" where the efficiency is greatest. This minimum is reached at different carrier gas velocities depending on the nature of the carrier gas. For example, speed of analysis must be sacrificed when nitrogen is used as a carrier gas. On the other hand, if one is willing to save time with slight loss of the efficiency, helium or hydrogen can be used. Additionally, efficiency varies slightly for hydrogen than helium as velocity increases. Finally, the use of hydrogen for any application in the laboratory always requires safety precautions in the event of leak:

- The flow rate, and consequently the linear velocity, through smaller columns is difficult to measure accurately and reproducibly by conventional apparatus. Linear velocity may be calculated, through a column of length L, by injecting a volatile, non-retained solute and noting its retention time t_M using this equation:

$$u\,(cm/s) = \frac{L}{t_M} \tag{3}$$

- In gas chromatography, when the temperature increases, linear velocity decreases because of increased viscosity of the carrier gas.
- Van Deemter curve is fitted under isothermal conditions.

4.3. Functionalization

It is quite straightforward to etch channels into silicon or glass chip. However, finding a comprehensive and reproducible method of fabrication enabling incorporation of a stationary phase inside the channel under conditions of extreme miniaturization, and production under clean room conditions, was a major challenge. This part covers various functionalization methods from classic coating to unusual MEMS-based techniques.

4.3.1. Polymer coating

In the beginning of the MEMS-based column era, researchers tried to adjust expertise gained from the preparation of conventional columns. Usually, columns are made by etching silica substrate followed by capping with Pyrex. Stationary phase application after sealing the channel was usually performed by liquid coating using static or dynamic method. These methods led to wall-coated open tubular MEMS (WCOT-MEMS) columns commonly named "open columns." The goal in coating is the uniform deposition of a thin film, typically ranging from 0.1 to 10 μm in thickness. To reach this, two varieties of coating exists: static and dynamic.

Polysiloxanes are the most widely used as stationary phases for both conventional and MEMS columns. They offer high solute diffusivities coupled with excellent chemical and thermal stabilities. Additionally, because a variety of functional groups can be incorporated into their structures, polysiloxanes exhibit a wide range of polarities. Since many polysiloxanes are viscous gums and, as such, coat well on MEMS columns. Polysiloxanes are easily cross-linked to be used as stationary phases. The basic structure of 100% dimethylpolysiloxane (PDMS) is depicted in **Figure 3**.

Lambertus et al. [18] reported a 3-m-long square-spiral MEMS column dynamically coated with PDMS achieving 8200 plates (**Figure 4**). Moreover, non-treated surface gave 1500 plates more than treated (CVD oxidation prior to bonding).

Nishino et al. [19] developed circular, 8.5–17.0-m-long MEMS columns to separate a mixture of 13 compounds which included polar and nonpolar compounds. Before coating with the liquid phase, deactivation treatment to reduce adsorption sites causing peak tailing or peak disappearance was completed. Stationary phase coating was performed by a static method with 5% phenyl 95% dimethylpolysiloxane to give a 0.25-μm-thick film.

Radadia et al. [20] improved separation of organophosphonate and organosulfur compounds by using a 3 m MEMS column coated with 0.25 μm OV-5 as stationary phase. To reduce Pyrex's active sites, they were deactivated by the use of a variety of agents. Organosilicon hydride deactivation reduced micro-column adsorption activity more than silazane and silane treatment, enabling baseline separation of nine compounds as peaks with very low asymmetry in 2 min and providing 5500 theoretical plates/m (**Figure 5**).

The most widely used non-silicon-containing stationary phases are the polyethylene glycols. They are commercially available in a wide range of molecular weights under several designations, such as Carbowax 20M and Superox-4. Unfortunately, their operational temperature is reduced compared to siloxane-based polymers. In addition, trace levels of oxygen and water from the sample or the carrier gas have adverse effects especially with Carbowax 20M leading to their fast degradation. An example of a MEMS-based column coated with Carbowax 20M was reporter by Lee et al. [21].

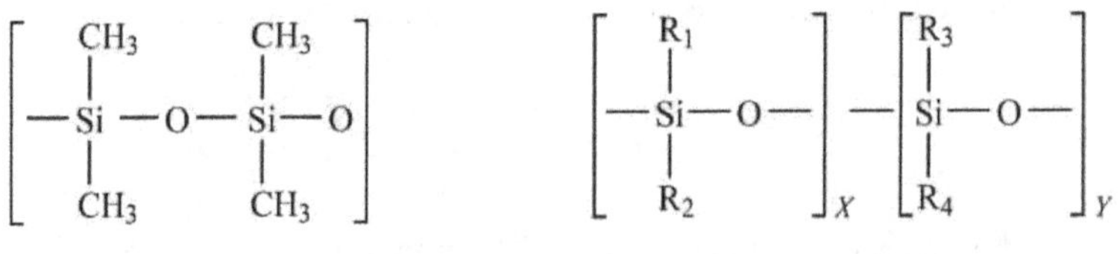

$$\left[-\underset{CH_3}{\overset{CH_3}{\mathrm{Si}}} - O - \underset{CH_3}{\overset{CH_3}{\mathrm{Si}}} - O \right] \qquad \left[-\underset{R_2}{\overset{R_1}{\mathrm{Si}}} - O - \right]_X - \left[\underset{R_4}{\overset{R_3}{\mathrm{Si}}} - O - \right]_Y$$

Figure 3. Chemical structure of basic dimethylpolysiloxane PDMS (left), and substituted polysiloxane.

4.3.2. Solid packing

A packed column refers to a column packed with either a solid adsorbent or solid support coated with a liquid phase. However, stable and reproducible performances depend mainly on the quality of packing. In conventional GC, this kind of column began to decline since 1979 by the apparition of capillary fused-silica columns. A packed column consists of three basic components: tubing in which packing material is placed (**Table 1**), packing retainers (such as glass wool plugs), and the packing material itself. In MEMS-based columns, tubes are replaced by MEMS channels and glass wool plugs by grids or meshes (**Figure 6**) [22].

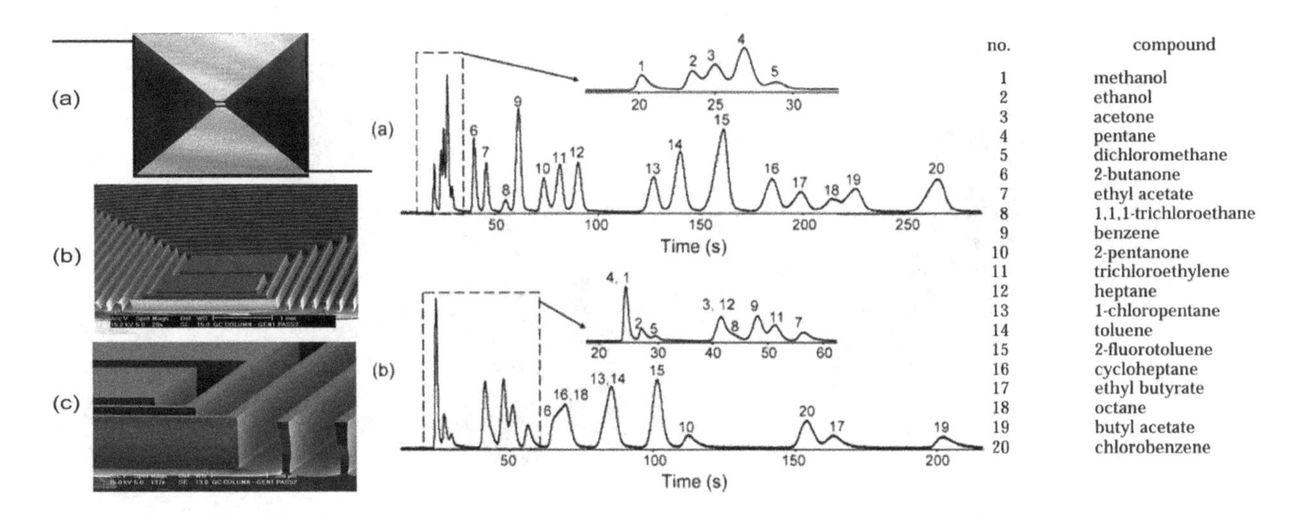

Figure 4. Left: (a) entire chip; (b) SEM image detail of gas flow (c) detail of etched-channel, right: Isothermal chromatograms at 22°C of the 20-component using channels coated with the nonpolar (a) and the moderately polar (b) stationary phases, reprinted with permission from [19].

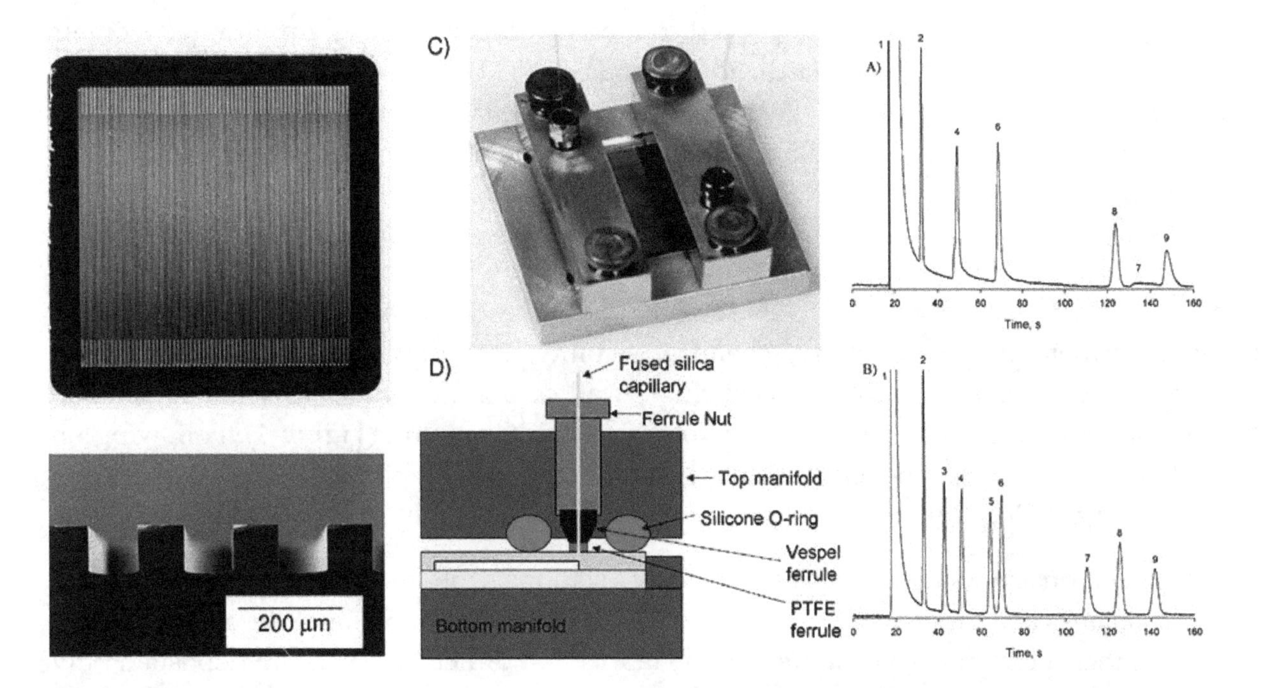

Figure 5. Left: (A) photograph of the MEMS column, (B) SEM of channels, (C) manifold packaging, and (D) connection to the micro-column. Right: Separation of test, reprinted with permission from [21].

Stationary phase	Usual applications
Alumina	Alkanes, alkenes, alkines, aromatic hydrocarbons (C1-C10)
Silica gel	Hydrocarbons (C1-C4), inorganic gases, volatile ethers
Carbon	Inorganic gases, hydrocarbons (C1-C5)
Carbon molecular sieves	Oxygenated compounds (C1-C6)
Molecular sieves (5X, 13 X)	Hydrogen, oxygen, methane, permanent gas, halocarbons

Table 1. Illustrative examples of some adsorbents and usual applications.

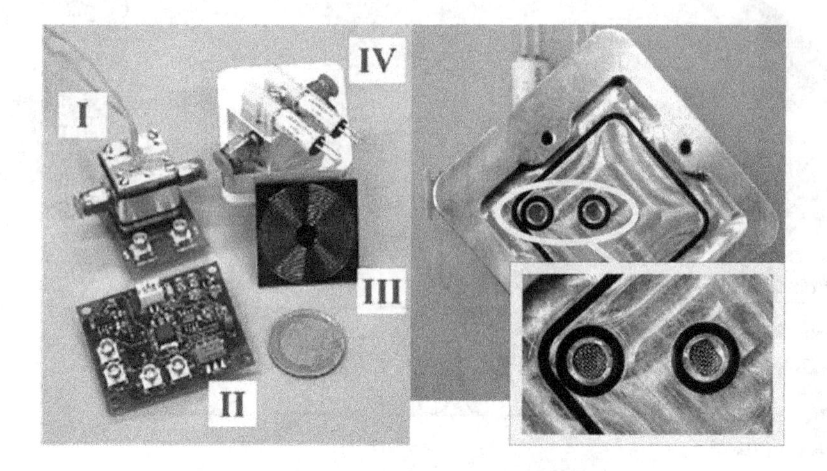

Figure 6. Left: Photograph of different components of miniaturized GC, right: Stainless steel meshes to keep the stationary phase particles in the column, adapted from [23].

Some separations require the use of packed columns: permanent gases, unsaturated isomers of light hydrocarbons, and standardized methods (ASTM E260, NF ISO 17494, etc.) [23]. Although these columns remain effective, their implementation in reduced sizes, low efficiency, and the pressure generated in the column are the main obstacles to their use.

4.3.3. *Carbon nanotubes*

Soon after their discovery in 1991 [24], carbon nanotubes (CNTs) received much attention because of their unique geometry, chemical stability, and high surface-to-volume ratio. Stadermann et al. [25] successfully used single-wall carbon nanotubes (SWCNTs) as a stationary phase by means of CVD in a microfabricated GC column (**Figure 7**). Following on from their study, the team developed a new process to produce a highly uniform mat of CNT stationary phase [26].

SWCNTs demonstrate a good ability to be used as stationary phase in gas chromatography to separate alkanes and other analytes. It can be used as is, and no functionalization is required. However, their performance is limited by the fabrication difficulty. CNTs are deposited only on three sides of the column's channel (silicon) leading to peaks broadening. Additionally, columns with CNTs suffer from poor separation of high-boiling compounds, which is often attributed to the thickness of the CNT layer.

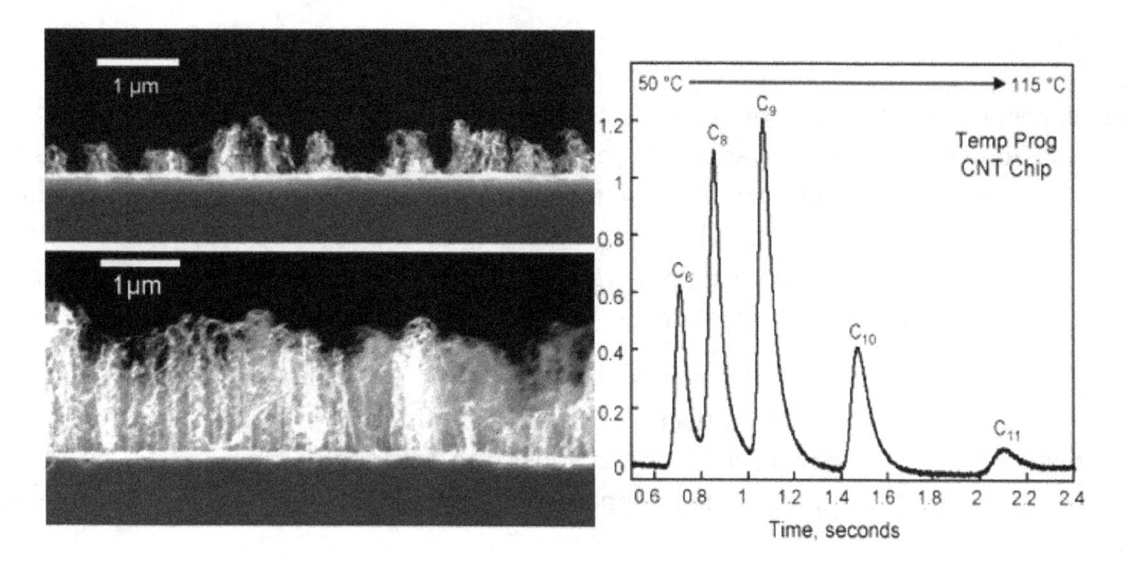

Figure 7. Left up: SEM image of the old CVD process to produce SWCNTS, left down: SEM image of the new CVD process lead to obtain a "mat" of SWCNTs, right: Separation chromatogram of n-alkanes with SWCNTs, reprinted with permission from [26, 27].

4.3.4. *Sputtering*

Sputtering is widely used in electronics for deposition of metals and dielectrics. Vial et al. [15] use this technique to provide solid and porous stationary phase. By varying the duration of the sputtering process, sputtered silica layers of different thicknesses were produced. For example, silica layer having 0.75 µm thickness produced 2500 theoretical plates for hydrocarbon separation (**Figure 8**). At the opposite, producing a thicker layer leads to loss separation efficiency (number of plates). To overcome this, the same group used a semi-packed column with high aspect-to-ratio pillars [13].

In that case separations were greatly improved because retention increased and efficiency was close to 5000 theoretical plates m^{-1}. The same group tested various targets such as graphite and alumina to separate light hydrocarbons [13, 16]. However, alumina requires a tedious activation step before using.

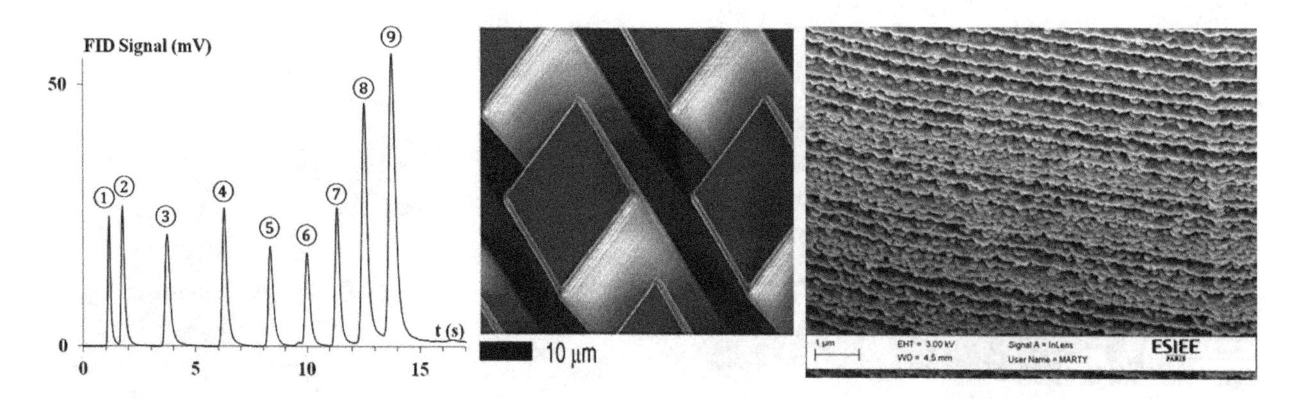

Figure 8. Left: Fast separation of linear hydrocarbons with a silica sputtered MEMS column, middle: Micro-pillars covered with sputtered silica, right: SEM picture of graphite-sputtered layer deposited on the inner wall of a micromachied column, reprinted with permission from [14, 17].

4.3.5. Gold layers

In the separation sciences, nanoparticles have been used as stationary phases to provide high separation efficiency for a variety of analytes. Because the nanoparticles are too small to be packed into the column, they are usually used as pseudo-stationary phase to enhance separation [27, 28]. Gold nanoparticles have become increasingly popular because of their long-term stability, high surface-to-volume ratio, and ease of chemical modification. The use of gold enables a variety of functional groups to be incorporated into the monolayer [29].

Agah's group introduced in 2010 a new stationary phase based on deposing gold by electroplating followed by its functionalization [30, 31]. The thickness and the regularity of the layer are customized by varying the current density. Additionally, they used a multi-capillary microfabricated 25 cm column to separate hydrocarbons yielding 20,000 plates m^{-1} (**Figure 9**).

Although such results were promising, a disadvantage is that nonselective deposition meant that the fabrication process required "mechanical" removal of gold from the upper surface. This step could damage the very thin fluidic channels. To resolve this problem, Shakeel et al. [32] proposed two different ways, highly reproducible, for the deposition of gold:

- Self-patterning gold on the vertical sidewalls only (varying electroplating conditions)
- Double-doped self-patterning to cover the interior surfaces of the channel (three silicon sides)

The use of gold stationary phases has furnished interesting results. However, uniformity and quality of deposition depend on the deposition conditions. Additionally, this stationary phase is not suitable for light hydrocarbons separation.

4.3.6. Ionic liquids

Ionic liquids constitute a group of organic salts with a particulate specification. They are liquid below 100°C and consequently liquids at room temperature. Ionic liquids are polar, nonflammable, chemically inert, thermally stable, easy to synthesize, and already used in conventional gas chromatography [33, 34]. Additionally, their selectivity can be tuned by altering the constituent cation or anion, and hence there is more than 300 commercially varieties.

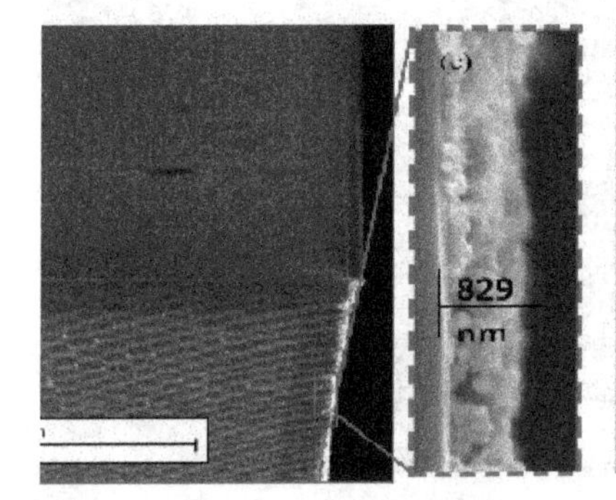

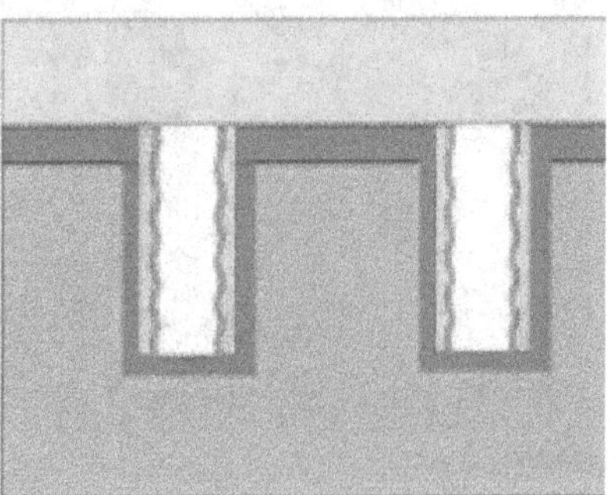
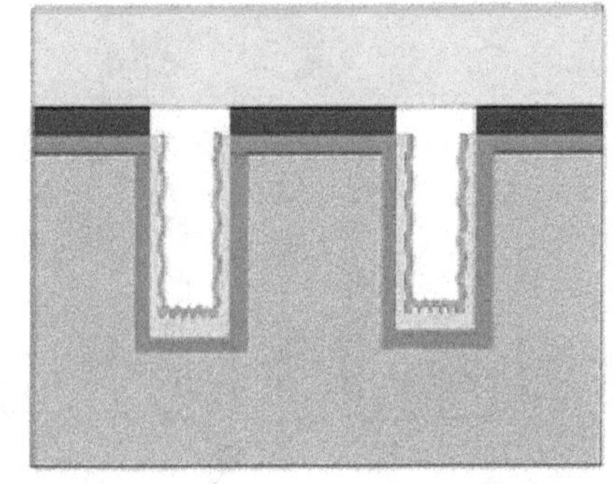

Figure 9. Left: Cross-section of a single side-wall with zoom (thickness of the gold layer, middle and right: Thiol deposition using single and double doping methods respectively, adapted from [33].

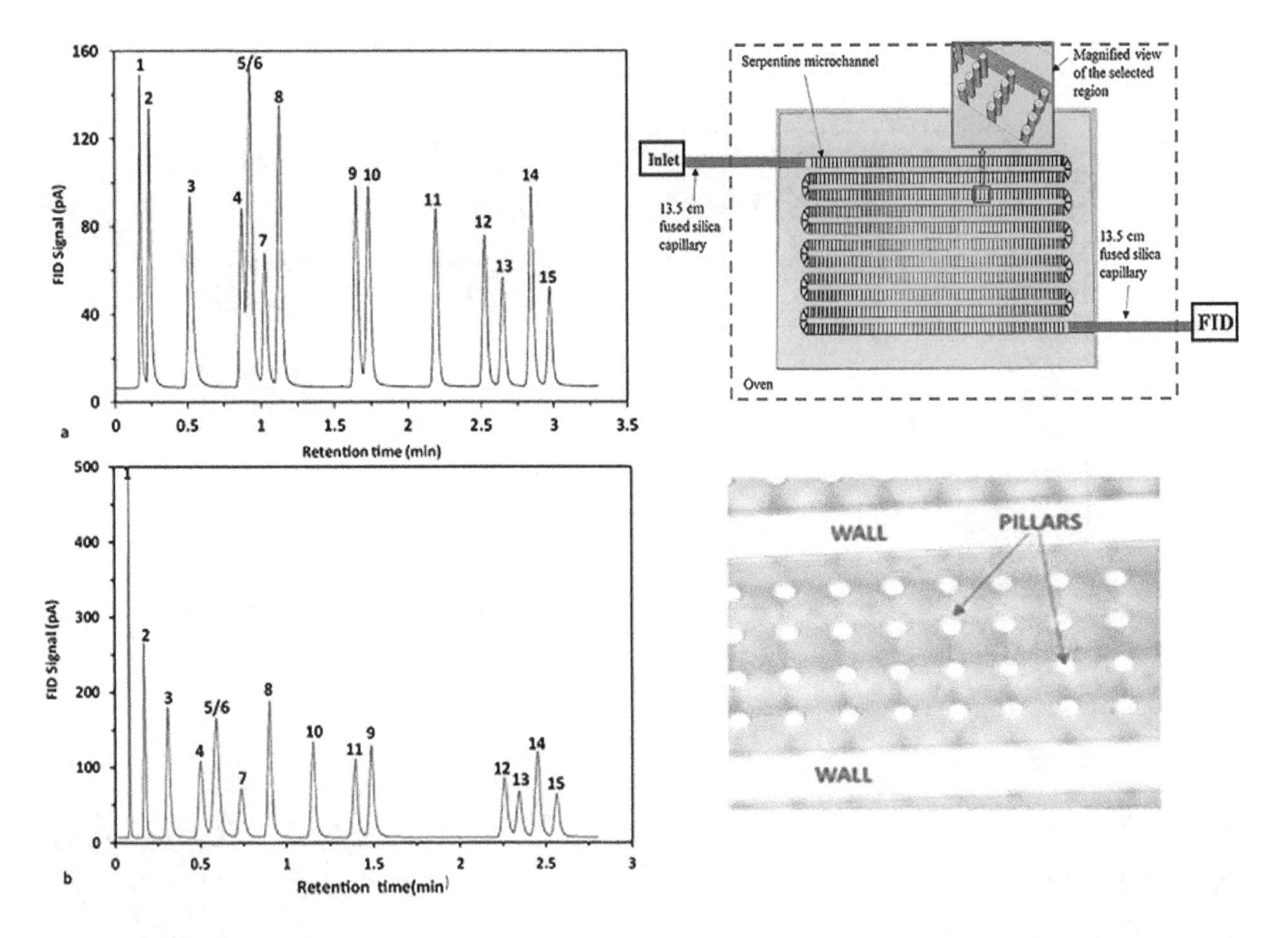

Figure 10. Left: Separation of a 15-compound mixture using (a) [P66614][NTf2]- and (b) [BPyr][NTf2]-coated columns, right: Up schematic diagram of the measurement setup, right down optical micrographs of the uncoated micro-column.

Zellers' group [35] was the first team to use ionic liquids in miniaturized gas chromatography by coating a rectangular column as a second dimension in a GC × GC system. Two years after, Agah's group [36] successes integration of ionic liquids for high-performance separation of complex chemical mixtures (**Figure 10**).

Ionic liquids can be easily statically or dynamically coated (immobilized). However, two points should be highlighted:

- Due to the vast number of ionic liquids, no correlation between the stationary phase and the group of analytes to be separated is known.
- Like normal polymer coating, homogeneity of the coating is not systematically reported. Moreover, no one can be sure that the coating thickness is homogeny along the column.

4.4. Geometry

In conventional gas chromatography, used columns are tubes functionalized by a stationary phase having length ranging from 10 to 100 m. To obtain an excellent column, few parameters can be optimized: length, inner diameter, film thickness, and the coiling radius [37]. Theory of chromatography predicts an increase of efficiency, while the diameter of a capillary column decreases. However, with the emerging of "planar columns," other parameters appear (**Figure 11**).

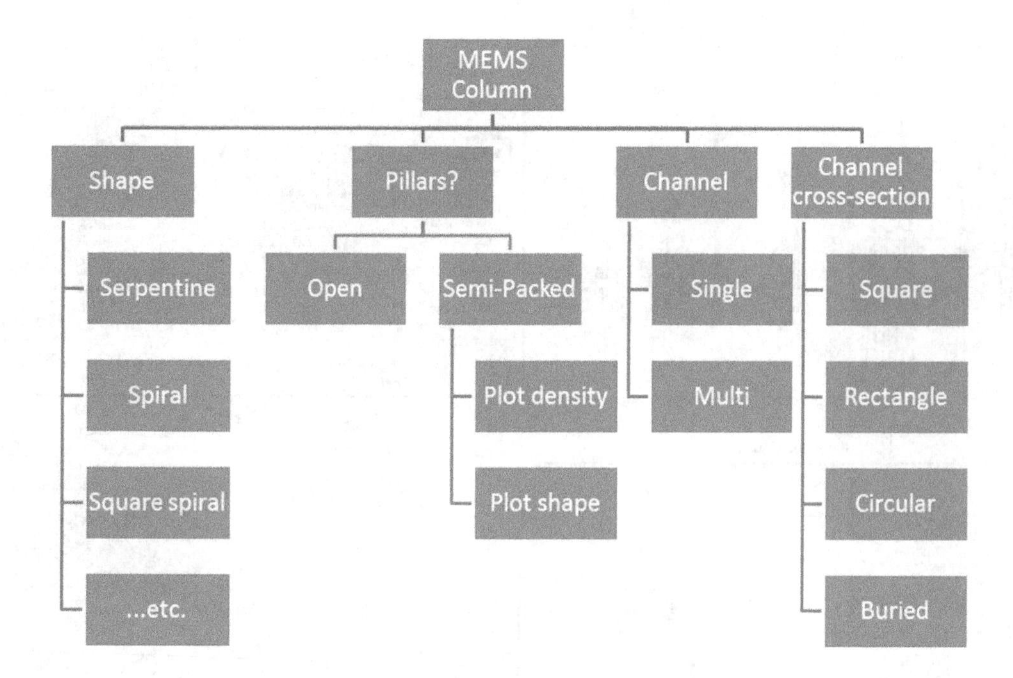

Figure 11. Illustration of some geometrical parameters related to MEMS columns.

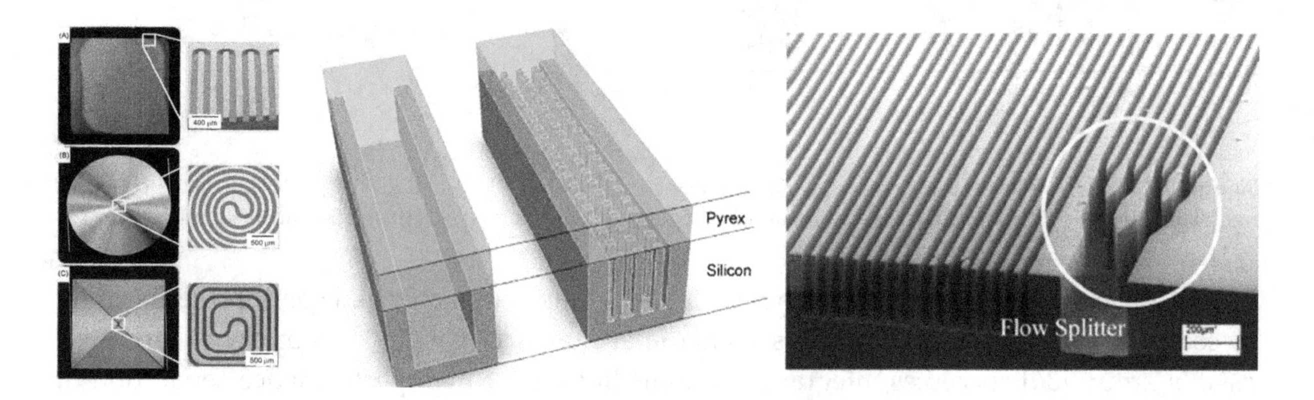

Figure 12. Left: Photograph showing three different micro-column (a) serpentine, (B) circular-spiral, and (C) square-spiral. Adapted from [39],middle: Section of an open and a semi-packed columns, adapted from [40], right: Multi-capillary MEMS column, adapted from [41].

The effect of microfabricated columns' geometries on separation performance was compared by Radadia et al. [38]. In fact, three configurations were tested under isothermal and temperature-programmed mode: serpentine, circular-spiral, and square-spiral (**Figure 12**). Although all the geometries have similar gas permeability, it is shown that the serpentine columns show higher separation plate numbers (lower band broadening) for retained solutes in isothermal mode of operation compared to circular- or square-spiral configurations. Additionally, in temperature-programmed mode of operation, the serpentine design yields higher separation numbers (peak-to-peak resolution) compared to spiral configurations. These performances were attributed to the more favorable hydrodynamic flow.

To increase the efficiency and the surface-to-volume and the loadability without scarifying inlet pressure, a new class of gas chromatographic column was introduced in 2009 by Agah's group [39]. This "semi-packed" column contains embedded 20 μm square posts along the length of the channel paced at 30 μm (**Figure 12**)." This novel configuration enhances both the sample capacity and the separation efficiency compared to the open rectangular columns. Furthermore, due to the uniform spacing and distribution of the posts, these columns have lower-pressure drops and eddy diffusion as compared to conventional packed columns.

Among the shape of the column and implemented pillars or none, some researchers tried different layouts including width modulation [40], multi-capillary [41], and partially buried channel [42].

4.5. Resistive heating

The major goal of GC method development is to minimize the analysis time with desired resolution for accurate qualitative and quantitative analysis. Additionally, fast analysis time means also bring the system back to its initial state (for another cycle). For conventional GC, the column is placed inside an oven and heated using a bare resistive metal wire positioned at the back of the oven. Heating rate for the entire GC analysis is between 30 and 60°C/min when cooling down after running a sample takes approximately 5 minutes. Slow heating and cooling are due mainly to the large total thermal mass of the oven making it unsuitable for separations in fast GC.

Microfabricated columns hold a promise for field applications, as they feature fast analysis time, low power consumption, and easy portability. Although the conventional oven has often been used to evaluate to MEMS column performance, heating element is directly incorporated on this plan columns. Because of the high thermal conductivity of silicon, localized heaters are usually deposited on that side to achieve reasonably uniform temperatures across the silicon chip (**Figure 13**).

Patterned resistive metal layers can be deposited on the surfaces of column substrates to form robust micro-heaters with good thermal conduction, wide temperature range, and extremely low thermal mass. Deposition is performed by various methods such as sputtering or CVD. The resistance temperature detector (RTD) is one of the most accurate temperature sensors. Not only does it provide good accuracy, but it also provides excellent stability and repeatability. Platinum (Pt) is often used in RTDs, and the thin metal film can also function as the heater and temperature sensor simultaneously, which is advantageous for system integration compared to external heaters [13, 43].

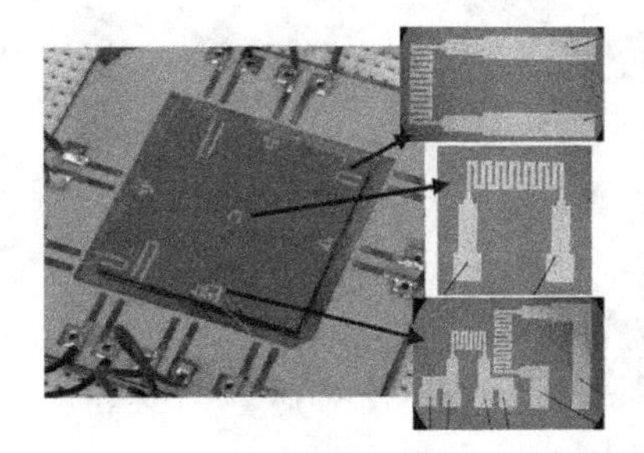

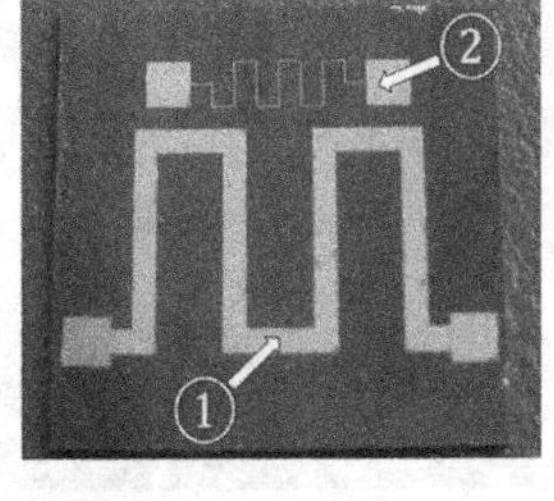

Figure 13. Examples of various deposited platinum resistance for sensing and heating the column, adapted from [14, 44].

Instead of using only platinum as metallic resistance, deposition of various metals was reported: chromium/gold film (Cr/Au) or titanium/platinum (Ti/Pt). Intimate contact between the heater and the column allows extremely high heating rates (1500°C/s) [44]. Depending on the thickness and size of the chip, a heating power consumption can be as low as 4 W/m.

In gas chromatography, separations are performed by temperature programming starting from lower to upper. For continuous monitoring or on-site analysis, GC system should be cooled to the initial state to start a new cycle. Peltier coolers are widely used in miniaturized GC. Also, the column can be set at sub-ambient temperature to retain volatile compounds, for example. This is an advantage compared to conventional GC systems which require liquid carbon dioxide or nitrogen.

5. Detectors

After separating the compounds, a detector is used to monitor the outlet stream from the column. Detection in analytical microsystems is a subject of paramount importance. Indeed, detection has been one of the main challenges for analytical microsystems, since very sensitive techniques are needed as a consequence of the ultrasmall sample volumes used in micron-sized environments.

The flame ionization detector (FID) is the most popular and widely used detector for the analysis of trace levels of organic compounds. Its success is based on outstanding properties, such as a very low minimum detectable limit, a high sensitivity, and a broad linear measurement range. Kuper's team works on miniaturized planar FID since 2000 where the oxygen-hydrogen flame burns inside a glass-silicon chip (**Figure 14**) [45].

At the opposite of FID, thermal conductivity detector (TCD) is a nondestructive system. It measures the difference in thermal conductivity between pure carrier gas and the carrier gas

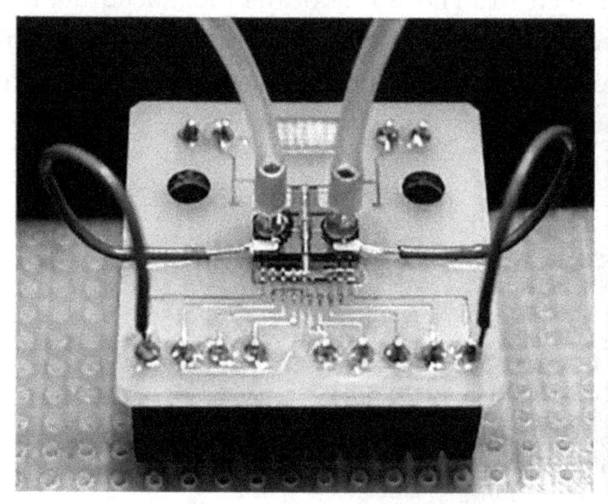

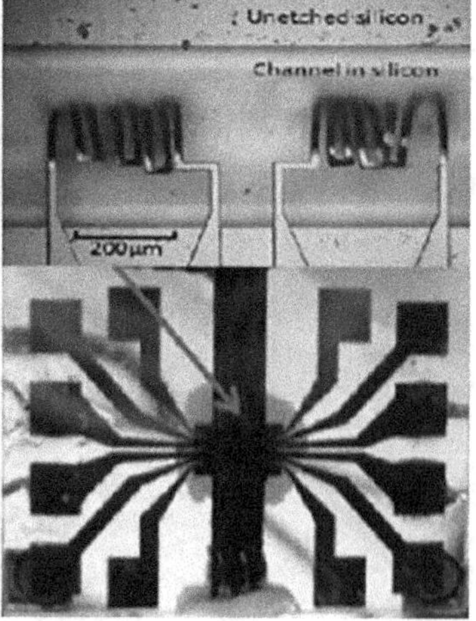

Figure 14. Left: Photography of a micro-FID on a PCB adapted from [47], right: A suspended micro-TCD, adapted from [48].

contaminated with the sample components. Miniaturization of TCD started with the first micro-GC in 1979, and since then several studies have been published in this area [46, 47].

Many sensors such as chemiresistor array and metal oxide (MOX) sensors have been reported for chip-based GC. The response mechanism of these sensors mainly relies on the impedance changes. Typically, a chemiresistor consists of a conductive or semiconductive polymer or emulsion and organometallic compounds [48, 49].

6. Integrated analytical systems

The development of MEMS gas chromatographic components is in progress at several laboratories and universities. Some characteristics of miniaturized GCs are listed in **Table 2**.

At this stage, microfabrication is an attractive option for the development of greatly improved instruments, and many investigations have been reported. However, there are no portable devices able to work anywhere, making accurate, automatic, and continuous analyses of gas samples.

Features→↓reference	Sampling and injection	Separation	Detection	Products to be separated
Sandia National Laboratories [51] (2004–2010)	MEMS cavities	MEMS spiral columns coated with PDMS, WAX, etc.	Chemiresistor, surface acoustic wave	C5–C12 polar and nonpolar compounds (hydrogen*)
μGC system CNR-IMM [52] (2009)	MEMS cavity filled with quinoxaline	0.5 m square-spiral column, packed with Carbograph	Metal-oxide semiconductor	Benzene, toluene, ethylbenzene, xylenes (ambient air*)
μGC system Arizona State University [53] (2013)	Stainless steel tube packed with Carbopack	2–19 m commercial columns	Quartz crystal fork detector with imprinted polymer (MIP)	Benzene, toluene, ethylbenzene, xylenes (ambient air*)
Intrepid GC University of Michigan WIMS [54] (2013)	Combination of stainless steel tube and MEMS elements	1 m MEMS column (PDMS coating)	Chemiresistor array	Explosives vapors (ambient air*)
Zebra GC system Virginia Tech [55] (2015)	MES cavity with embedded pillars (Tenax TA coating)	2 m MEMS column (PDMS coating)	μTCD embedded in the column	Benzene, toluene, tetrachloroethylene, chlorobenzene (helium*)
μGC system University of Michigan WIMS [56] (2016)	MEMS cavity filled with Carbopack	10 m commercial column	Homemade PID	26 VOCs (helium cartridge*)

*carrier gas.

Table 2. Comparison of some portable GC systems.

C2V Company released its first commercial micro-GC product (model C2V-200) in 2010. The platform, based on cartridges, allows hybrid integration of components such as sensors and valves to be assembled together with reduced dead volume. Integrated microchip technology combined with narrow bore capillary GC columns results in a higher performance for lower costs. The C2V-200 micro-GC is designed for ease of use, reduced maintenance, and low gas consumption. Exchangeable column cartridges, with integrated heating zones, can be easily installed (up to 4) with a different column and detection method. This modular setup allows the ability to monitor a wider spectrum of gas components in the same timeframe of 10 to 60 seconds [50].

7. Conclusion

The concept of micro-gas chromatographs demonstrates the potential of mobile devices in various fields related to analytical chemistry such as oil and gas, air analyses, defense, food processing industry, etc. New instrument designs and component manufacturing methods are coming on line that will result in the development of a new generation of high-performance, moveable, and miniaturized instruments for high-performance gas chromatography (HP-GC). The use of microelectromechanical system technologies for the manufacturing of microfabricated gas chromatographic components results in very small, autonomous, and low-cost instruments. Completely autonomous GC instruments require no daily maintenance and can be placed in remote locations for long-term service. This requires battery operation, wireless communications, and freedom from tanks of compressed gases. To this end, work is in progress to develop a high-performance micro-GC that will have acceptable volume. To achieve complete autonomy, vacuum outlet GC should be used with ambient air as carrier gas. In addition, remote battery charging with radiofrequency transmission will be feasible. The use of ambient air as a carrier gas poses several challenges. First, some stationary phases rapidly decompose in air. Poly(ethylene glycol) is a good example. In addition, particulate material and water vapor may need to be removed. Sensor array detection also is needed because these devices can be microfabricated with very low dead volumes; they require no support gases for their operation, and they can be fabricated with a variety of selectivities, which can be used for vapor recognition and for the deconvolution of overlapping peaks. This can reduce the resolution requirements for the column. Sensors and detectors usually have lower sensitivity than detectors incorporated in laboratory gas chromatographic instruments. Low detector sensitivity, coupled with the very low concentrations often associated with air monitoring, requires the use of a preconcentrator for sample enrichment prior to separation and detection. More energy is then needed to heat the preconcentrator to release the adsorbed sample. It seems that one way to solve these problems is a future integration of the instrument on a single chip, focusing each device to one field of application instead to make universal apparatus emulating conventional gas chromatographs.

Conflict of interest

The authors declare no conflict of interest.

Author details

Imadeddine Azzouz[1,2*] and Khaldoun Bachari[1]

*Address all correspondence to: imadeddine.azzouz@esiee.fr

1 Research Center in Analytical Chemistry and Physics (CRAPC), Algiers, Algeria

2 ESYCOM/ESIEE Paris, University of Paris-Est, Noisy-le-Grand, France

References

[1] Brown PR. Advances in Chromatography. New York: CRC press; 1993. 296 p

[2] Terry SC, Jerman JH, Angell JB. A gas chromatographic air analyzer fabricated on a silicon wafer. IEEE Transactions on Electron Devices. 1979;**26**(12):1880-1886

[3] Wang J, Nuñovero N, Lin Z, Nidetz R, Buggaveeti S, Zhan C, et al. A wearable MEMS gas chromatograph for multi-vapor determinations. Procedia Engineering. 2016; **168**(Supplement C):1398-1401

[4] Alfeeli B, Hogg D, Agah M. Solid-phase microextraction using silica fibers coated with tenax-TA films. Procedia Engineering. 2010;**5**(Supplement C):1152-1155

[5] Nachef K, Bourouina T, Marty F, Danaie K, Bourlon B, Donzier E. Microvalves for natural-gas analysis with poly ether ether ketone membranes. Journal of Microelectromechanical Systems. 2010;**19**(4):973-981

[6] Nachef K, Marty F, Donzier E, Bourlon B, Danaie K, Bourouina T. Micro gas chromatography sample injector for the analysis of natural gas. Journal of Microelectromechanical Systems. 2012;**21**(3):730-738

[7] Holland PM, Chutjian A, Darrach MR, Orient OJ. Miniaturized GC/MS instrumentation for in situ measurements: micro gas chromatography coupled with miniature quadrupole array and Paul ion trap. Mass spectrometers. 2003

[8] Voiculescu I, Zaghloul M, Narasimhan N. Microfabricated chemical preconcentrators for gas-phase microanalytical detection systems. TrAC Trends in Analytical Chemistry. 2008;**27**(4):327-343

[9] Frye-Mason GC, Manginell RP, Heller EJ, Matzke CM, Casalnuovo SA, Hietala VM, et al., editors. Microfabricated gas phase chemical analysis systems. Microprocesses and Nanotechnology Conference, 1999 Digest of Papers Microprocesses and Nanotechnology '99 1999 International; 1999 6-8 July 1999

[10] Lahlou H, Vilanova X, Correig X. Gas phase micro-preconcentrators for benzene monitoring: A review. Sensors and Actuators B: Chemical. 2013;**176**(Supplement C):198-210

[11] Bryant-Genevier J, Zellers ET. Toward a microfabricated preconcentrator-focuser for a wearable micro-scale gas chromatograph. Journal of Chromatography A. 2015; **1422**(Supplement C):299-309

[12] Junghoon Y, Christopher RF, Byunghoon B, Richard IM, Mark AS. The design, fabrication and characterization of a silicon microheater for an integrated MEMS gas preconcentrator. Journal of Micromechanics and Microengineering. 2008;**18**(12):125001

[13] Haudebourg R, Vial J, Thiebaut D, Danaie K, Breviere J, Sassiat P, et al. Temperature-programmed sputtered micromachined gas chromatography columns: An approach to fast separations in oilfield applications. Analytical Chemistry. 2013;**85**(1):114-120

[14] Robertson JK. A vertical micromachined resistive heater for a micro-gas separation column. Sensors and Actuators A: Physical. 2001;**91**(3):333-339

[15] Vial J, Thiébaut D, Marty F, Guibal P, Haudebourg R, Nachef K, et al. Silica sputtering as a novel collective stationary phase deposition for microelectromechanical system gas chromatography column: Feasibility and first separations. Journal of Chromatography A. 2011;**1218**(21):3262-3266

[16] Haudebourg R, Matouk Z, Zoghlami E, Azzouz I, Danaie K, Sassiat P, et al. Sputtered alumina as a novel stationary phase for micro machined gas chromatography columns. Analytical and Bioanalytical Chemistry. 2014;**406**(4):1245-1247

[17] Golay MJE. In: Coates VJ, Noebels HJ, Fagerson IS, editors. Gas Chromatography (East Lansing Symposium). New York: Academic Press; 1958

[18] Lambertus G, Elstro A, Sensenig K, Potkay J, Agah M, Scheuering S, et al. Design, fabrication, and evaluation of microfabricated columns for gas chromatography. Analytical Chemistry. 2004;**76**(9):2629-2637

[19] Nishino M, Takemori Y, Matsuoka S, Kanai M, Nishimoto T, Ueda M, et al. Development of μGC (micro gas chromatography) with high performance micromachined chip column. IEEJ Transactions on Electrical and Electronic Engineering. 2009;**4**(3):358-364

[20] Radadia AD, Masel RI, Shannon MA, Jerrell JP, Cadwallader KR. Micromachined GC columns for fast separation of Organophosphonate and Organosulfur compounds. Analytical Chemistry. 2008;**80**(11):4087-4094

[21] Lee C-Y, Liu C-C, Chen S-C, Chiang C-M, Su Y-H, Kuo W-C. High-performance MEMS-based gas chromatography column with integrated micro heater. Microsystem Technologies. 2011;**17**(4):523-531

[22] Sklorz A, Janßen S, Lang W. Application of a miniaturised packed gas chromatography column and a SnO_2 gas detector for analysis of low molecular weight hydrocarbons with focus on ethylene detection. Sensors and Actuators B: Chemical. 2013;**180**(Supplement C):43-49

[23] Sun JH, Guan FY, Zhu XF, Ning ZW, Ma TJ, Liu JH, et al. Micro-fabricated packed gas chromatography column based on laser etching technology. Journal of Chromatography A. 2016;**1429**(Supplement C):311-316

[24] Iijima S. Helical microtubules of graphitic carbon. Nature. 1991;**354**:56

[25] Stadermann M, McBrady AD, Dick B, Reid VR, Noy A, Synovec RE, et al. Ultrafast gas

chromatography on single-wall carbon nanotube stationary phases in microfabricated channels. Analytical Chemistry. 2006;**78**(16):5639-5644

[26] Reid VR, Stadermann M, Bakajin O, Synovec RE. High-speed, temperature programmable gas chromatography utilizing a microfabricated chip with an improved carbon nanotube stationary phase. Talanta. 2009;**77**(4):1420-1425

[27] Bächmann K, Göttlicher B. New particles as pseudostationary phase for electrokinetic chromatography. Chromatographia. 1997;**45**(1):249-254

[28] Yu C-J, Su C-L, Tseng W-L. Separation of acidic and basic proteins by nanoparticle-filled capillary electrophoresis. Analytical Chemistry. 2006;**78**(23):8004-8010

[29] Ventra M, Evoy S, Heflin JR. Introduction to Nanoscale Science and Technology. Boston: Springer US; 2004 611 p

[30] Zareian-Jahromi MA, Agah M. Microfabricated gas chromatography columns with monolayer-protected gold stationary phases. Journal of Microelectromechanical Systems. 2010;**19**(2):294-304

[31] Zareie H, Alfeeli B, Zareian-Jahromi MA, Agah M, editors. Self-patterned gold electroplated multicapillary separation columns. 2010 IEEE Sensors; 2010 1-4 Nov. 2010

[32] Shakeel H, Agah M. Self-patterned gold-electroplated multicapillary gas separation columns with MPG stationary phases. Journal of Microelectromechanical Systems. 2013;**22**(1):62-70

[33] Poole SK, Shetty PH, Poole CF. Chromatographic and spectroscopic studies of the solvent properties of a new series of room-temperature liquid tetraalkylammonium sulfonates. Analytica Chimica Acta. 1989;**218**(Supplement C):241-264

[34] Anderson JL, Armstrong DW. High-stability ionic liquids. A new class of stationary phases for gas chromatography. Analytical Chemistry. 2003;**75**(18):4851-4858

[35] Collin WR, Bondy A, Paul D, Kurabayashi K, Zellers ET. μGC × μGC: Comprehensive two-dimensional gas chromatographic separations with microfabricated components. Analytical Chemistry. 2015;**87**(3):1630-1637

[36] Regmi BP, Chan R, Agah M. Ionic liquid functionalization of semi-packed columns for high-performance gas chromatographic separations. Journal of Chromatography A. 2017;**1510**(Supplement C):66-72

[37] Sumpter SR, Lee ML. Enhanced radial dispersion in open tubular column chromatography. Journal of Microcolumn Separations. 1991;**3**(2):91-113

[38] Radadia AD, Salehi-Khojin A, Masel RI, Shannon MA. The effect of microcolumn geometry on the performance of micro-gas chromatography columns for chip scale gas analyzers. Sensors and Actuators B: Chemical. 2010;**150**(1):456-464

[39] Ali S, Ashraf-Khorassani M, Taylor LT, Agah M. MEMS-based semi-packed gas chromatography columns. Sensors and Actuators B: Chemical. 2009;**141**(1):309-315

[40] Shakeel H, Wang D, Heflin JR, Agah M. Width-modulated microfluidic columns for gas separations. IEEE Sensors Journal. 2014;**14**(10):3352-3357

[41] Zareian-Jahromi MA, Ashraf-Khorassani M, Taylor LT, Agah M. Design, Modeling, and fabrication of MEMS-based multicapillary gas chromatographic columns. Journal of Microelectromechanical Systems. 2009;**18**(1):28-37

[42] Radadia AD, Morgan RD, Masel RI, Shannon MA. Partially buried microcolumns for micro gas Analyzers. Analytical Chemistry. 2009;**81**(9):3471-3477

[43] Reidy S, George D, Agah M, Sacks R. Temperature-programmed GC using silicon microfabricated columns with integrated heaters and temperature sensors. Analytical Chemistry. 2007;**79**(7):2911-2917

[44] Kim S-J, Reidy SM, Block BP, Wise KD, Zellers ET, Kurabayashi K. Microfabricated thermal modulator for comprehensive two-dimensional micro gas chromatography: Design, thermal modeling, and preliminary testing. Lab on a Chip. 2010;**10**(13):1647-1654

[45] Zimmermann S, Wischhusen S, Müller J. Micro flame ionization detector and micro flame spectrometer. Sensors and Actuators B: Chemical. 2000;**63**(3):159-166

[46] Narayanan S, Agah M. Fabrication and characterization of a suspended TCD integrated with a gas separation column. Journal of Microelectromechanical Systems. 2013;**22**(5): 1166-1173

[47] Kaanta BC, Chen H, Zhang X. Novel device for calibration-free flow rate measurements in micro gas chromatographic systems. Journal of Micromechanics and Microengineering. 2010;**20**(9) 095034

[48] Lorenzelli L, Benvenuto A, Adami A, Guarnieri V, Margesin B, Mulloni V, et al. Development of a gas chromatography silicon-based microsystem in clinical diagnostics. Biosensors and Bioelectronics. 2005;**20**(10):1968-1976

[49] Yang J-C, Dutta PK. High temperature amperometric total NOx sensors with platinum-loaded zeolite Y electrodes. Sensors and Actuators B: Chemical. 2007;**123**(2):929-936

[50] Thermo Scientific C2V-200 Micro GC. Available from: https://static.thermoscientific.com/images/D01461~.pdf

[51] Bhushan A, Yemane D, Trudell D, Overton EB, Goettert J: Fabrication of micro-gas chromatograph columns for fast chromatography. Microsystem Technologies. 2007;**13**:361-368. DOI: 10.1007/s00542-006-0210-3

[52] Zampolli S, Elmi I, Mancarella F, Betti P, Dalcanale E, Cardinali GC, Severi M. Real-time monitoring of sub-ppb concentrations of aromatic volatiles with a MEMS-enabled miniaturized gas-chromatograph. Sensors and Actuators B: Chemical. 2009;**141**:322-328

[53] Chen C, Tsow F, Campbell KD, Iglesias R, Forzani E, Tao N. A wireless hybrid chemical sensor for detection of environmental volatile organic compounds. IEEE Sensors Journal. 2013;**13**:1748-1755. DOI: 10.1109/JSEN.2013.2239472

[54] Serrano G, Chang H, Zellers E T. A micro gas chromatograph for high-speed determinations of explosive vapors. In: Proceedings of the Solid-State Sensors, Actuators and Microsystems Conference (TRANSDUCERS 09); 21-25 June 2009; Denver, CO, USA: IEEE; 2009. p. 1654-1657

[55] Garg A. Zebra GC: A Fully Integrated Micro Gas Chromatography System [thesis]. Blacksburg: Virginia Polytechnic Institute; 2014

[56] Zhou M, Lee J, Zhu H, Nidetz R, Kurabayashi K, Fan X: A fully automated portable gas chromatography system for sensitive and rapid quantification of volatile organic compounds in water. RSC Advances. 2016;**6**:49416-49424. DOI: 10.1039/C6RA09131H

Permissions

All chapters in this book were first published in MEMSSDA, by InTech Open; hereby published with permission under the Creative Commons Attribution License or equivalent. Every chapter published in this book has been scrutinized by our experts. Their significance has been extensively debated. The topics covered herein carry significant findings which will fuel the growth of the discipline. They may even be implemented as practical applications or may be referred to as a beginning point for another development.

The contributors of this book come from diverse backgrounds, making this book a truly international effort. This book will bring forth new frontiers with its revolutionizing research information and detailed analysis of the nascent developments around the world.

We would like to thank all the contributing authors for lending their expertise to make the book truly unique. They have played a crucial role in the development of this book. Without their invaluable contributions this book wouldn't have been possible. They have made vital efforts to compile up to date information on the varied aspects of this subject to make this book a valuable addition to the collection of many professionals and students.

This book was conceptualized with the vision of imparting up-to-date information and advanced data in this field. To ensure the same, a matchless editorial board was set up. Every individual on the board went through rigorous rounds of assessment to prove their worth. After which they invested a large part of their time researching and compiling the most relevant data for our readers.

The editorial board has been involved in producing this book since its inception. They have spent rigorous hours researching and exploring the diverse topics which have resulted in the successful publishing of this book. They have passed on their knowledge of decades through this book. To expedite this challenging task, the publisher supported the team at every step. A small team of assistant editors was also appointed to further simplify the editing procedure and attain best results for the readers.

Apart from the editorial board, the designing team has also invested a significant amount of their time in understanding the subject and creating the most relevant covers. They scrutinized every image to scout for the most suitable representation of the subject and create an appropriate cover for the book.

The publishing team has been an ardent support to the editorial, designing and production team. Their endless efforts to recruit the best for this project, has resulted in the accomplishment of this book. They are a veteran in the field of academics and their pool of knowledge is as vast as their experience in printing. Their expertise and guidance has proved useful at every step. Their uncompromising quality standards have made this book an exceptional effort. Their encouragement from time to time has been an inspiration for everyone.

The publisher and the editorial board hope that this book will prove to be a valuable piece of knowledge for researchers, students, practitioners and scholars across the globe.

List of Contributors

Lluis Pradell, Marco Antonio Llamas and Julio Heredia
Department of TSC, Technical University of Catalonia (UPC), Barcelona, Catalonia, Spain

David Girbau, Antonio Lázaro and Adrián Contreras
Department of EEEA, Rovira i Virgili University, Tarragona, Catalonia, Spain

Miquel Ribó
Department of ET, La Salle-Ramon Llull University, Barcelona, Catalonia, Spain

Jasmina Casals-Terré
Department of EM, UPC, Barcelona, Catalonia, Spain

Flavio Giacomozzi and Benno Margesin
Fondazione Bruno Kessler, Trento, Italy

Minami Kaneko, Ken Saito and Fumio Uchikoba
Department of Precision Machinery Engineering, College of Science and Technology, Nihon University, Chiba, Japan

Bassem Jmai and Ali Gharsallah
Department of Physics, FST, Unit of Research in High Frequency Electronic Circuit and System, University Tunis El Manar, Tunis, Tunisia

Adnen Rajhi
Department Electrical Engineering, National School of Engineering Carthage, Tunis, Tunisia
Laboratory of Physics Soft Materials and EM Modelisation, FST, University Tunis El Manar, Tunis, Tunisia

Paulo Mendes
Department of Industrial Electronics, Microelectromechanical Systems Research Center, University of Minho, Guimarães, Portugal

Bahadir Tunaboylu
Istanbul Sehir University, Department of Industrial Engineering, Istanbul, Turkey

Ali M. Soydan
Gebze Technical University, Institute of Energy Technologies, Gebze-Kocaeli, Turkey

Rajagopal Kumar and Fenil Chetankumar Panwala
National Institute of Technology Nagaland Chumukedima, Nagaland, India

Huiliang Cao
Science and Technology on Electronic Test and Measurement Laboratory, North University of China, Tai Yuan, China

Jianhua Li
National Key Laboratory of Science and Technology on Electromechanical Dynamic Control, Beijing Institute of Technology, Beijing, China

Hai-peng Liu and Shi-qiao Gao
State Key Laboratory of Explosion Science and Technology, Beijing Institute of Technology, Beijing, China

Lei Jin and Shao-hua Niu
School of Mechatronical Engineering, Beijing Institute of Technology, Beijing, China

Chunxiu Liu, Chenghua Xu, Ning Xue, Jian Hai Sun, Haoyuan Cai and Tong Li
State Key Laboratory of Tranducer Technology, Institute of Electronics, Chinese Academy of Sciences, Beijing, China

Yuanyuan Liu
Institute of Semiconductors, Chinese Academy of Sciences, Beijing, China

Jun Wang
School of Materials Science and Engineering, Beijing Institute of Technology, Beijing, China

Imadeddine Azzouz and Khaldoun Bachari
Research Center in Analytical Chemistry and Physics (CRAPC), Algiers, Algeria
ESYCOM/ESIEE Paris, University of Paris-Est, Noisy-le-Grand, France

Index

R

S

T

U

W

CPSIA information can be obtained
at www.ICGtesting.com
Printed in the USA
LVHW062100111022
730499LV00004B/41